Issues in Acoustic Signal – Image Processing and Recognition

NATO ASI Series

(Advanced Science Institutes Series)

A series presenting the results of activities sponsored by the NATO Science Committee, which aims at the dessemination of advanced scientific and technological knowledge, with a view to strengthening links between scientific communities.

The Series is published by an international board of publishers in conjunction with the NATO Scientific Affairs Division

A Life Sciences B Physics	Plenum Publishing Corporation London and New York
C Mathematical and Physical Sciences	D. Reidel Publishing Company Dordrecht and Boston
D Behavioural and Social Sciences E Applied Sciences	Martinus Nijhoff Publishers The Hague, The Netherlands
F Computer and Systems Sciences G Ecological Sciences	Springer-Verlag Berlin Heidelberg New York

Series F: Computer and System Sciences No. 1

Issues in Acoustic Signal – Image Processing and Recognition

Edited by

C. H. Chen

Springer-Verlag Berlin Heidelberg New York Tokyo 1983

Proceedings of the NATO Advanced Research Workshop on Issues in Acoustic Signal – Image Processing and Recognition

Professor Dr. C. H. Chen
Southeastern Massachusetts University, Department of Electrical and Computer Engineering, North Dartmouth, MA 02747, USA

Published in corporation with NATO Scientific Affaires Division

ISBN 3-540-12192-7 Springer-Verlag Berlin Heidelberg New York Tokyo
ISBN 0-387-12192-7 Springer-Verlag New York Heidelberg Berlin Tokyo

Library of Congress Cataloging in Publication Data.
Issues in acoustic signal — Image processing and recognition. (NATO ASI series. Series F, Computer and system sciences; no. 1)
"Proceedings of the NATO Advanced Workshop Research on Issues in Acoustic Signal — Image Processing and Recognition" — T.p. verso.
"Held August 5–9, 1982 ... in San Miniato, Italy" — Pref.
"Published in cooperation with NATO Scientific Affaires Division."
1. Underwater acoustics — Congresses. 2. Image processing – Congresses. 3. Signal processing — Congresses. I. Chen, C. H. (Chi-hua), 1937. II. North Atlantic Treaty Organization. Scientific Affairs Division. III. NATO Advanced Study Institute on Issues in Acoustic Signal — Image Processing and Recognition (1982: San Miniato, Italy). IV. Series.
QC242.I77. 1983. 620.2'5. 83-547
ISBN 0-387-12192-7 (U.S.)

Printed in Germany

Printing: Beltz Offsetdruck, Hemsbach; Bookbinding: J. Schäffer OHG, Grünstadt
2145/3140-543210

Preface

The NATO Advanced Research Workshop on Issues in Acoustic Signal/Image Processing and Recognition was held August 5-9, 1982 at the Cappuccini complex in San Miniato Italy. The Workshop was primarily concerned with the underwater acoustic signal processing and seismic signal analysis and a major effort was made to link these topics with pattern recognition, image processing and artificial intelligence. Major issues and new approaches in these interrelated areas were closely examined in the Workshop. In addition to paper presentations three discussion sessions were held on: (1) spectral analysis in underwater acoustics, (2) seismic wave propagation, seismic imaging and migration, and seismic inversion, and (3) unresolved issues and future directions.

This Proceedings volume includes most presentations made at the Workshop. The publication, like the meeting itself, is unique in the sense that it provides extensive interactions among the closely related areas stated above. Such interactions which usually result in the integration of different systems or approaches are certainly much needed to achieve some performance breakthrough while individual systems or approaches reach their performance limit.

I am grateful to all participants for their active participation that makes the Workshop very productive, and to Dr. Lewis J. Lloyd and Dr. Ralph Goodman for their help to arrange an informative visit to the SACLANT ASW Research Centre for the Workshop participants. I am confident that this publication will be equally productive to report important current research results and near-future research activity particularly in underwater acoustic signal processing.

C.H. Chen
Workshop Director

Table of Contents

A COMMON FRAMEWORK FOR ACOUSTIC SIGNAL ANALYSIS IN THE OCEAN ENVIRONMENT

Dr. J. C. Hassab
Naval Underwater Systems Center
Newport, Rhode Island 02840

ABSTRACT. Estimates of arrival time intervals and their variation in time constitute basic measurement in a signal processor for undersea applications. Those measurements are then mapped in the data processor into the desired estimate of emitting source range, direction, depth and velocity. The integration of the signal and data processing stages within the signal analysis function (see Fig. 1) form the subject of this paper. Figure 2 gives the three basic types of channels with the pertinent time intervals and processing functions. The processor structure has a spectral type formulation for its stationary elements and a state space formulation for its nonstationary ones.

Though varied, the basic signal processing elements have a common set of issues that can be addressed simultaneously (Fig. 3,4). Each signal processing stage has a set of linear and non linear operations, i.e. Fourier transform, absolute value, log, $\tan^{-1}$...that map the embedded time interval in the input function into a recognizable clue denoting the value of that time interval. The clue may be a dominant peak or a slope. To enhance a deteriorating clue in the presence of varied interferences, windowing has been incorporated into the basic signal processing stage. Gating is added to limit the search for the clue to the most probable region in the signal processor output. The detected time intervals could then be passed through a linear weighed least square filter to improve and assess the quality of the estimates. Those estimates are fed back in a loose couple to adapt the parameters of the gate and in turn those of the window, for instance. The same estimates are also fed forward through a linearized or nonlinear weighted least square filter to map the time interval estimates into smoothed source position and velocity estimates (Fig. 5). The source parameter estimates are also fed back for use in adaptive design of gate, window...

An interactive system is depicted in Figure (6) that utilizes to best advantage all available information in a synergistic way. The motivation for the described approach lies in the desire to accommodate the widest variety of practical situations with a priori information generally lacking. Too often sub-systems are

NATO ASI Series, Vol. F1
Issues in Acoustic Signal/Image Processing and Recognition
Edited by C. H. Chen

compartmented and there is little or haphazard communication from one to another. Each sub-system by itself has natural performance limits which can be enhanced through sub-system interactions. The coupling provides guidance for localized search and adaptation, not direction to the local processor.
Directive information may lead to system instability.

1. INTRODUCTION

This paper addresses the problem of estimating source location and velocity through passive observations of the source signal. Active observation of a scatterer with the a-priori information on the probing signal is considered as a subset of the passive observation problem. The total gain in a localization and motion analysis system is derived from spatial or array gain (size, sensors number and placement etc.), from inclusion of propagation channel features and from temporal or processing gain (signal, data). More attention by system designers should be drawn to the simultaneous optimization along both spatial and temporal dimensions.

Practical considerations such as array stabilization, cost, spatial coherence, platform dimension ... limit the achievable spatial gain, while the temporal dimension is limited by the requirement to yield an acceptable solution within a given time lapse. When the elements of the problem are stationary, the temporal processing reduces to a continual integration over the observed source clues until desired accuracy in the solution is obtained. When the elements of the problem are nonstationary i.e. moving source/receiver, changing channel characteristics ... the source observation is limited to a basic observation time interval, over which the process may be considered locally stationary. The system gives then short memory estimates i.e. its lowest level of performance, and yields a fuzzy idea on the location and motion of the source. With the succession of basic time observation intervals, the temporal processor basically removes the non-stationarity in the problem and superimposes the repeated fuzzy pictures to reinforce the invariant parameters in the problem and to develop a well-defined picture of source location and motion.

In its totality, the estimation process constitutes a non-linear and non-stationary problem that is not amenable to optimum global system synthesis. The state-of-the-art would allow for subsystem (local) optimization and for integration into a self-adjusting/self-organizing global system that achieves the desired performance. The overall design guideline would be a robust performance under the varied realistic conditions that the system is expected to encounter. Those conditions should accommodate non-gaussian noise, non-stationary data due to relative source/receiver motion or to varying signal/noise characteristics, channel distortion and dispersion, sensor position instabilities, off-broadside source directions, recursive short observation times of source, tonals in otherwise broadband signals, interference. When the conditions permit idealization, i.e., gaussian signal and noise, very long observation time of source, stationary source/receiver, the system should self-adjust to yield a performance that approximates the optimal performance criterion given by the Cramer-Rao bound.

This paper is not intended to be a list of contributors and contributions. It is an attempt at merger of those contributions into a design assessment and into candidate structures. Hopefully it serves to highlight a system design approach in underwater acoustics. Section (2) describes the elements of the signal, the basic types of channels through which the signal would propagate, and the appropriate processors to the encountered signals and channels. Section (3) discusses two multiplicative operations common to all cases. Windowing is executed in the frequency domain and gating in the time domain. Section (4) assesses the various processing schemes appropriate to channels with intersecting ray paths. Section (5) evaluates the processors for non-intersecting ray path channels. Section (6) merges the varied time delay measurements into source position and velocity estimates. Section (7) formalizes the discussions in the preceding sections and presents an interactive source tracking approach. Section (8) contains summary comments.

2. SOURCE, CHANNEL, PROCESSOR

Figure (1) describes a basic system and its function. The source signature may include wideband, narrowband and transient elements. The signature traverses non-intersecting or single paths and/or intersecting or multipath. The channel may be distortionless or display distortion, dispersion and may support noise and interference. The receiving sensors, after beamforming, are effectively concentrated into a single point or spatially distributed points. The traditional signal analysis of underwater systems have separated the signal from the data processing functions and have marginally included channel characteristics in the analysis process.

The two basic ocean channels are viewed, that of horizontal and of multipath propagation. For each type of channel, recent signal and data processing methods have been evaluated analytically and experimentally, then contrasted to existing techniques. The traditional separation in the design of signal and data processing should be replaced by an integrated approach that can lead to a measurable improvement in system performance. Since both types of channels may occur alternately or simultaneously in practice, a joint treatment should be given to both types where much of the structure and concepts are shared.

In general, the acoustic energy emitted by the source reaches the receiver through different paths: as direct, reflected and/or diffracted waves. Reception of the source signal is picked up at a single or at multiple spatial points. The processing techniques depend on the number of receiving points and the number of arrivals at each point (Figure 2). In one basic type, multiple propagation paths become scattered by the obstacles in the channel which lead to their intersection at a sensing point. To dissect the resulting composite function, generalized cepstrum, autocorrelation or complex demodulation have been applied (Figure 3). In the other basic type, the propagation paths do not intersect at the sensing points; then spectral estimation may be carried out at each sensing point, and generalized cross-correlation across sensing points (Figure 4). Using recursive least-squares filters, the pertinent time intervals from each channel may be smoothed and mapped under a weighted least square error criterion into the desired source range, depth, course, and speed.

In general, localization system designs have been optimized to some average and idealized operating conditions in the source and channel which have caused long interruption in useful data and a wide gap between detection and source tracking. Weak links result in the system chain from such a design. Sensors, channel, signal and data processors are then disjointed in the localization and tracking chain. The realization of potential gains in localization and source motion analysis performance can result from functionally cooperative, mutually supportive subsystems having real-time access to a shared information base. Figures 3, 4 and 5 give a sampling of such systems. At times independent or dedicated designs have been developed for a sub-set sensor configuration, single source characteristic i.e. wideband, and a fixed and simple window characteristic. This prevents the sharing/selecton of appropriate sub-arrays out of the total sensor suites as determined by source signature and range, array speed, noise distribution over arrays, channel conditions, and the adaptation of the window to more complex situations.

3. WINDOWING AND GATING

Windowing and gating have been the subject of extensive studies for both types of channels. Techniques for the measurement of time intervals are varied because of their differing limitations and their unacceptable performance under general conditions. Windowing lowers the operational threshold of the basic processor (Fig.

7,8). The windows are designed to remedy or compensate for physical conditions that affect unfavorably the performance of a given signal processor. They are dependent on signal spectra, noise spectra and channel parameters (Fig. 9-11). Unfailingly, addition of a properly designed window has extended the region of satisfactory performance of a given conventional signal processor. Guidance for the design and implementation of such windows has been documented in a series of papers cited in the bibliography (I,II). It should be stressed that the windows must be designed to suit the processor at hand and the situation under consideration. Hence, analysis should precede any design in order to determine the cause of deficiency. Otherwise, improper windowing may be applied that would have the reverse effect of deteriorating the performance instead of improving it. Often the input parameters for the design of the window are not known a-priori, and adaptive techniques are used to execute the implementation. The ultimate guide to the functioning of the window is in the added enhancement to one of two basic clues denoting the value of the time interval parameter to be estimated. For correlation and cepstrum the basic clue is a peak in the output, and for complex demodulation it is a slope.

The simplistic scheme of independently selecting the dominant clue from the total signal processor output can deliver erratic time interval estimates, whenever adverse but temporary conditions exist at the input. For initialization, an ensemble average over a number of signal processor outputs are taken to enhance the clue against the mean background noise. Where the clue is identified as having sufficient signal power over noise power, a gate is centered at the corresponding output region. The characteristics of the gate may be provided by a Kalman filter operating on the raw time interval estimates. Such gating enhances the robustness of the processor against fades and limits the clue search to the most probable region in the processor output. A successful stabilization process of the estimates allows for automatic and quasi-optimal processing of the data to estimate source location and motion. Furthermore, the linear Kalman filter can detect easily a source maneuver through a jump in the time delay rate and pass on this information to the linearized Kalman filter that is estimating the source state dynamics.

4. MULTIPATH AND SINGLE SENSING POINT SIGNAL PROCESSING (I)

In a multipath situation with an effective single sensing point, the estimated time intervals are the time delays among the channel paths (Fig 3). Homomorphic deconvolution can determine, under certain conditions, the difference in arrival times or time delays among these paths and estimate both the impulse response of that channel, and the signal propagating through it. This is a very powerful processor, considering little a-priori information is required. Certain assumptions are presumably obeyed; otherwise deterioration and failure of the estimation process occur. Failure occurs principally in the phase unwrapping stage due to excessive interference noise or differential dispersion among the paths. This preempts the estimation of signal or channel impulse response and limits the estimation process to that of time delay.

Time delay estimation relies, then, on the real part, i.e., amplitude cepstrum of the complex cepstrum in the homomorphic filter. Noise interference with the amplitude or power cepstrum may be minimized through proper design of windowing functions. Distortion and dispersion in the channel would continue to deteriorate the performance of the power cepstrum (Fig. 12).

A variant on complex demodulation has been successfully used to avoid the detrimental effect of distortion and dispersion in the channel (Fig. 13). This is conditioned on the absence of zero spectral content in the considered spectrum and on successful phase unwrapping (Fig. 14). For the power cepstrum, such a zero singularity deteriorates, similar to interference noise, the performance but would not disrupt the processing. For the auto-correlation, such singularity problems do not arise, but problems occur if windowing is not applied in the presence of highly peaked spectral content. Not-withstanding, the cepstrum has displayed a better capability to measure decreasingly small time delays, under low S/N and in the presence of interference sinusoids. For the power cepstrum, its weakness and sensitivity to certain traits in the processed composite function can be shielded through proper design of log spectrum windows. For the cepstrum, the windowing function is to enhance and limit consideration to high signal spectrum to noise spectrum over the observation interval of the composite function.

Over successive observation intervals, procedures for detection, estimation and tracking have been considered in the literature. They include moving averages, gating, editing and filtering. Those processing steps enhance the quality of time delay estimates by controlling the false assignment of peaks and thus reducing the bias and variance. The reduction in bias renders the data compatible with the direct processing requirements by a Kalman filter. The cited data processing steps are repeated for the single path situation with multiple sensing points.

5. SINGLE PATH WITH MULTIPLE SENSING POINTS SIGNAL PROCESSING (II)

Now considerations are given to multiple sensing points with a single transmission path at each. Processing at an effective single sensing point yields the spectral content. Smoothing is applied only for spectral estimation and not for estimation and tracking of a sinusoidal frequency value. Detection, estimation and tracking of a sinusoidal frequency follow those described for time delay.

Across sensing points, the generalized correlator yields the difference in arrival time or time dalay in a wideband noise source. Interference noise with the correct peak or slope denoting the time delay value is combatted through inclusion of a windowing function. Application of windowing in this context is well-rooted and several windows are found in the literature.

An analytical and experimental comparison of several windows has been carried out. The Hassab-Boucher W_{HB-II} window is recommended for peak enhancement. This window accommodates an enlarged set of realistic conditions, i.e. tonals in signals, while reducing to the maximum likelihood or Eckart window under specialized conditions(Fig. 15). As in the multipath case, any selected window requires a-priori information about the signal and noise characteristics. Also, the windows do change in shape with varying effective parameters such as SNR and bandwidths. Such information is not presumed available and its exact estimation is made difficult by a dynamic environment. Consequently, some adaptive variation of the window parameters would be needed until detection is confirmed. In this regard, similar concepts to those used in multipath are adopted.

6. ESTIMATION OF SOURCE LOCATION AND VELOCITY (III)

Reduction of the bias and variance in source location and motion estimation can be accomplished by effective use of statistical estimation techniques over sequential and finite observation of the source signal. The estimator is an expanding memory filter that uses the imperfect estimates of time intervals and ultimately maps them onto invariant source trajectory parameters over which smoothing is performed to reduce, jointly, the variance and the bias in the estimate of source kinematic parameters (Fig. 16). The coupling of this data processor with the generalized signal processor improves substantially on the techniques that process inappropriately mapped time delays or those that transform directly the best time delays available into source motion estimates. The latter approach can only be optimum when stationarity of all the elements in the problem can be presumed. For this limiting case, the present approach converges to the optimum estimates. For generalized cases, the present approach remains viable for moving sources at long range, off-broadside source directions, high time delay variances.

Implicit in this discussion is a requirement for correct statistical descriptors of the processes at hand. The recovery from an incorrect description is aided by use of the coupling loops for detection of such an event, for by-passing the ensuing divergence and by routing the processes in a degraded mode until system recovery. The traditionally separated signal and data processing stages have to be integrated in order to increase the performance of a given system. Total system gain is derived from spatial or array gain (size, sensors placement, etc.) and temporal or processing gain (signal, data). Practical considerations eventually limit the achievable spatial gain. The usefulness of such a system can still be enhanced through inclusion of temporal processing techniques along the lines described above.

7. INTERACTIVE SOURCE TRACKING FUNCTION

The variously discussed concepts are now merged into an interactive source tracking function. A formalized description is given (Fig. 6). The function basically accepts varying levels of information reflecting on the source dynamic state, maps them through an adaptive and expanding memory processor into firm estimates of source position (including depth), velocity and solution quality. The information may be primitive, such as source signature, or at a high level such as a source direction, a speed gate, a range gate. Speed gate may be an aural estimate by an operator. Range gate(s) may reflect what the propagation channel can support, given source temporal characteristic i.e. S/N, bandwidth ...Figure (6) gives a block diagram of the process.

7.1 INPUT DATA

The integrated source tracking system would receive information from the acoustic performance prediction function and the focused beamformer. The beamformer would pass, for the selected source of interest, its signature along with higher characteristics (bandwidth, SNR...) within depression/elevation and azimuthal sectors, sensors location vector, and "short memory" processing estimates of the source range and direction.

The acoustic performance prediction would develop metrics to the spatial cells apportioning the channel that links the sensors to the surrounding environment. The feature vector attached to each cell may comprise such items as: ray-paths types and number, expected types of differential distortion and dispersion, predicted difference of arrival times (time delays), predicted relative signal strengths. Any environmental highlights that are recognized from a cluster of cells, such as convergence zone/front size or sharpness, would be provided.

7.2 PROCESSING ELEMENTS

The signal processing, data processing, and environmental processing stages should be synergistic through cooperative couplings that will enable sharing the extracted information within each stage for on line adaptations and consistency checks of the global system. The shared information provides guidance vice direction for neighborhood search within a processing stage. The signal processor provides, over short observation intervals of the source, fuzzy snapshots of the dynamic scene and yields lowest level performance of the system. The quality of these independent snapshots do not improve in time. The data processor superimposes the invariant parameters and filters the fuzzy snapshots to produce a clear dynamic scene. The channel provides the interpretation of the noisy observations into noisy source dynamics descriptors.

7.3 BASIC CLUE EXTRACTOR

The basic clue extractor will provide measurements of time intervals (such as period or frequency, time delay) from the processing of observed source signature. The elements should include spectral analysis, signal comparison (like auto-and cross-correlation), homomorphic filtering and complex demodulation. These elements will operate on the appropriate outputs from the beamformer. They map embedded measures in the signature of some indirect aspect (like time interval) of the source dynamic state into recognizable clues (minimum, maximum or slope) and into measurable clues.

7.4 ENHANCED CLUE EXTRACTOR

The mapping process leads to a varied level of accuracies depending on the received source signature, the existing environmental conditions and the time allowed for observation. Based on channel characteristics and signal and noise descriptors, parametric and non-parametric windows/gates would be adapted on line and exercised in parallel to remedy or compensate for unfavorable conditions that would otherwise deteriorate clue recognition/measurement process.

7.5 CLUE TRACK EXTRACTOR

Data non-stationarity dictate its observation and processing in short time frames to avoid smearing of the clue and spoiling of its interpretation. The clue track extractor will align recursive snap shots of the clues relying on parameter invariants and using a linear statistical pattern recognizer (based on weighted least square or maximum likelihood criterion as appropriate), that yield improved estimates of the clues, their doppler variation, and quality measures. The clue track extractor outputs are fed forward to the source track extractor and back to the window/gate design controller, and beamformer. A maneuver by the source track extractor is reflected into a sudden change in the clue doppler variation; the change detection is reported to the source track extractor. The doppler estimate may be fed back to form a loose frequency, time delay, or phase lock loops that minimize the dynamic nonstationarity in the signature and allows a longer observation (coherent integration) over the source signature.

7.6 SOURCE TRACK EXTRACTOR

The source track extractor aligns all relevant profiles from the acoustic, environmental, a-priori information...clue extractors to develop a firm and clear picture of the source dynamic track. The source track extractor is an expanding and flexible memory estimator. It would use recursive and sparse clue estimates (phase, time delay, frequency, doppler, direction, range, speed...) their measures of quality if available, channel cells characteristics, receiver/sensor motion estimates...and map them onto invariant source trajectory parameters over which long temporal smoothing is performed.

The source extractor would match/correlate (deterministic first, statistical if necessary) the set of clues from the short memory clue track extractor and the beamformer with the set of clues attached to channel cells in order to develop a most probable cluster of possible source locations. Outlier cells in the cluster will be rejected through primitive consistency checks with the stored source dynamics class capabilities. For rejected cells, the gating mechanism in the enhanced clue extractor will be adjusted accordingly whenever unambiguous mapping exists between the basic clue field and source location field.

Tracking the evolution of the cluster(s) due to non-stationary features (like relative source/receiver motion) and more importantly the reduction of bias and variance of source track estimates may be achieved in a statistical estimator which constrains the extracted clue track to point to invariants (like initial range, constant source velocity). This flexible and expanding memory system will yield an enhanced estimation of the source track.

Hypothesis testing of the postulated invariants are formulated. Detection of variants will signal an ensuing divergence. Until system recovery, process routing in a degraded mode will use the clue track extractor output. In parallel, a parametric "learning" process could be developed to estimate the variation.

Finally, the source track extractor performs consistency checks and sensitivity analysis prior to confirmation of an extracted source track. Machine generated and <u>operator</u> introduced source track templates could be used to test the uniqueness of the extracted source track. The confirmed source track is fed to the windowing/gating controller and the beamformer for aid during signal fading.

8. SUMMARY COMMENTS

A few summary comments highlight the need for bringing new issues into active consideration of the design process:

1. There is a fixation with signal to noise ratio parameter description of system performance while defering other major, and at times more dominant, descriptions of system performance. Performance of a nonlinear signal processor depends on some function of signal spectra and noise spectra describable in terms of effective parameters, such as signal to noise ratio, center frequency, relative bandwidth. The bandwidth issue, for instance, is often a neglected parameter. Furthermore, since a data processor can deal more directly with unbiased lower SNR data than with biased high SNR data, systems should not stress SNR gain without examining the side effects developed during the acquisition of higher signal to noise ratio to the remaining estimation processes.

2. The system chain is not evenly reinforced. Weak links exist. Proper windowing, for instance, is often a neglected area where a noticeable improvement in the system performance may be accrued at a relatively low cost. For a given relative signal to noise bandwidth, proper windowing can lower the threshold by an average 6 dB, and this improvement process is often a neglected avenue.

3. System design has relied on average operating conditions causing wide interruption in useful localization data. For instance, a design may be optimized for a distortionless channel and a wideband noise source. When channel dispersion occurs, biased estimate in the optimized peak detection processor results; a slope processor should then be used. If a narrowband signal arises, system failure through ambiguity results. Variant windowing should be offered. Exclusion of the troublesome narrowband signal through reduction in signal bandwidth yields a poorer system performance. Linear analysis shows that an increase of both bandwidth/center frequency and observation time by multiples has similar effect on variance of range estimates, to effective array length increase, to source range decrease. System

design should be robust with ability to accommodate a range of conditions, and to transition gracefully from average to specialized conditions.

4. Sensor, channel description, signal and data processing continue to be designed as disjointed processes in the localization and motion analysis chain. Various indicators developed through analysis and simulation reinforce the contention that an integrated system with cooperative couplings among its elements could reduce the localization and motion analysis time by one half or more over an uncoupled system and bridge the wide gap between source detection and tracking.

5. Processing approaches ignore often the nonlinearity of the operations within the localization and tracking system. This can lead to substantial residual bias in the source state estimates. The source dynamics are nonlinearly embedded in the received signals which are corrupted by additive noise. The system should be designed so that the simple rule, stating that non-linear and expectation operators be applied over small perturbation, is obeyed throughout the system.

List of References

Four sections are given. For each section, the references are listed chronologically.

I. Multipath With Single Sensing Point Signal Processing

1. B. P. Bogert, J. J. Healy, and J. W. Tukey, "The Frequency Analysis of Time Series for Echoes; Cepstrum Pseudo-Autocovariance, Cross-Cepstrum, and Saphe Cracking," in Time Series Analysis, M. Rosenblatt, Ed. Chap. 15 pp. 209-243, New York: Wiley, 1963.

2. A. Noll, "Short-Time Spectrum and 'Cepstrum' Techniques for Vocal-Pitch Detection," J. Acoust. Soc. Amer., Vol. 36, pp. 296-302, Feb. 1964.

3. B. P. Bogert and J. F. Ossanna, "The Heuristics of Cepstrum Analysis of a Stationary Complex Echoed Gaussian Signal in Stationary Gaussian Noise," IEEE Trans. Inform. Theory, Vol. IT-12, pp. 373-380, July 1966.

4. A. V. Oppenheim, "Generalized Superposition," Inform. Contr., Vol. 11, pp. 528-536, Nov. - Dec. 1967.

5. A. V. Oppenheim, R. W. Schafer, and T. G. Stockham, Jr., "Nonlinear Filtering of Multiplied and Convolved Signals," Proc. IEEE, Vol. 56, pp. 1264-1291, Aug. 1968.

6. J. Prabhakar and S. C. Gupta, "Separation of Rayleigh and Poisson Density Functions Through Homomorphic Filtering," Nat. Elec. Conf., pp. 605-610, Dec. 1970.

7. T. Cohen, "Source-Depth Determination Using Spectral, Pseudo-Autocorrelation and Cepstral Analysis," Geophys. J. Roy. Astron. Soc., Vol. 20, pp. 223-231, 1970.

8. T. Ulrych, "Application of Homomorphic Deconvolution to Seismology," Geophysics, Vol. 36, No. 4, pp. 650-660, Aug. 1971.

9. S. Senmoto and D. G. Childers, "Decomposition of a Composite Signal of Unknown Wavelets in Noise," in Int. Conf. Comm. (ICC), 71C 28-COM, Montreal, Canada, pp. 5-14-5-19, 1971.

10. R. Kemerait and D. G. Childers, "Signal Detection and Extraction by Cepstrum Techniques," IEEE Trans. on Inform. Theory, Vol. IT-18, pp. 745-759, Nov 1972.

11. T. Ulrych, O. G. Jensen, R. M. Ellis, and P. G. Sommervile, "Homomorphic Deconvolution of Some Teleseismic Events," Bull. Seismological Soc. Amer., Vol. 62, No. 5, pp. 1253-1265, Mar. 1972.

12. P. Buhl, P. L. Stoffa, and G. M. Bryan, "The Application of Homomorphic Deconvolution to Shallow Water Marine Seismology - Part II; Real Data," Geophyiscs, Vol. 39, No. 4, pp. 417-426, Aug. 1974.

13. P. L. Stoffa, P. Buhl, and G. M. Bryan, "The Application of Homomorphic Deconvolution to Shallow Water Marine Seismology - Part I: Models," Geophysics, Vol. 39, No. 4, pp. 401-416, Aug. 1974.

14. J. F. Bohme, "The Cepstrum as a Generalized Function," IEEE Trans. on Inform. Theory, Vol. IT-20, pp. 650-653, Sept. 1974.

15. J. C. Hassab, "Time Delay Processing Near the Ocean Surface," Journal of Sound Vibration, Vol. 35, (4), pp. 489-501, 1974.

16. J. C. Hassab, "On the Convergence Interval of the Power Cepstrum," IEEE Transaction on Information Theory, Vol. IT-20, No. 1, pp. 111-112, 1974.

17. P. O. Fjell, "Decomposition of Signal Arrival Times Due to Multipath Conditions in Shallow Waters, Norwegian Defense Research Establishment Report N-U-319, 1975.

18. R. Rom, "On the Cepstrum of Two-Dimensional Functions," IEEE Trans. on Inform. Theory, pp. 214-217, Mar. 1975.

19. T. G. Stockham, Jr., T. M. Cannon, R. B. Ingebretsen, "Blind Deconvolution Through Digital Signal Processing," Proc. IEEE, vol. 63, pp. 678-692, Apr. 1975.

20. H. Schwazlander, B. Silverman, R. Hohensee, "Use of the Cepstrum for Processing Multipath Signals - II," Rome Air Development Center TR-75-117, May 1975.

21. R. G. Smith, "Cepstrum Discrimination Functions," IEEE Trans. Inform. Theory, Vol. IT-21, pp. 332-334, May 1975.

22. J. C. Hassab and R. E. Boucher, "Analysis of Signal Extraction, Echo Detection and Removal by Complex Cepstrum in Presence of Distortion and Noise," Journal of Sound and Vibration, Vol. 40, No. 3, pp. 321-335, 1975.

23. J. C. Hassab, "Network Function Theory and Complex Cepstrum," Journal of Sound and Vibration, Vol. 41, No. 1, 1975.

24. D. P. Skinner and D. G. Childers, "Real-time Composite Signal Decomposition," IEEE Trans. on Acoust., Speech, and Signal Processing, Vol. ASSP-24, pp. 267-270, June 1976.

25. J. C. Hassab, "The Probability Density Functions at the Output of the Complex Cepstrum," Proceedings IEEE International Symposium on Information Theory, Ronneby, Sweeden, June 1976.

26. J. C. Hassab, R. E. Boucher, "A Probalistic Analysis of Time Delay Extraction in Stationary Gaussian Noise," IEEE Transactions on Information Theory, Vol. IT22, pp. 444-454, Jul. 1976.

27. J. K. Hammond and L. G. Peardon, "The Power Cepstrum Applied to Multi-Peaked Wavelets," J. Sound and Vibration, Vol. 48, pp. 537-541, 1976.

28. J. C. Hassab, "Further Statistical Measures in the Cepstrum," IEEE Transactions on Information Theory, Vol. IT23, No. 4, pp. 540-544, July 1977.

29. L. B. Poche, Jr., "Complex Cepstrum Processing of Digitized Transient Calibration Data for Removal of Echos," Naval Research Lab Report 8143, Sept. 1977.

30. D. G. Childers, D. P. Skinner, and R. C. Kemerait, "The Cepstrum; a Guide to Processing," Proc. IEEE. Vol. 65, No. 10. pp. 1428-1443, Oct. 1977.

31. M. H. Loew, R. Shankar, A. Mucciardi, "Experiments with Echo Detection in the Presence of Noise Using the Power Cepstrum and a Modification," IEEE Int'l Conf. on Acoust., Speech, and Sig. Processing, pp. 307-310, 1977.

32. J. M. Tribolet, "Seismic Applications of Homomorphic Signal Processing," Ph.D. Dissertation, Massachusetts Institute of Technology, Cambridge, MA, 1977.

33. J. C. Hassab, "The Smoothing of 'Zero' Singularities in the Cepstrum," Journal of Sound and Vibration, Vol. 57, No. 2, pp. 299-301, 1978.

34. J. C. Hassab and R. E. Boucher, "Improved Cepstrum Performance Through Windowing of Log Spectrum," Journal of Sound and Vibration, Vol. 58, No. 4, pp. 597-598, 1978.

35. J. C. Hassab and R. E. Boucher, "Improved Time Delay Estimation Given a Composite Signal in Noise," Proceedings IEEE Int'l Conference on Communications, pp. 16.6.1-16.6.7, Toronto, Canada, 1978.

36. J. C. Hassab, "Homomorphic Deconvolution in Reverberant and Distortional Channels: An Analysis," Journal of Sound and Vibration, Vol. 58 (2), pp. 215-231, 1978.

37. J. C. Hassab and R. E. Boucher, "Further Comments on Windowing in the Power Cepstrum," Proceedings of the IEEE, Vol. 66, No. 10, pp. 1290-1291, October 1978.

38. J. C. Hassab and R. E. Boucher, "The Effect of Dispersive and Non-Dispersive Channels on Time Delay Estimation," Journal of Sound Vibration, Vol. 66 (2), pp. 247-253, 1979.

II. Single Path With Multiple Sensing Points Signal Processing

1. V. H. MacDonald and P. M. Schultheiss, "Optimum Passive Bearing Estimation," J. Acoust. Soc. Am., 46, 37-43, 1969.

2. H. J. Hannan and P. J. Thomson, "The Estimation of Coherence and Group Delay," Biometrika, 58, 469-481, 1971.

3. Melvin J. Hinich and Paul Shaman, "Parameter Estimation for an R-Dimensional Plane Wave Observed with Additive Independent Gaussian Errors," The Annals of Mathematical Statistics, Vol. 43, No. 1, 153-169, 1972.

4. W. R. Hahn and S. A. Tretter, "Optimum Processing for Delay-Vector Estimation in Passive Signal Arrays," IEEE Trans. Inf. Theory IT-19, 608-614, September 1973.

5. C. N. Pryor, "Minimum Detectable Signal for Spectrum Analyzer Systems," Signal Processing, Academic Press London and New York, 1973.

6. E. J. Hannan and P. J. Thomson, "Estimating Group Delay," Biometrika, Vol. 60, No. 2 pp. 241-253, 1973.

7. B. V. Hamon and E. J. Hannan, "Spectral Estimation of Time Delay for Dispersive and Non-Dispersive Systems," Appl. Statist., Vol. 23, No. 2, pp. 134-14

8. C. H. Knapp and G. C. Carter, "The Generalized Correlation Method for Estimation of Time Delay," IEEE Trans. Acoust., Speech, Signal Process., ASSP-24, pp. 320-327, Aug. 1976

9. C. H. Knapp and G. C. Carter, "Estimation of Time Delay in the Presence of Source or Receiver Motion," J. Acoust. Soc. Am., 61, 1545-1549, 1977.

10. R. Kirlin, "Augmenting the Maximum Likelihood Delay Estimator to Give Maximum Likelihood Direction," IEEE, Vol. ASSP-26, No. 1, February 1978.

11. Y. T. Chan, R. V. Hattin and J. B. Plant, "The Least Squares Estimation of Time Delay and Its Use in Signal Detection", IEEE Trans Acoust., Speech Signal Processing, Vol. ASSP-26, No. 3, pp. 217-222, 1978.

12. J. T. Patzewitsch, M. D. Srinath, and C. I. Black, "Nearfield Performance of Passive Correlation Processing Sonars," J. Acoust. Soc. Am., Vol. 64, No. 5, pp. 1412-1423, 1978.

13. R. Kirlin, "Improvement of Delay Measurements from Sonar Arrays via Sequential State Estimation," Univ. of Wyoming, Laramie, WY, May 1979.

14. J. C. Hassab and R. E. Boucher, "Optimum Estimation of Time Delay by a Generalized Correlator," IEEE Trans. on Acoustics, Speech, and Signal Processing, Vol. 27, No. 4, pp. 373-380, August 1979.

15. J. C. Hassab and R. E. Boucher, "A Quantitative Study of Optimum and Sub-Optimum Filters in the Generalized Correlator," ICASS P-79 Conference Record, Catalog No. 79 CH 1379-7 ASSP, IEEE Press, Piscataway, NJ. pp. 124-127, 1979.

16. W. B. Adams, J. P. Kuhn, and W. P. Whyland, "Correlator Compensation Requirements for Passive Time Delay Estimation with Moving Source or Receivers," IEEE Trans, Acoust,. Speech, Signal Processing, Vol. ASSP-28, No. 2, pp. 158-168, 1980.

17. J. C. Hassab, R. E. Boucher, "An Experimental Comparison of Optimum and Sub-Optimum Filters Effectiveness in the Generalized Correlator," Journal of Sound and Vibration, Vol. 76, pp. 117-128, May 1981.

18. G. L. Sackman and S. C. Shelef, "The Use of Phase Difference Trace Functions for Bearing Estimation with Small Circular Arrays," IEEE Transactions on Acoust., Speech, and Signal Processing, Vol. ASSP-29, No. 3, pp. 501-507, June 1981.

19. R. L. Kirlin, D. F. Moore and R. F. Kubicheck, "Improvement of Delay Measurements from Sonar Arrays Via Sequential State Estimation," IEEE Transactions on Acoust., Speech and Signal Processing, Vol. ASSP-29, No. 3, pp. 514-519, June 1981.

20. D. M. Etter and S. D. Stearns, "Adaptive Estimation of Time Delays in Sampled Data Systems," IEEE Transactions on Acoust., Speech and Signal Processing, Vol. ASSP-29, No. 3, pp. 582-587, June 1981.

21. R. E. Boucher, J. C. Hassab, "Analysis of Discrete Implementation of Generalized Cross Correlator," IEEE Trans. Acoustics, Speech, and Signal Processing, Vol. 29, No. 3, pp. 609-611, June 1981.

22. J. C. Hassab, R. E. Boucher, "Performance of the Generalized Cross Correlator in the Presence of a Strong Spectral Peak in the Signal," IEEE Trans. Acoustics, Speech, and Signal Processing, Vol. 29, No. 3, pp. 549-555, June 1981.

23. J. C. Hassab, "Signal Processing in Difficult Channels", in Int'l Symposium on Underwater Acoustics, Tel Aviv, Israel, June 15-18, 1981.

24. J. P. Ianniello, "Time Delay Estimation via Cross-Correlation in the Presence of Large Estimation Errors," Oceans 81 Conference Record, Vol. 2, pp. 998-1001, Sept. 1981.

25. J. C. Hassab, "Generalized Window Synthesis for Optimum Time Delay Detection," J. Acoustical Society of America, Supp. 1, Vol. 70, pp. 516, Dec 1981.

III. Estimation of Source Location and Velocity

1. D. J. Murphy, "Noisy Bearings-Only Target Motion Analysis," Ph.D. thesis, Department of Electrical Engineering, Northeastern Univ., 1970.

2. Per Heimdal and Finn Bryn, Passive Ranging Techniques, Signal Processing, edited by J. W. Griffiths, P. L. Stocklin, and C. Van Schoonveld Academic Press London and New York, 1973.

3. Norman L. Owsley, A Recent Trend in Adaptive Spatial Processing for Sensor Arrays: Constrained Adaptation, Signal Processing, Academic Press London and New York, 1973.

4. Loren W. Nolte, Adaptive Processing: Time varying Parameters, Signal Processing, edited by J. W. Griffiths, P. L. Stocklin, and C. Van Schoonveld, Academic Press London and New York, 1973.

5. W. J. Bangs and P. M. Schultheiss, "Space-Time Processing for Optimal Parameter Estimation," in Signal Processing, edited by J. W. Griffiths, P. L. Stocklin, and C. Van Schoonveld (Academic, New York), 1973.

6. J. C. Hassab and D. Watson, "Positioning and Navigation Under the Sea: An Acoustic Method," Proceedings IEEE International Conference on Engineering in Ocean Environment, pp. 145-149, August 1974.

7. B. Widrow et al., "Adaptive Noise Cancelling: Principles and Applications,: Proc. IEEE, Vol. 63. pp. 1692-1716, Dec. 1975.

8. W. R. Hahn, "Optimum Signal Processing for Passive Sonar Range and Bearing Estimation," J. Acoust. Soc. Amer., Vol. 58, No. 1, pp. 201-207, 1975.

9. J. C. Hassab, "Passive Tracking of a Moving Source by Single Observer in Shallow Water," Journal of Sound and Vibration, Vol. 44, No. 1, pp. 127-145, 1975.

10. N. H. Gholson and R. L. Moose, "Maneuvering Target Tracking Using Adaptive State Estimation," IEEE Trans. Aerosp. Electron. Syst., Vol. AES-13, May 1977.

11. G. C. Carter, "Variance Bounds for Passively Locating an Acoustic Source with a Symmetric Line Array," J. Acoust. Soc. AM 62, 922-926, (1977).

12. J. S. Davis and K. F. Gong, "Adaptive Filtering via Maximization of Residual Joint Density Functions," Proc. IEEE Conf. on Decision and Control, New Orleans, LA, December 1977.

13. J. Billingsley, A Comparison of the Source Location Techniques of the Acoustic Telescope and Polar Correlation, Journal of Sound and Vibration 61(3), 419-425, 1978.

14. J. M. F. Moura and A. B. Baggeroer, "Passive Systems Theory with Narrow-Band and Linear Constraints: Part I-Spatial Diversity," IEEE J. Oceanic Eng., Vol OE-3, p. 5, 1978.

15. P. M. Shultheiss and E. Weinstein, "Passive Localization of a Moving Source," Eascon 78 Conference Record, No. CH1352, pp. 258-266, IEEE, Piscataway, NJ, 1978.

16. R. Moose, H. F. Vanlandingham and D. H. McCabe, "Modeling and Estimation of Tracking Maneuvering Targets," IEEE Trans, Aerosp. Electron Syst., Vol 15, p.p. 448-456, 1979.

17. P. M. Schultheiss, "Locating a Passive Source with Array Measurements: A Summary of Results", ICASSP-79 Conference Record, Catalog No. 79CH1379-7ASSP, IEEE Press, Piscataway, NJ, pp. 967-970, 1979.

18. J. M. F. Moura, "Passive Systems Theory with Narrow-Band and Linear Constraints: Part II-Temporal Diversity," IEEE J. Oceanic Eng., Vol. OE-4, p. 19, 1979.

19. J. M. F. Moura, "Passive Systems Theory with Narrow-Band and Linear Constraints: Part III-Spatial/Temporal Diversity," IEEE J. Oceanic Eng., Vol. OE-4, p. 113, 1979.

20. J. C. Hassab and R. E. Boucher, "Passive Ranging Estimation from an Array of Sensors," Journal of Sound and Vibration, Vol. 67 (1), 1979.

21. E. J. Hilliard, Jr. and R. F. Pinkos, "An Analysis of Triangulation Ranging Using Beta Density Angular Errors," J. Acoust. Soc. Amer., Vol. 65, pp. 1218-1228, May 1979.

22. B. W. Guimond, "Joint Estimation and Adaptive Identification for Systems with Unknown Inputs," 13th Asilomar Conf. on Circuits, Systems, and Computers, Monterey, CA, November 1979.

23. G. C. Carter and P. B. Abraham, "Estimation of Source Motion From Time Delay and Time Compression Measurements," J. Acoust. Soc. Amer., Vol 67, No. 3, pp. 830-832, 1980.

24. J. C. Hassab, B. W. Guimond, and S. C. Nardone, "A Structure for the Combined Reduction of Bias and Variance in Estimating Source Location and Motion," Int. Conf. on Acoust. Speech and Signal Processing, Denver, CO, April 1980.

25. Peter M. Schultheiss, John P. Ianniello, "Optimum Range and Bearing Estimation with Randomly Perturbed Arrays," J. Acoust. Soc. AM 68(1), July 1980.

26. N. L. Owsley, Model Decomposition of Data Adaptive Spectral Estimates, NUSC Technical Document 6479, p.p. 27-29, May 1981.

27. P. M. Schultheiss and E. Weinstein, "Lower Bounds on the Localization Errors of a Moving Source Observed by a Passive Array," IEEE Transactions on Acoust., Speech, and Signal Processing, Vol. ASSP-29, No. 3, pp. 600-607, June 1981.

28. D. H. McCabe and R. L. Moose, "Passive Source Tracking Using Sonar Time Delay Data," IEEE Transactions on Acoust., Speech, and Signal Processing, Vol. ASSP-29, No. 3, pp. 614-617, June 1981.

29. J. C. Hassab, B. Guimond, S. Nardone, "Estimation of Location and Motion Parameters of a Moving Source Observed from a Linear Array," J. Acoustical Society of America, Vol. 70, No. 4, pp. 1054-1061, Oct 1981.

30. M. J. Hinich and M. C. Bloom, "Statistical Approach to Passive Target Tracking," J. Acoust. Soc. Am,. Vol. 69., p.p. 738-743, 1981.

31. J. C. Hassab, B. Guimond, S. Nardone, "Comments on Inherent Bias in Wavefront Curvature Ranging, IEEE Transactions on Acoustic Speech and Signal Processing, Vol. ASSP 30, No. 1, p. 99, Feb 1982.

IV. General References

1. E. J. Hannan, Mulitple Time Series, New York: Wiley 1970.

2. C. H. Chen, Statistical Pattern Recognition, Hayden Book Company, Inc, Rochelle Park, New Jersey, 1973.

3. J. W. R. Griffiths, P. L. Stocklin and C. Van Schooneweld, Editors, Signal Processing, Academic Press London and New York, 1973.

4. J. C. Hassab, Editor, "Proceedings of the Time Delay Estimation and Applications Conference," Vol. 1, DDC No. 9312068, Naval Postgraduate School, Monterey, CA July 1978.

5. J. C. Hassab, "Problems, Obstacles and Approaches in Naval Signal Analysis: A Sampling," presented to the U.S. National Academy of Sciences Panel on Applied Mathematics Research Alternatives for the Navy, Washington, DC, November 2, 1979.

6. D. G. Childers and A. E. Durling, Digital Filtering and Signal Processing, St. Paul, MN: West Publishing, 1975.

7. G. C. Carter, Editor, IEEE Special Issue on Time Delay Estimation, IEEE Trans. Acoustics, Speech, and Signal Processing, Vol 29, No. 3, June 1981.

8. C. H. Chen, Editor, Digital Waveform Processing and Recognition, CRC Press, Inc, Boca Raton, Florida, 1982.

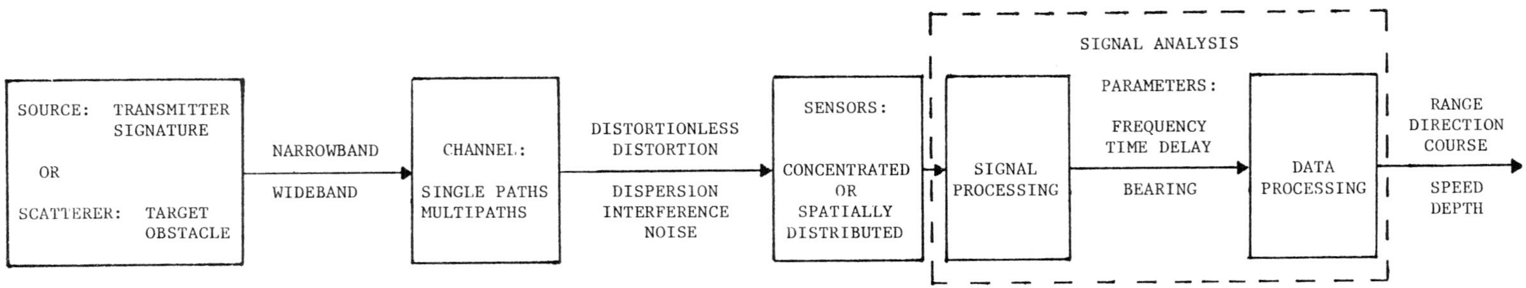

Fig. 1. Characterization of source localization and tracking problem

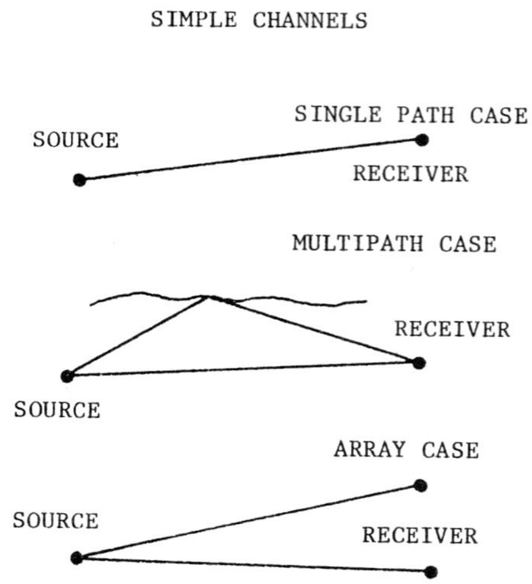

SIMPLE CHANNELS	PARAMETERS	PROCESSING STRUCTURES
SINGLE PATH CASE	FREQUENCY	FOURIER COMPLEX DEMOD. PLL, KF
MULTIPATH CASE	TIME DELAY	GENERALIZED AUTOCORRELATION CEPSTRUM COMPLEX DEMOD. LEAST SQUARES
ARRAY CASE	TIME DELAY	GENERALIZED CROSS CORRELATION LEAST SQUARES

COMMON CLUES IN SIGNAL ANALYSIS

- MAXIMUM—MINIMUM
- SLOPE

COMMON QUESTIONS:

- HOW TO ENHANCE THE CLUE IN PRESENCE OF CONTAMINANTS?
- HOW TO MAP A CLUE ONTO DESIRED FUNCTIONS?

Fig. 2. A sampling of channels, parameters, processors, issues.

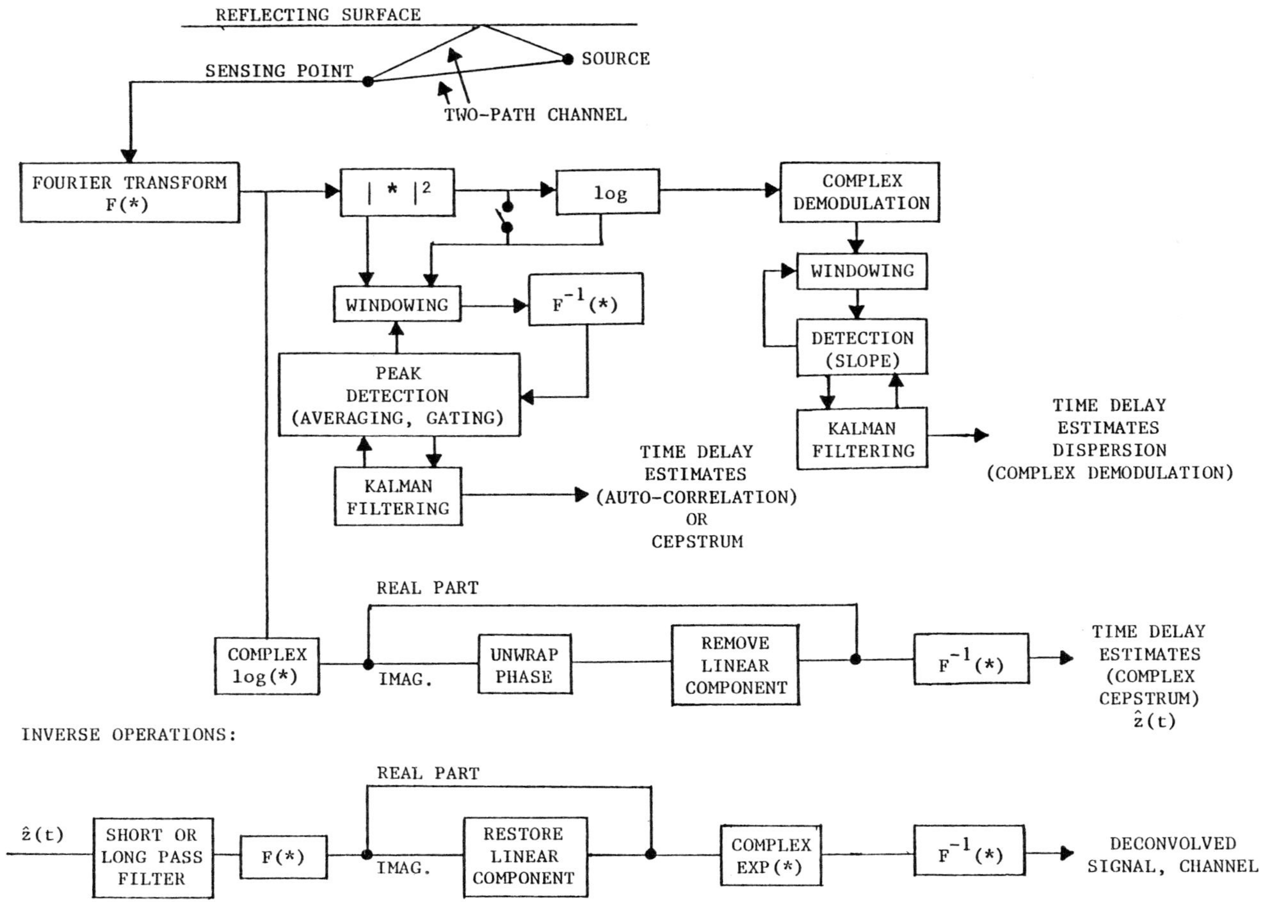

Fig. 3. Signal processing in a multipath channel.

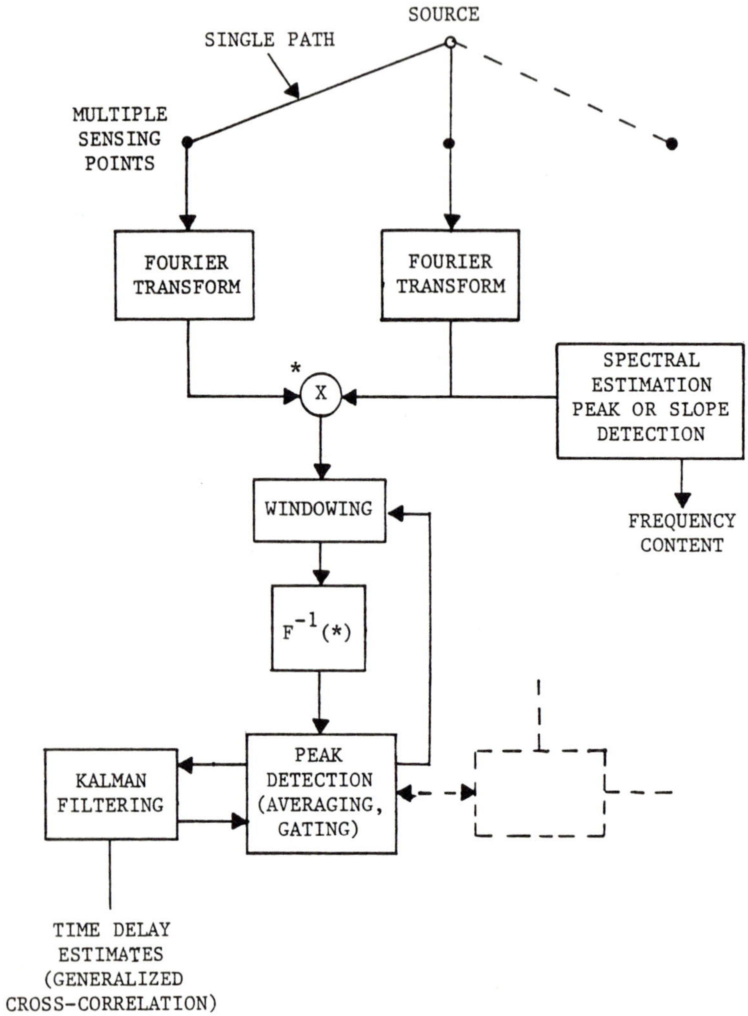

Fig. 4. Signal processing for separated sensors. Frequency tracking follows steps in time delay tracking. Application of complex demodulation on the Fourier transform outputs follows that of Fig. 3

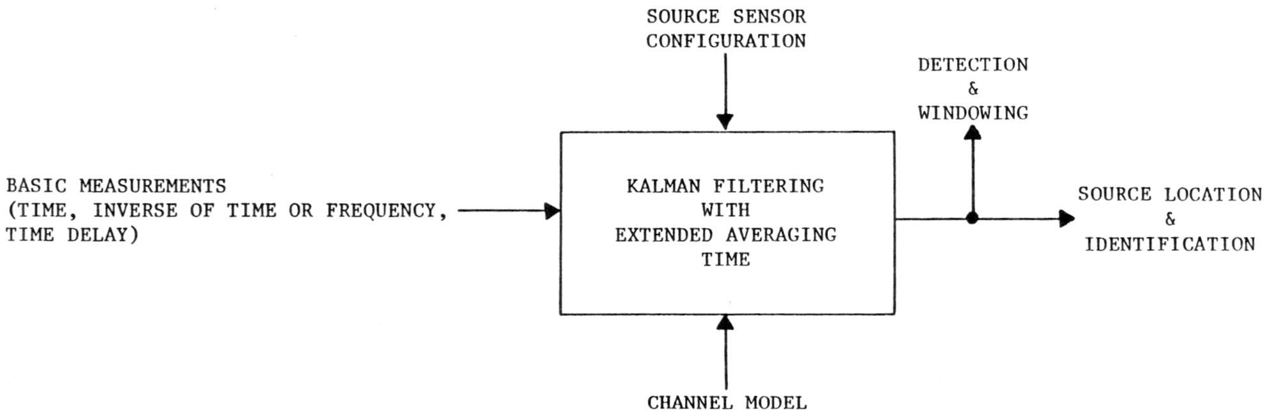

Fig. 5. Data processing for the three basic types of channels. A Kalman filtering implementation is given as an example

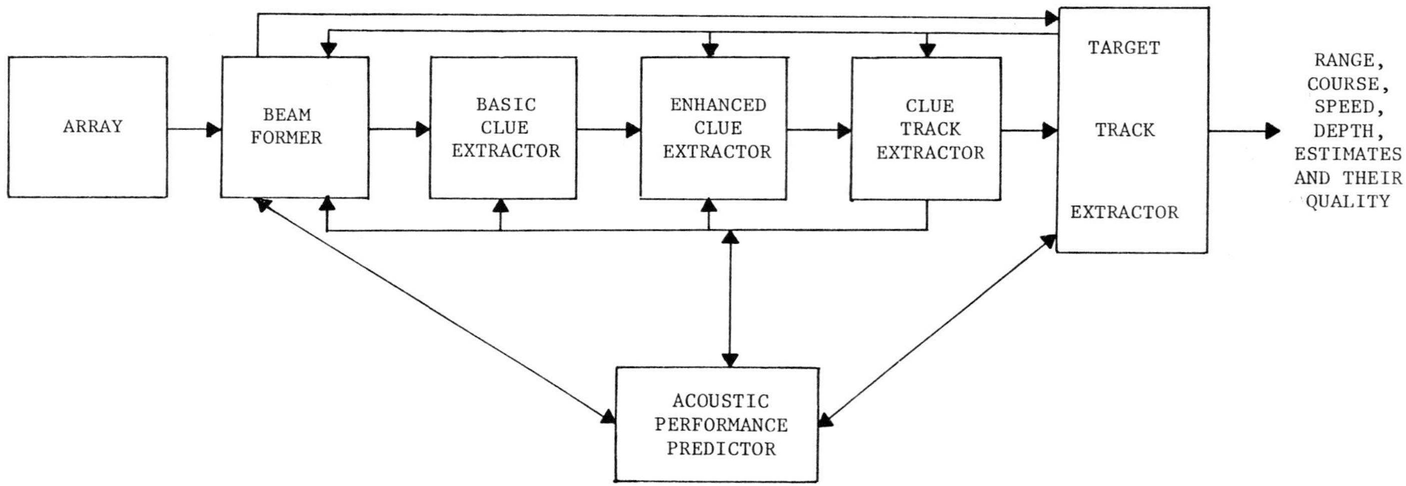

Fig. 6. An interactive system for signal, data and channel processing with loose couplings. Structure provides incremental approach to improved performance through synergistic interactions

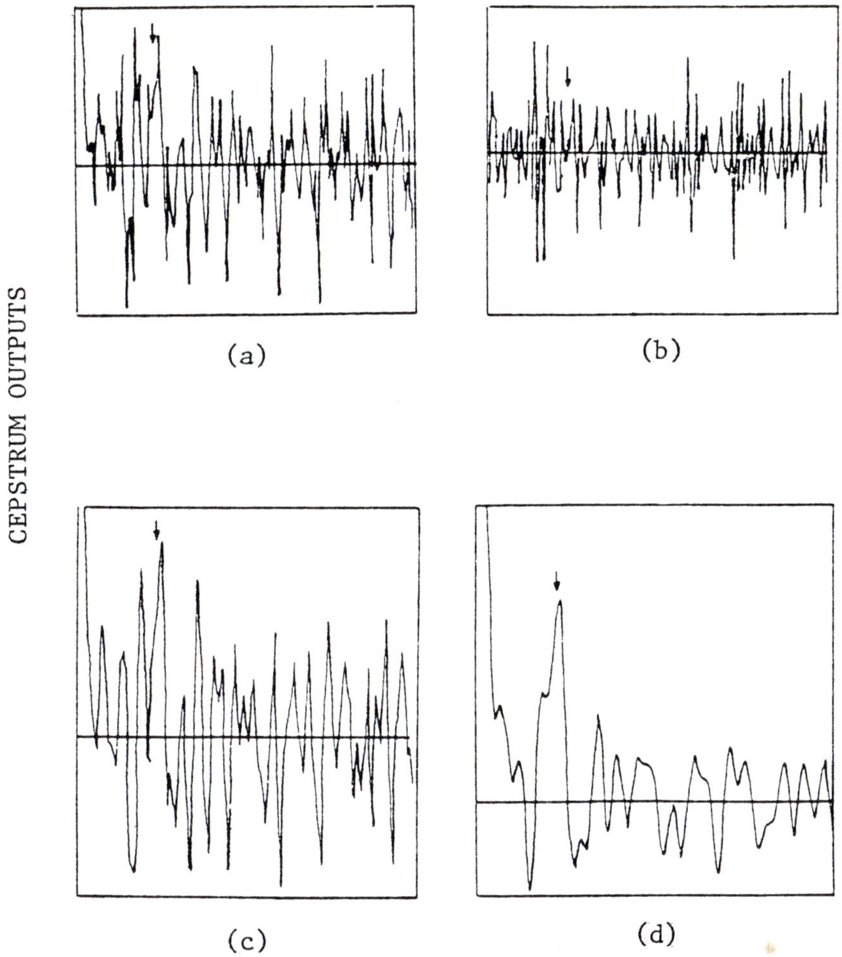

Fig. 7. Effect of properly designed window on processor performance. (a) Failure of unwindowed case. (b) Failure of inappropriate Hamming window. (c) Success of shifted Hamming window. (d) Success of Gaussian window denotes correct time delay (I-34; I-37)

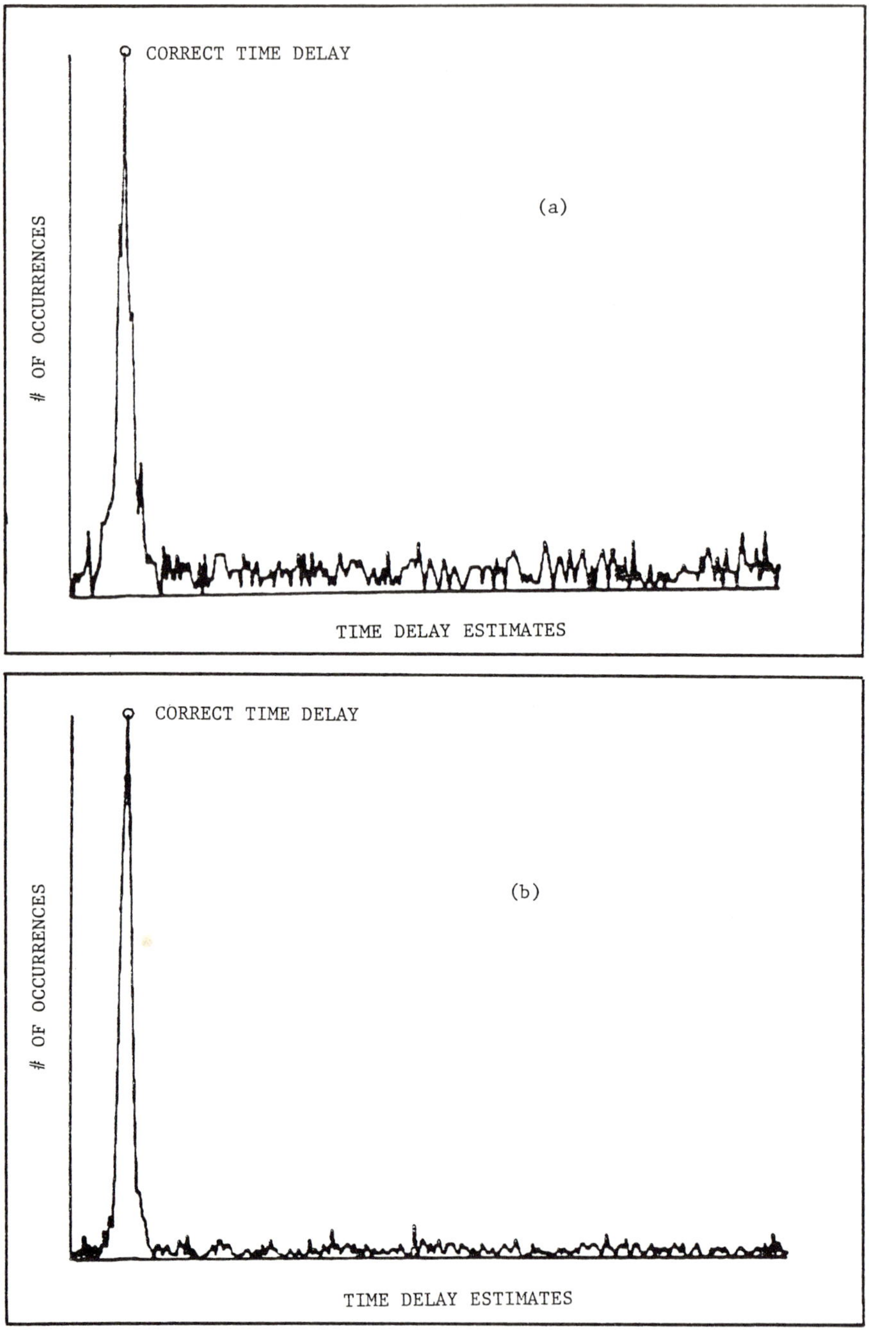

Fig. 8. Effect of windowing on improved time delay estimation. Histograms for white noise, SNR = -10 dB and signal spectrum $\alpha^2/(\alpha^2+\omega^2)^2$ α=0.333; (a) unwindowed; (b) W_I Maximum Likelihood window or energy detection window (II-14; IV-4)

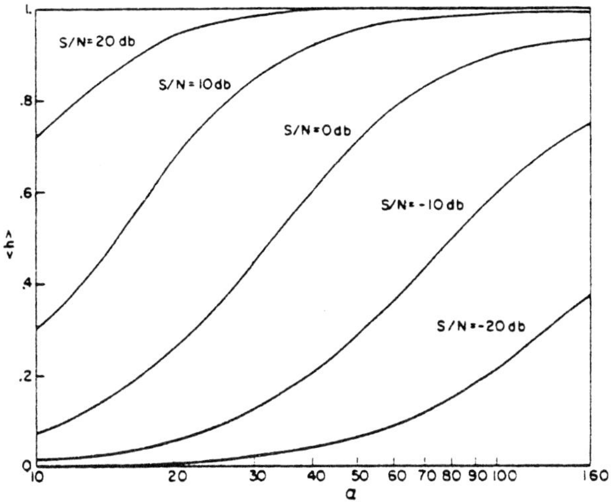

Fig. 9. Effect of bandwidth on performance of a nonlinear processor. Mean reduction <h> in peak denoting time delay for cepstrum processor as a function of signal bandwidth α and S/N; signal spectrum $\Phi_y(\omega)=Y_o\alpha^2/(\alpha^2+\omega^2)^2$ noise spectrum $\Phi_n(\omega)=N_o(40)^4/(40^2+\omega^2)^2$ (I-26)

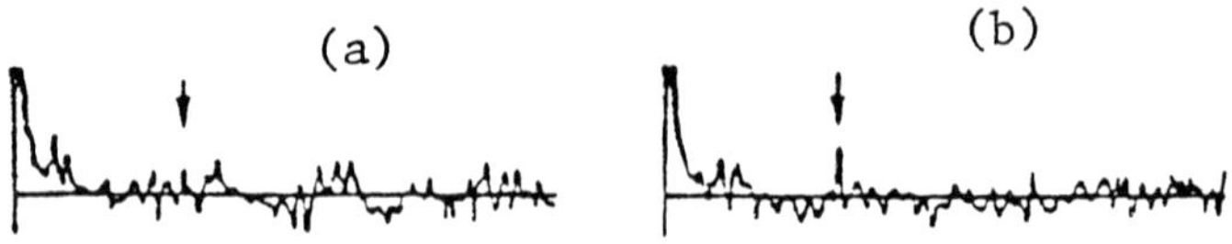

Fig. 10. Effect of Signal Bandwidth α, versus Noise Bandwidth β on Cepstrum, SNR = -2 dB (a) $\alpha = \beta$; (b) $\alpha = 2\beta$; (I-35)

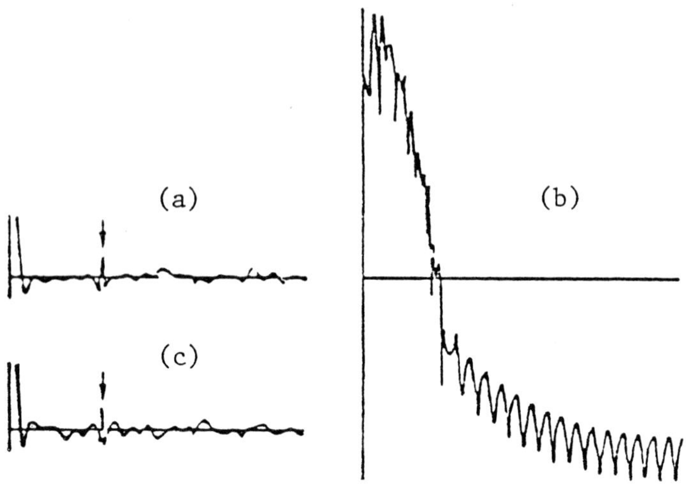

Fig. 11. Effect of spectrum Rolloff on Cepstrum. Noise Spectrum has Gaussian Shape. (a) SNR = -16 dB; (b) Log Spectrum of Case (α); (c) SNR = -30 dB; (I-35)

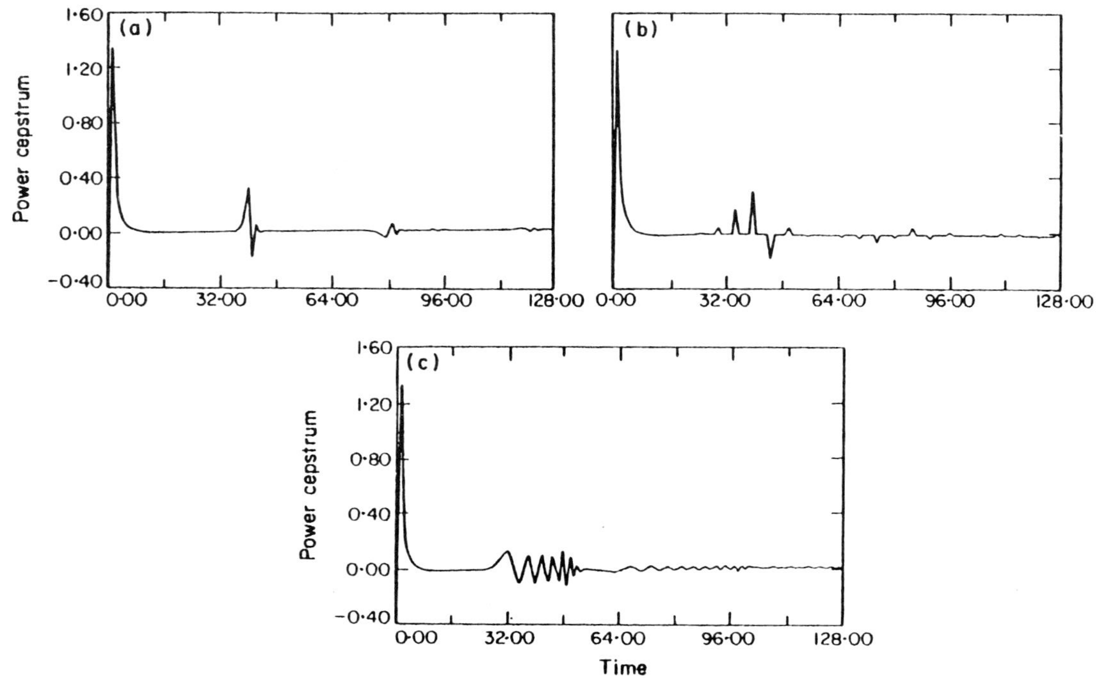

Fig. 12. Effect of channel dispersion on peak detectors gradual deterioration of power cepstrum with increase in phase distortion $[\beta_h = \omega\tau - b_\rho \sin \omega\tau]$, $a = 0.4$. (a) $\tau = 40$, $b_\rho = 1$, $t_\rho = 1$, then $\tau_e = 39.7$; (b) for $\tau = 40$, $b_\rho = 1$, $t_\rho = 5$, then $\tau_e = 39.99$; (c) for $\tau = 40$, $b_\rho = 10$, $\tau_\rho = 1$, failure of cepstrum to detect correct time delay (I-38)

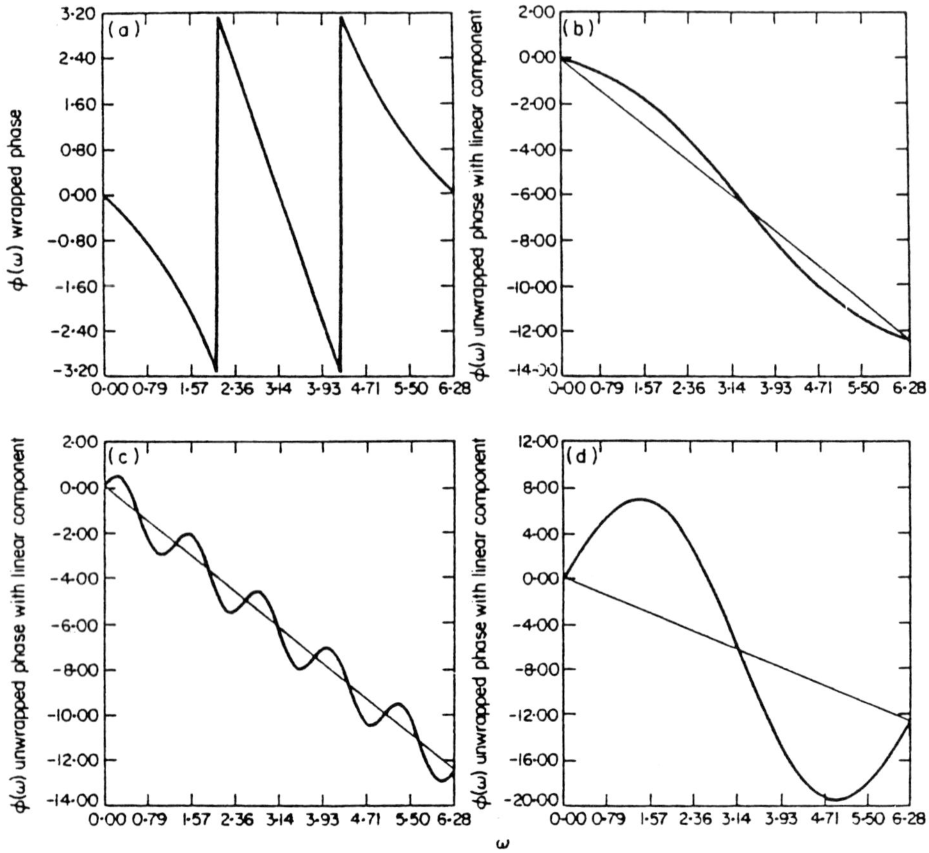

Fig. 13. Effect of dispersive channel on complex demodulation processor. Complex demodulation in the presence of phase distortion; $\beta_h = \omega\tau - b_\rho$, $\sin \omega t_\rho$, $\tau = 40$, $a = 0\cdot4$. (a) Biased time delay estimate for wrapped phase, slope = -1 and $\tau_e = 39$, $b_\rho = 1$, $t_\rho = 1$; (b) successful time delay estimate for unwrapped phase, slope = -2, $\tau_e = 40$, $b_\rho = 1$, $t_\rho = 1$; (c) success of unwrapped phase for case of Fig. 12(b), $b_\rho = 1$, $t_\rho = 5$; (d) success of unwrapped phase for case of Fig. 12(c), $b_\rho = 10$, $t_\rho = 1$ (I-38)

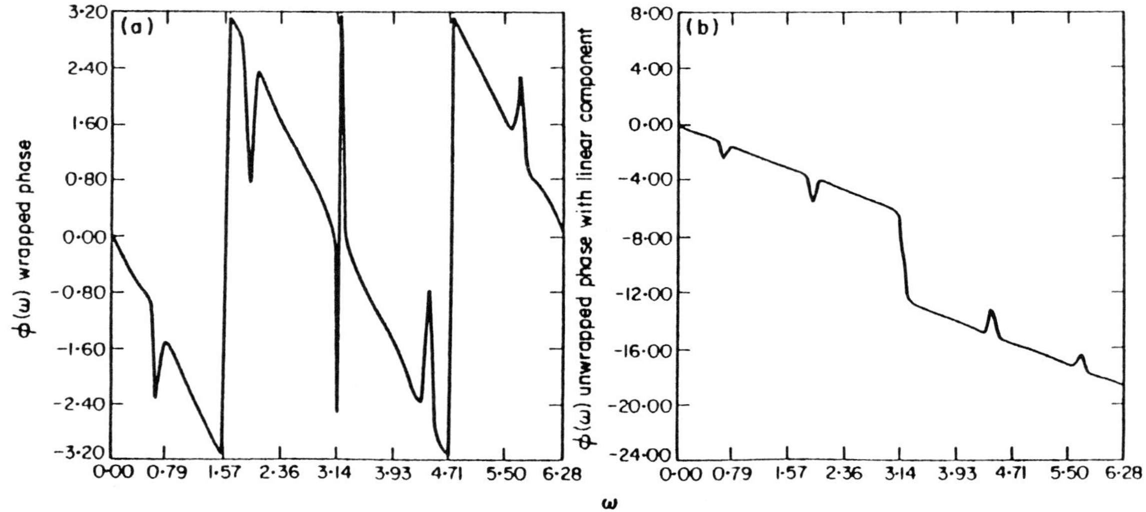

Fig. 14. Effect of singularities in complex demodulation. Complex demodulation and singularities; $\tau = 40$, $\hat{\tau} = 38$, $|H(\omega)| = 0 \cdot 41(1 + \cos 5\omega)$. (a) Wrapped phase output, slope = $-2 \cdot 11$, $\tau_e = 40 \cdot 11$; (b) biased result with unwrapped phase, slope = -3, $\tau_e = 41$ (I-38)

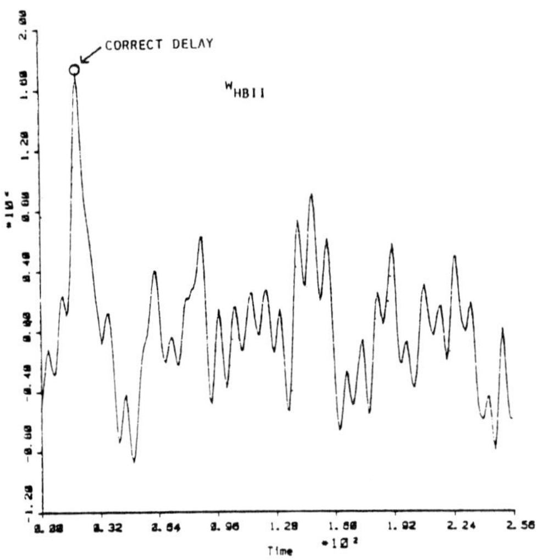

(a)

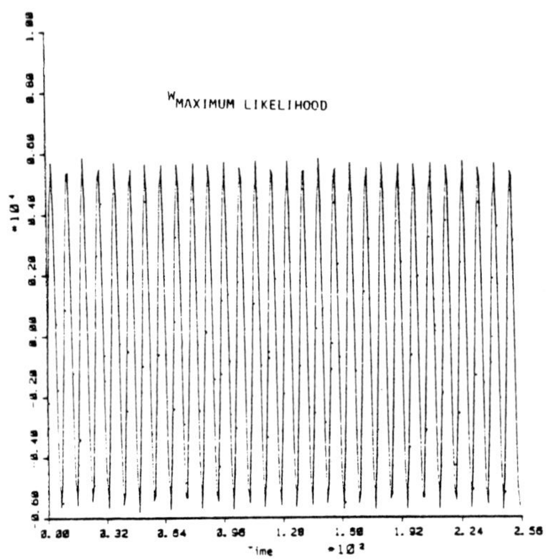

(b)

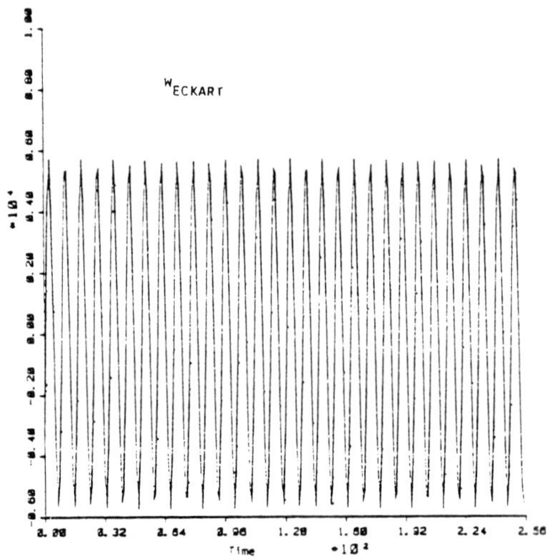

(c)

Fig. 15. Effect of a well optimized window to signals at hand. Generalized correlator output with an optimized window, W_{HBII}, to composite signals (wideband plus sinusoid); optimized windows, W_{ML} and W_E to wideband element only. Only W_{HBII} avoids the ambiguity problem (II-22; II-17)

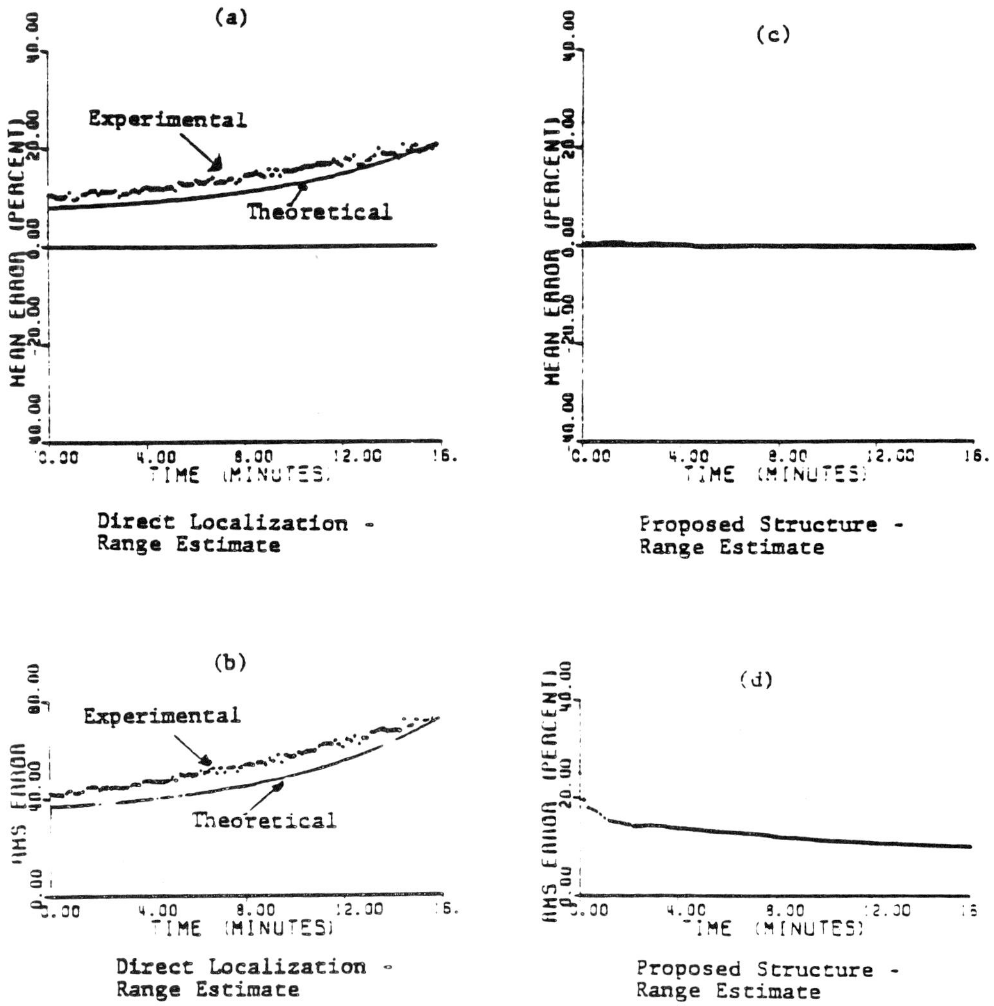

Fig. 16. Effect of snapshot alignments in data processor. (a), (b) performance of signal processor giving low quality of snapshots; (c), (d) improvement through appropriate data processing (III-29)

APPLICATION OF SIGNAL PROCESSING AND PATTERN RECOGNITION TO UNDERWATER ACOUSTICS

C. H. Chen
Electrical Engineering Department
Southeastern Massachusetts University
N. Dartmouth, Mass. 02747 U.S.A.

Abstract

This paper consists of four inter-related topics on the application of signal processing and pattern recognition to underwater acoustics. Part I is on pattern recognition processing for target motion analysis. The three-dimensional target motion analysis (TMA) in a multipath environment with bearing and multipath delay information has been examined and the extended Kalman filter was used to estimate the target parameters for localization and tracking of a moving underwater target. In this paper, pattern recognition is considered in the context of TMA. Two approaches are examined. The first approach uses the classification method and the pattern matching idea to identify the target location, without employing the conventional optimal filtering. The second approach uses the extended Kalman filtering but with initial parameters determined by the classification method. The simulation results clearly demonstrate a major improvement in the TMA capability via pattern-recognition processing.

Part II is on time dealy estimation from the generalized correlator. The performance improvement available from using the Hassab-Boucher window is closely examined for both known and unknown signals in white or colored noise. Part III is on the parameter estimation of a moving source observed from a linear array. Three simulation examples are presented that use time delays directly in the extended Kalman filter to obtain the localization and motion solutions. Part IV is on multipath ranging pattern recognition. The fundamental issues on this relatively unexplored area are examined. The results presented in this paper clearly illustrate that the study of underwater acoustics can benefit considerably from the rapid progress in signal processing and pattern recognition.

NATO ASI Series, Vol. F1
Issues in Acoustic Signal/Image Processing and Recognition
Edited by C. H. Chen

Part I

Pattern Recognition Processing for Target Motion Analysis

1.1 Introduction

In the passive sonar systems, one is principally concerned with estimating both the temporal and spatial structure of an observed signal field. The passive sonar bases its detection and estimation on sounds which emanate from the target itself, including machinery noise, flow noise and transmissions from its active sonar, etc. An excellent survey of the passive target tracking in the undersea environment is given in Ref. 1. In addition to the direct path, there are multipaths in the sound-wave propagation. Two typical propagation paths along with the target/observer geometry are shown in Fig. 1. With the use of the signal processing algorithms, the travel time differences between various paths can be measured. The multipath delays can be used to augment bearing and other sensor information to enable a three-dimensional TMA. The doppler shift measurement when available can also be processed to improve target localization

As the target signal is detected, a combat system requires two processors to obtain the desired information. First, the signal processor is used to extract all possible information such as bearing, time delay, frequency, etc. Secondly, the data processor is used to extract from the signal processor output all information relating to target motion, target characteristics, etc. The desired target information includes range, bearing, speed, course and depth.

In this paper, the classification method of pattern recognition is utilized in data processing. Two approaches of pattern recognition processing to the three-dimensional TMA are presented. The first approach generates a reference map of the pitch plane by partitioning it into a number of small cells. The feature set consists of multipath delays and the bearing measurement. Each input feature set is identified to the most likely cell with the smallest Euclidean distance. The second approach uses the extended Kalman filter and the initial parameter values determined from the recognition procedure. Both approaches are shown to provide a major improvement over the existing methods. Furthermore, it is demonstrated that the approaches are fairly insensitive to noises in the measurements. The first approach is particularly simple and effective. Any additional information can be easily integrated into the feature measurement in the pattern recognition processing.

The idea of generating a reference map was suggested in Ref. 2. Related pattern recognition study for the passive ranging is not available in open literature [3][4]. The simulation results presented in this paper suggests that the pattern recognition processing is very useful in TMA. As the interest in developing a fully automatic passive ranging system increases, the use of pattern recognition is expected to become essential to the implementation of such a system.

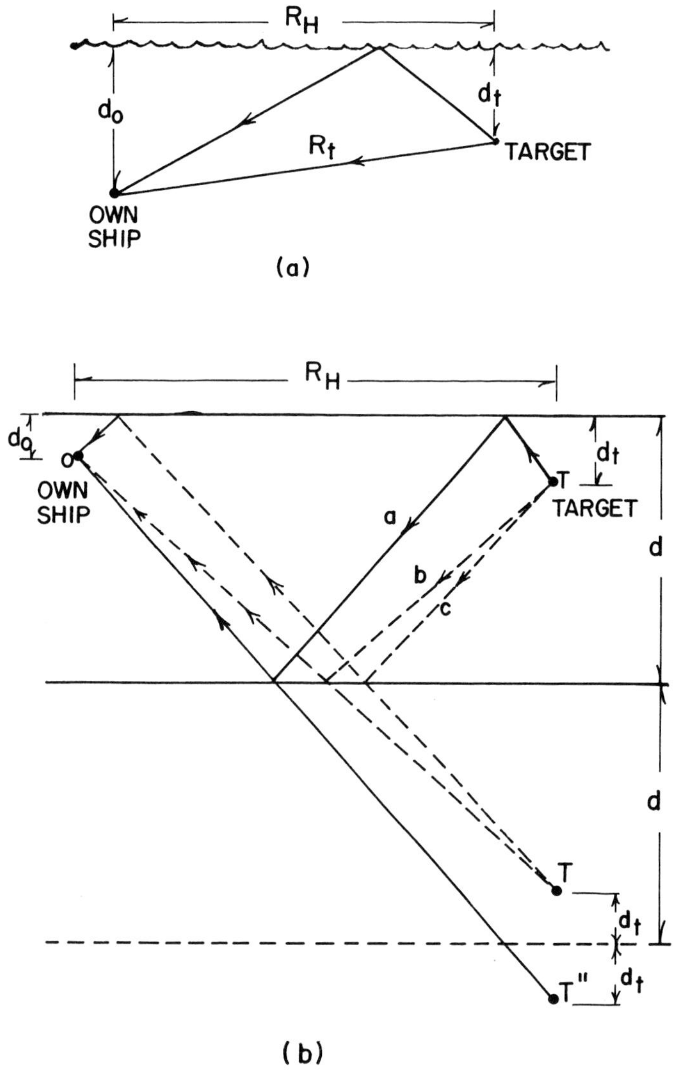

Figure 1. Typical multipath propagation paths

1.2 Computational Procedures

Sonar detection and classification are closely related to the general problem of pattern recognition [5] as illustrated in Fig. 2. The function of the structure shown here is to associate the observed multipath data vector with a specific pattern class. The pattern classes are generated by dividing the pitch plane into a number of square cells as shown in Fig. 3a. Four multipath delays are computed as features for each cell. The corresponding R_H and d_t are computed by the measurement equations which are described as follows.

Figures 3b and c depict the own ship/target geometry in both the azimuth and pitch planes. From Fig. 3b, the true bearing $\beta(t)$ is related to the state vector by nonlinear expression.

$$\beta(t) = \text{arc tan } [R_x(t)/R_y(t)] \tag{1}$$

For the pitch plane geometry shown in Fig. 3c, d_o is the depth of the receiver (own ship), d_t is the depth of the target, R_H is the azimuth range, and R_t is the slant range from the target to the receiver. Sound energy travels directly from the assumed point source to the receiver and mathematically from the image of the source as a consequence of reflection. In the absence of refraction it can be shown (see Ref. 6) that the differences in travel time between the direct and reflected paths of sea surface and bottom, expressed as τ_1 and τ_2 respectively are

$$\tau_1 = \frac{1}{c} \{- R_t + [R_t^2 + 4d_o d_t]^{1/2}\} \tag{2a}$$

$$\tau_2 = \frac{1}{c} \{- R_t + [R_t^2 + 4(d - d_o)(d - d_t)]^{1/2}\} \tag{2b}$$

For the case of the deep ocean multipath propagation in which three paths are available (Fig. 1), the multipath delays are

$$\tau_{bc} = \frac{1}{c} \{- R_p + [R_p^2 + 4d_o(2d - d_t)]^{1/2}\} \tag{2c}$$

$$\tau_{ab} = \frac{1}{c} \{- R_p + [R_p^2 + 4d_t(2d - 2d_o)]^{1/2}\} \tag{2d}$$

where $R_H^2 = R_x^2 + R_y^2$

$$R_t^2 = R_x^2 + R_y^2 + R_z^2 = R_H^2 + R_z^2$$

$$R_p^2 = R_x^2 + R_y^2 + [2(d - d_o) - R_z]^2$$

where c is the speed of underwater sound and positive z is taken to be down. If range is much larger than the ocean depth, Eq. (2) may be approximated as

$$\tau_1 \approx \frac{2d_o d_t}{R_t C} \tag{3a}$$

$$\tau_2 \approx \frac{2(d - d_o)d - d_t)}{R_t C} \tag{3b}$$

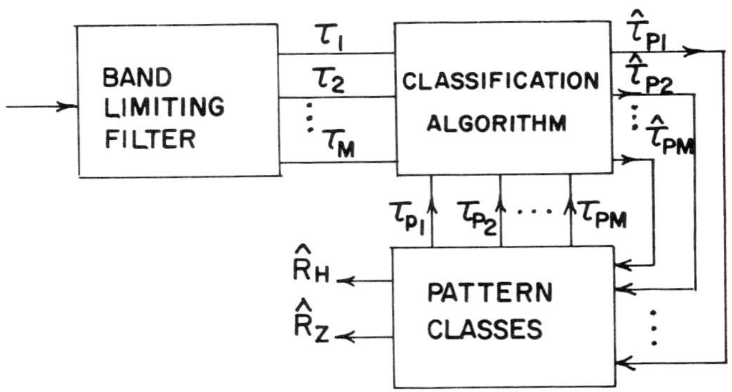

CLASSIFICATION ALGORITHM:

$$D_P = \frac{1}{M} \sum_{i=1}^{M} | \tau_{Pi} - \tau_i |$$

$$\hat{\tau}_{Pi} = \min_{P \in N} D_P = \{ \tau_{P1}, \tau_{P2}, \cdots, \tau_{PM} \}$$

WHERE $P = 1, 2, \ldots, N$

$i = 1, 2, \ldots, M$

HERE $N = 25 \times 100$

$M = 4$

Figure 2. Structure of pattern recognition processing

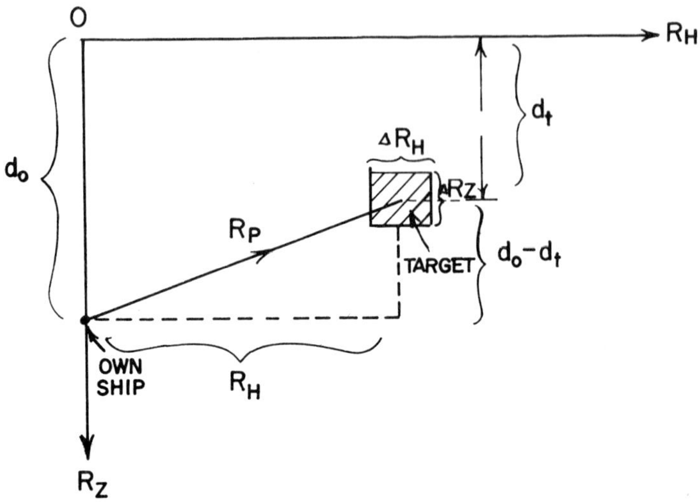

Figure 3(a). Configuration of reference map

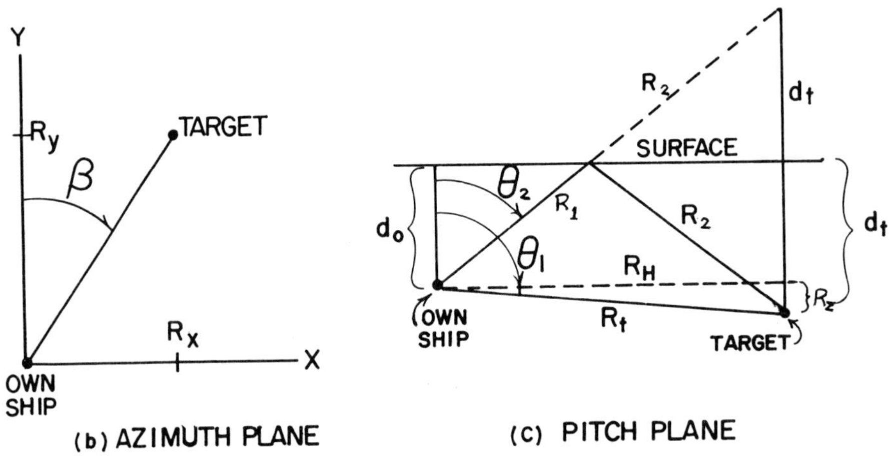

Figures 3(b) & 3(c). Geometric representation of (b) azimuth and (c) pitch planes of own ship/target encounter

$$\tau_{bc} \approx \frac{4d_o d}{R_p C} \tag{3c}$$

$$\tau_{ab} \approx \frac{4d_t d}{R_p C} \tag{3d}$$

Equation (3) is used for the multipath-delays computation. The cell identified by the pattern recognition processing provides estimates, $\hat{R}_H$ and $\hat{R}_z$, from which we can calculate*

$$\hat{R}_x = \hat{R}_H \sin\beta(t) \tag{4a}$$

$$\hat{R}_y = \hat{R}_H \cos\beta(t) \tag{4b}$$

Thus the target location is completly determined.

For the second approach which utilizes the extended Kalman filter, the computational procedures follow exactly those described in Ref. 6. The dynamic system is described in general by a nonlinear stochastic vector differential state equation and a nonlinear measurement vector equation. Let's consider first the state equation with the assumption of the linear target motion. The relative parameters that describe the plant dynamics can be defined as follows:

Azimuth plane description

$$\dot{V}_x(t) = \dot{V}_{xt}(t) - \dot{V}_{xo}(t) = W_x(t) \tag{5a}$$

$$\dot{V}_y(t) = \dot{V}_{yt}(t) - \dot{V}_{yo}(t) = W_y(t) \tag{5b}$$

$$\dot{R}_x(t) = \dot{R}_{xt}(t) - \dot{R}_{xo}(t) = V_x(t) \tag{5c}$$

$$\dot{R}_y(t) = \dot{R}_{yt}(t) - \dot{R}_{yo}(t) = V_y(t) \tag{5d}$$

Pitch plane description

$$\dot{V}_z(t) = \dot{V}_{zt}(t) - \dot{V}_{zo}(t) = W_z(t) \tag{5f}$$

$$\dot{R}_z(t) = \dot{R}_{zt}(t) - \dot{R}_{zo}(t) = V_z(t) \tag{5g}$$

where $V_x(t)$ and $V_z(t)$ are the relative velocity components between the target and and own ship; $R_x(t)$, $R_y(t)$, and $R_z(t)$ are the relative range components; $W_x(t)$, $W_y(t)$ and $W_z(t)$ are zero-mean white noise components with variances σ_x^2, σ_y^2, σ_z^2

*In the classification method, bearing measurement is not filtered. Although R_H may be determined accurately, variances of R_x, R_y are roughly proportional to the range square multiplied by variance of bearing measurement and can be very significant even for small deviation of bearing. However the percentage change in range depends only on deviation of bearing. This explains why the range error is small in simulation.

respectively. These random perturbations are included to account for errors resulting from own ship's measurement of its own velocity and/or the occurrence of the target maneuvers. The state vector equation which is now linear can be written as,

$$\dot{X}(t) = A(t)X(t) + \Gamma(t)W(t) \tag{6}$$

where the state vector X(t) is given by

$$X^T(t) = [V_x(t)V_y(t)R_x(t)R_y(t)V_z(t)R_z(t)];$$

the plant matrix A(t) is

$$A(t) = \begin{bmatrix} 0 & 0 & 0 & 0 & 0 & 0 \\ 0 & 0 & 0 & 0 & 0 & 0 \\ 1 & 0 & 0 & 0 & 0 & 0 \\ 0 & 1 & 0 & 0 & 0 & 0 \\ 0 & 0 & 0 & 0 & 0 & 0 \\ 0 & 0 & 0 & 0 & 1 & 0 \end{bmatrix};$$

the plant noise vector is

$$W^T(t) = [W_x(t)W_y(t)W_z(t)]$$

and the plant noise coefficient matrix is given by

$$\Gamma(t) = \begin{bmatrix} 1 & 0 & 0 \\ 0 & 1 & 0 \\ 0 & 0 & 0 \\ 0 & 0 & 0 \\ 0 & 0 & 1 \\ 0 & 0 & 0 \end{bmatrix}$$

Here W(t) is a vector zero-mean Gaussian white noise process with covariance Q(t) $\delta(t - t_i)$. The measurement equation given by

$$Y(t) = h(x, t) + V(t) \tag{7}$$

is nonlinear where*

$$Y^T(t) = [\beta(t)\tau(t)]$$

$$V^T(t) = [V_\beta(t)V_\tau(t)]$$

$$h(x,t) = \begin{bmatrix} \arctan\,[R_x(t)/R_y(t)] \\ \frac{1}{c}\,\{-[R_x^2(t) + R_y^2(t) + R_z^2(t)]^{1/2} + [R_x^2(t) + R_y^2(t) \\ + R_z^2(t) + 4d_o(d_o + R_z(t))]^{1/2}\} \end{bmatrix}$$

*In the simulation experiment, all four delays are used in the classification method, while only the surface bounced delay is used, i.e., $\tau(t) = \tau_1(t)$ in the extended Kalman filtering method.

Here V(t) is a vector zero-mean Gaussian white noise process with covariance R(t) $\delta(t - t_i)$. To apply the extended Kalman filter, the measurement matrix is linearized such that

$$Y(t) \approx \hat{Y}(t) + M(\hat{X}, t)\,[X(t) - \hat{X}(t)] + V(t) \tag{8}$$

where

$$M(\hat{X}, t) = \frac{\partial}{\partial x}\,[h(X, t)]\Big|_{x=\hat{x}} \tag{9}$$

Here $\hat{X}$ is the nominal trajectory about which successive linearization is performed. The details of the calculation of the terms in M(X, t) are given in Appendix D of Ref. 6. Further details about hte extended Kalman filter are well documented in the literature (see e.g., [7]). Only the outline of the computational procedures used in the discrete filter simulation is presented here.

System dynamic description

$$x(k + 1) = \Phi(k + 1, k)x(k) + G(k)W(k) \tag{10}$$

$$y(k + 1) = M(k + 1)x(k + 1) + V(k + 1) \tag{11}$$

where

$\Phi(k + 1, k)$ = the state transition matrix

$E[W(k)] = E[V(k)] = 0$

$E[W(k)W^T(\ell)] = Q(k)\delta_{K\ell}$

$E[V(k)V^T(\ell)] = R(k)\delta_{K\ell}$

G(k) = the plant noise covariance matrix

M(k) = the linearized measurement matrix

State prediction

$$\hat{X}(k + 1/k) = \Phi(k + 1, k)\hat{X}(k/k) \tag{12}$$

Covariance matrix prediction

$$P(k + 1/k) = \Phi(k + 1, k)P(k/k)\Phi^T(k + 1, k) + G(k)Q(k + 1)G^T(k) \tag{13}$$

Kalman gain calculation

$$K(k + 1) = P(k + 1/k)M^T(k + 1)\,[M(k + 1)P(k + 1/k)M^T(k + 1) + R(k + 1)]^{-1} \tag{14}$$

State estimate

$$\hat{X}(k + 1/k + 1) = \hat{X}(k + 1/k) + K(k + 1)\,[Y(k + 1) - \hat{Y}(k + 1)] \tag{15}$$

where $\hat{Y}(k + 1) = h[\hat{X}(k + 1/k); k + 1]$

Covariance estimate

$$P(k + 1/k + 1) = [I - K(k + 1)M(k + 1)]\,P(k + 1/k)\,[I - K(k + 1)M(k + 1)]^T + K(k + 1)R(k + 1)K^T(k + 1) \tag{16}$$

It is noted that the filter requires the computation of the inverse of the residual covariance matrix

$$M(k+1)P(k+1/k)M^T(k+1) + R(k+1)$$

As the system is often faced with the task of processing multiple input data such as bearing, frequency, multipath time dealy, D/E angle, and others, the computational burden may be substantial. If it is assumed that the various sensor data are uncorrelated, the simultaneous measurement data can be processed sequentially within the sampling interval as the scalar quantities. The computational requirement can then be significantly reduced.

Now consider the plant noise covariance matrix Q(k). It is assumed that the existence of Q(k) is due to the target acceleration perturbations in a constant velocity model; that is the target is assumed to travel in a constant trajectory whose states V_{xt}, V_{yt} and V_{zt} are contaminated by the white Gaussian noise perturbations introduced by the random target accelerations. This noise is assumed to have zero mean and respective variances σ_x^2, σ_y^2 and σ_z^2. Recall that the state equation may be written as

$$X(k+1) = \Phi(k+1, k)X(k) + U(k)$$

where U(k) is the driving input with covariance matrix

$$L(k) = E[U(k)U^T)] = G(k)Q(k)G^T(k).$$

It can be shown that (see Appendix E, Ref. 6),

$$L(k) = \begin{bmatrix} \sigma_x^2 T & 0 & \sigma_x^2 T^2 & 0 & 0 & 0 \\ 0 & \sigma_y^2 T & 0 & 0 & 0 & 0 \\ \sigma_x^2 T^2/2 & 0 & \sigma_x^2 T^3/3 & 0 & 0 & 0 \\ 0 & \sigma_y^2 T^2/2 & 0 & \sigma_y^2 T^3/3 & 0 & 0 \\ 0 & 0 & 0 & 0 & \sigma_z^2 T & \sigma_z^2 T^2/2 \\ 0 & 0 & 0 & 0 & \sigma_z^2 T^2/2 & \sigma_z^2 T^3/3 \end{bmatrix} \qquad (17)$$

where T is the sampling interval. Equations (12) to (17) constitute the discrete extended Kalman filter employed in the simulation.

1.3 Simulation Experiments and Results

The principal objectives of the simulation experiments are:

(1) To assess the efficiency of the pattern recognition processing.

(2) To examine the improvement in covergent rate of the extended Kalman filter algorithm by using the initial values of $\hat{R}_H(0/0)$ and $\hat{R}_z(0/0)$ determined from the classification method.

In the simulation, specific consideration was given to modeling the real world properties of the measured data. In each experiment, the initial data point always contains both bearing and time-delay information. The laboratory simulation used for all the experiments was run on the PDP 11/45 computer. Sampling rate is 1 sample every 20 seconds. Two geometries considered are shown in Figs. 4 and 5. For the first encounter the initial range is 2617 yards and the target is a straigt-running vehicle with a speed of 5 knots and a course of 169.7^{o}. Own ship runs three 240-second legs at a constant speed of 8 knots and respective courses of 140°, 186.8^{o} and 125.6^{o}. A 0.4^{o}/second turning rate is employed for all own ship maneuvers. Also own ship and the target operate at the constant depths of 294 and 618 feet respectively.

For the second encounter, the initial range is 5000 yards and the target is a constant velocity vehicle with a speed of 5 knots and a course of 0°. Own ship runs four 240-second legs at a constant speed of 5 knots and respective courses of 60°, 147°, 6°, and 126°. A 3°/second turning rate is employed for all own ship maneuvers. Own ship and the target are both operating at the same depth of 1000 feet.

In generating the reference maps for the classification method, the size of the cell is selected as 40 x 40 yards. There are 25 x 100 cells. To improve the classification accuracy, each cell is subdividied into 8 x 8 subcells of the size 5 x 5 yards each. The classification is performed in two steps. In the first step, a cell is identified; then in the second step, a subcell of the cell is identified. The use of subcells greatly improves the resulting classification accuracy.

The processing results and the errors of the azimuth range and velocity are illustrated in Figs. 6-15 and Tables 1-10 respectively. For Figs. 6, 7, 12, and 13, the standard deviations of the time-delay and bearing noises are $\sigma_{\tau} = 0.001$ sec., $\sigma_{\beta} = 0.5^{o}$. The corresponding azimuth range errors are listed in Tables 1, 3, 7 and 8. For Figs. 8, 9, 14 and 15, the standard deviations are $\sigma_{\tau} = 0.005$ sec., $\sigma_{\beta} = 0.5^{o}$ and Tables 3, 4, 9 and 10 are the corresponding azimuth range. For Figs. 10 and 11, the standard deviations are $\sigma_{\tau} = 0.005$, $\sigma_{\beta} = 2.5^{o}$ with corresponding azimuth range errors listed in Tables 5 and 6.

By comparing the errors of the azimuth range in both methods of approach, we note that the classification method which has more input information performs better and requires less computation. The range errors in both methods are well within the normal requirements in TMA. The initial values determined by the classification method and used in the extended Kalman filter have considerably improved the filter covergence. Thus the simulation results clearly demonstrate that the pattern recognition processing is very effective in TMA.

1.4 Concluding Remarks

Many problem areas remain to be examined. When a number of measurements are available, the effective feature selection method must be considered. The automatic

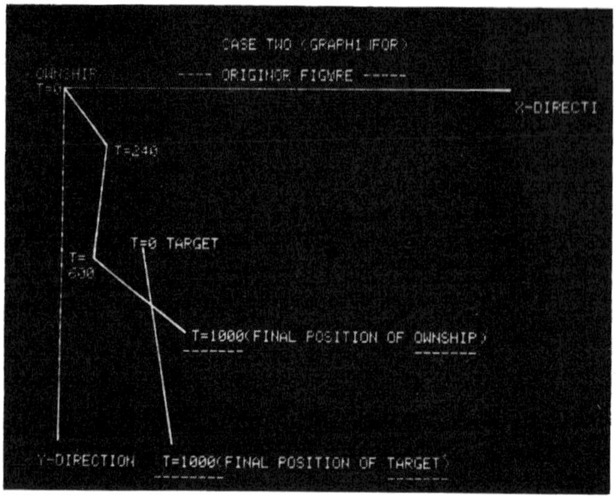

Figure 4. First geometry considered

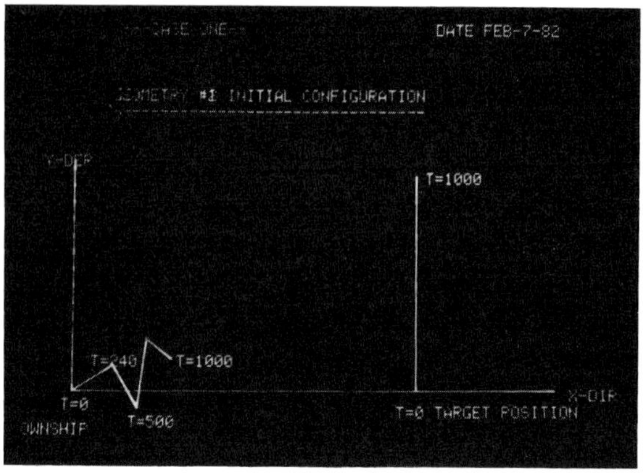

Figure 5. Second geometry considered

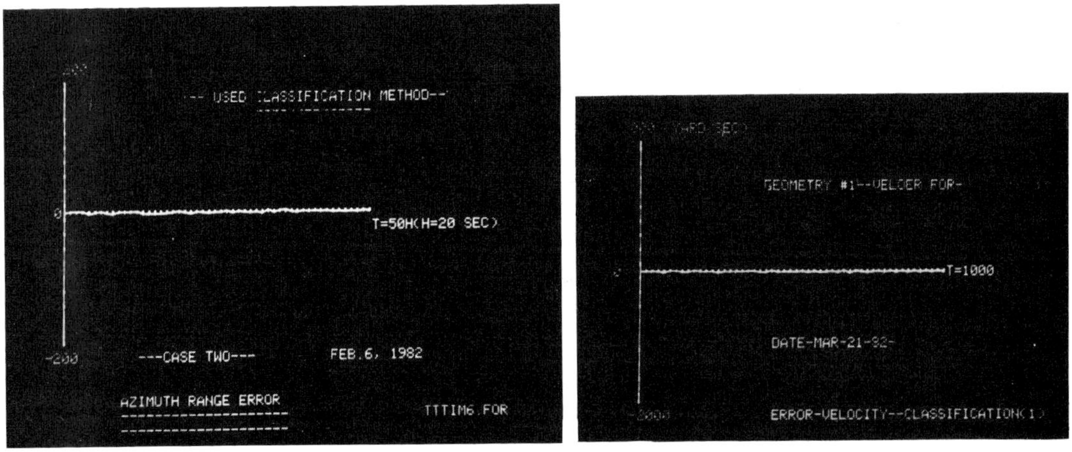

(a) Range error (b) Velocity error

Figure 6. Results of classification method with σ_τ = 1 m sec, σ_β = 0.5^o for first geometry

TABLE 1

Time index	ERROR of RH					ERROR OF VELOCITY				
1-5	2	2	2	-2	0	0	0	0	-.2	.1
6-10	2	-1	-2	1	1	.1	-.1	-.1	.2	.0
11-15	1	2	-2	-2	-2	.0	.0	-.2	.0	.0
16-20	-2	-1	0	2	0	.0	.0	.1	.1	-.1
21-25	-2	1	0	-1	0	-.1	.2	-.1	.0	.0
26-30	0	2	-1	1	0	.0	.1	-.1	.1	-.1
31-35	-1	0	-2	3	-1	.0	.1	-.1	.2	-.2
36-40	2	3	0	-1	1	.2	.0	-.1	.0	.1
41-45	0	1	0	1	-1	-.1	.1	-.1	.1	-.1
46-50	0	-1	0	-1	1	.1	-.1	.1	-.1	.1
51	0					-.1				

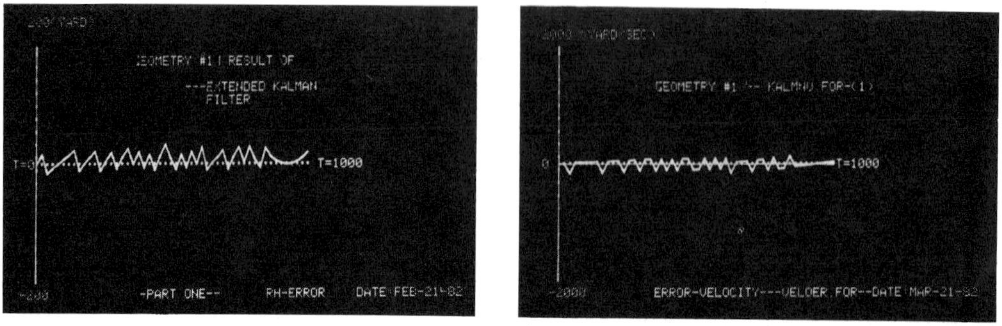

(a) Range error

(b) Velocity error

Figure 7. Results of EKF with σ_τ = 1 m sec, σ_β = 0.5^o for first geometry

Table 2

Time Index	ERROR of RH					ERROR of velocity				
1-5	17	-16	- 6	1	9	.0	-1.7	.3	.3	.3
6-10	16	24	-11	2	12	.3	.4	-1.6	.4	.5
11-15	22	-11	5	20	- 8	.5	-1.4	.6	.7	-1.2
16-20	12	27	0	21	- 6	.8	.8	-1.2	.8	-1.1
21-25	17	-10	15	34	9	+ .8	-1.0	.9	.9	-1.
26-30	- 9	17	- 6	22	0	- .9	1.0	- .9	1.	- .8
31-35	30	-10	3	12	24	1.	-1.6	.4	.4	.5
36-40	- 9	13	29	3	30	-1.3	.7	.8	-1.	1.1
41-45	8	- 6	28	13	6	- .8	- .7	1.4	- .4	- .3
46-50	2	0	1	5	11	- .2	- .1	.1	.2	.3
51	20					.4				

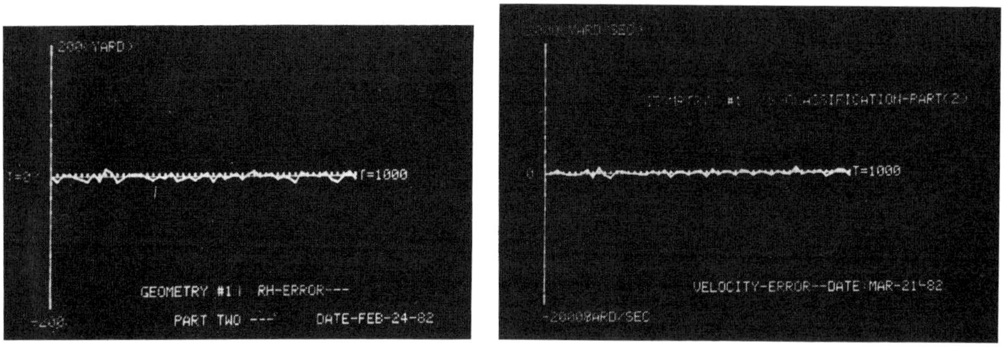

(a) Range error

(b) Velocity error

Figure 8. Results of classification method with σ_τ = 5 m sec, σ_β = 0.5^o for first geometry

Table 3

Time index	ERROR of RH					ERROR of velocity				
1-5	- 8	2	2	- 2	-5	0.1	.5	.0	-.2	-.2
6-10	- 8	4	- 7	11	6	- .1	.6	-.6	.9	-.3
11-15	- 9	-3	3	3	3	- .7	.3	.3	.0	.0
16-20	- 2	-6	0	- 3	5	- .2	-.2	.3	-.2	.4
21-25	- 7	-4	0	- 6	0	-0.6	.2	.2	-.3	.2
26-30	5	-8	4	- 4	0	.3	-.6	.5	-.3	.2
31-35	4	-5	3	8	-1	.2	-.4	.4	.2	-.4
36-40	2	3	- 5	- 1	-4	.2	.0	.4	.2	-.2
41-45	-10	6	0	1	-1	- .3	.8	-.3	.1	-.1
46-50	5	-6	-10	- 1	6	.3	-.6	-.2	.4	.3
51	- 1					-.5				

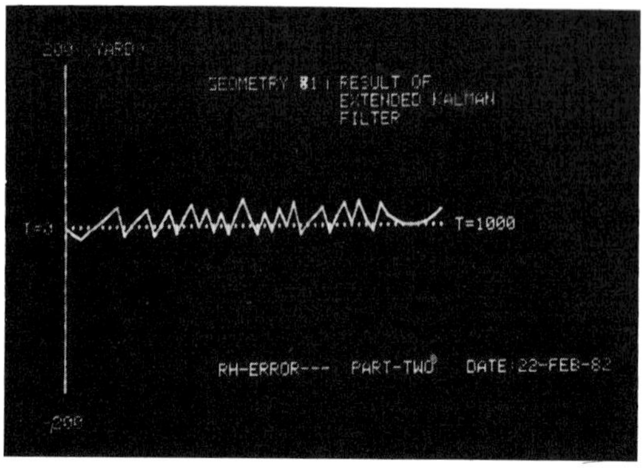

(a) Range error

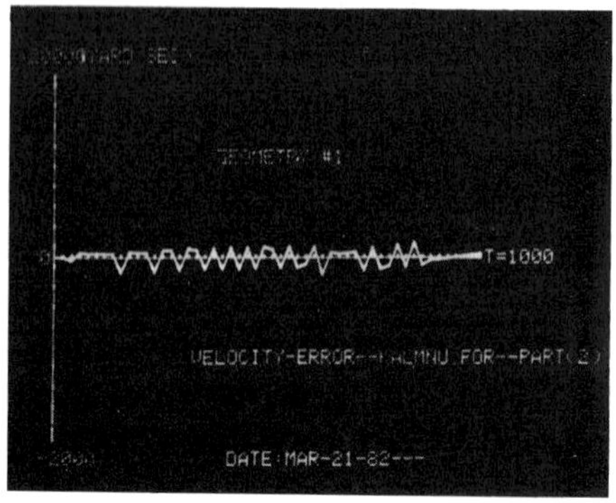

(b) Velocity error

Figure 9. Results of EKF with σ_τ = 5 m sec, σ_β = 0.5° for first geometry

Table 4

Time index	ERROR of RH					ERROR of velocity				
1-5	- 8	-14	- 5	2	9	0.	- .4	.5	.4	.4
6-10	17	25	-10	3	12	.4	.4	-1.7	.6	.5
11-15	22	-11	5	21	- 8	.5	-1.6	.8	.8	-1.4
16-20	12	28	0	22	- 6	1.	.8	-1.3	1.	-1.3
21-25	17	- 9	15	34	10	1.1	-1.3	1.1	-1.2	1.2
26-30	- 9	17	- 6	22	0	.9	-1.1	1.2	-1.1	- .7
31-35	30	-10	4	13	24	1.2	-1.9	.6	.4	.5
36-40	- 8	13	29	3	31	.6	-1.4	1.1	-1.3	.9
41-45	8	- 6	28	13	7	1.4	-1.0	1.7	- .8	- .3
46-50	2	1	2	5	11	- .2	- .1	.1	.2	.3
51	20					.4				

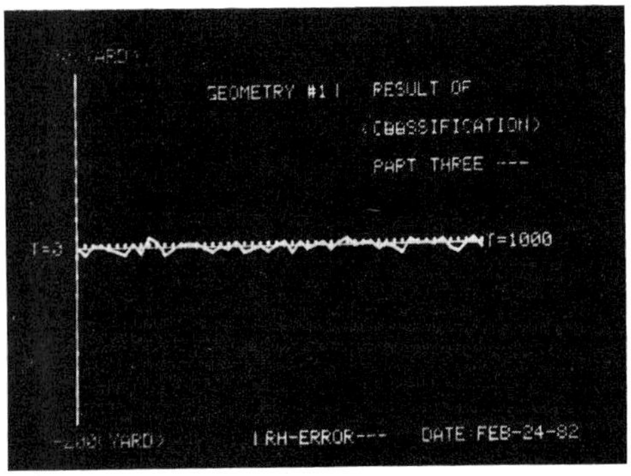

(a) Range error

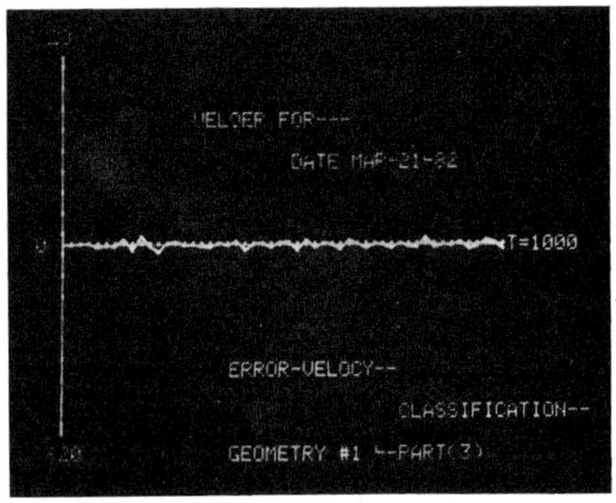

(b) Velocity error

Figure 10. Results of classification method with σ_τ = 5 m sec, $\sigma_\beta = 2.5^o$ for first geometry

Table 5

Time Index	ERROR of RH					ERROR of velocity				
1-5	- 8	2	2	- 2	-5	0.	.1	.1	-.3	-.1
6-10	- 8	4	-7	11	6	- .2	.6	-.6	.9	-.2
11-15	- 9	-3	3	3	3	- .8	.2	.3	-.1	.0
16-20	- 2	-6	0	- 3	5	- .2	-.3	.2	-.1	.4
21-25	- 7	-4	0	- 6	0	- .6	.1	.2	-.3	.3
26-30	5	-8	4	- 4	0	.3	-.6	.5	-.3	.2
31-35	4	-5	3	8	-1	.0	-.5	.5	.1	-.4
36-40	2	3	-5	- 1	-4	.1	.0	-.4	.2	-.2
41-45	-10	6	0	1	-1	- .3	.8	-.3	.0	-.1
46-50	5	-6	-10	- 1	6	.3	-.5	-.2	.5	.2
51	- 5					- .5				

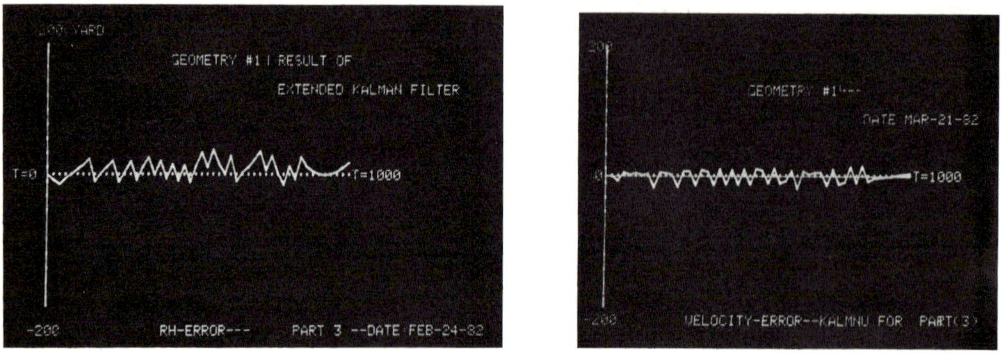

(a) Range error (b) Velocity error

Figure 11. Results of EKF with $\sigma_\tau = 5$ m sec, $\sigma_\beta = 2.5^\circ$ for first geometry

Table 6

Time index	ERROR of RH					ERROR OF VELOCITY				
1-5	- 8	-15	- 5	2	9	.0	- .7	.6	.2	.5
6-10	16	25	-10	3	12	.2	.4	-1.8	.6	.5
11-15	22	-11	5	21	- 8	.3	-1.7	.8	.8	-1.4
16-20	12	27	0	22	- 6	1.	.7	-1.4	1.1	-1.3
21-35	17	-9	15	-11	15	1.	-1.3	1.1	-1.3	1.2
26-30	35	12	40	17	1	.9	-1.1	1.2	-1.1	- .7
31-35	30	-10	4	12	24	.9	-2.	.8	.3	.6
36-40	37	7	30	3	-15	.6	-1.5	1.1	-1.3	- .9
41-45	14	- 6	28	13	6	1.5	-1.	1.6	- .8	- .3
46-50	2	0	2	5	12	- .2	- .1	.0	.2	.3
51	19					.4				

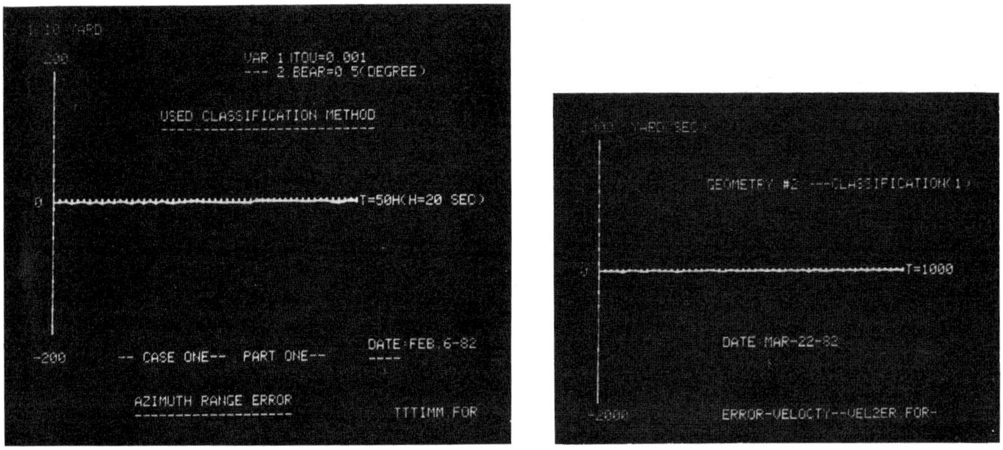

(a) Range error (b) Velocity error

Figure 12. Results of classification method with σ_τ = 1 m sec, $\sigma_\beta = 0.5^o$ for second geometry

Table 7

Time index	ERROR of RH					ERROR of velocity				
1-5	0	1	-2	0	1	.0	.0	-.0	-.2	.1
6-10	-2	1	-2	1	-1	.1	-.1	-.1	.2	.0
11-15	2	0	-2	2	-2	.0	.0	-.2	.0	.0
16-20	2	-2	2	-1	-2	.0	.0	-.1	.1	-.1
21-25	0	0	2	2	0	-.1	.2	.1	.0	.0
26-30	0	0	-1	-1	-1	.0	.1	-.1	.1	-.1
31-25	-1	-2	-2	3	2	.0	.1	-.1	.2	-.2
36-40	2	2	1	1	1	.2	.0	-.1	.0	.1
41-45	1	0	0	0	2	-.1	.1	-.1	.1	-.1
46-50	1	-2	-3	-1	3	.1	-.1	.1	-.1	.1
51	-1					-.1				

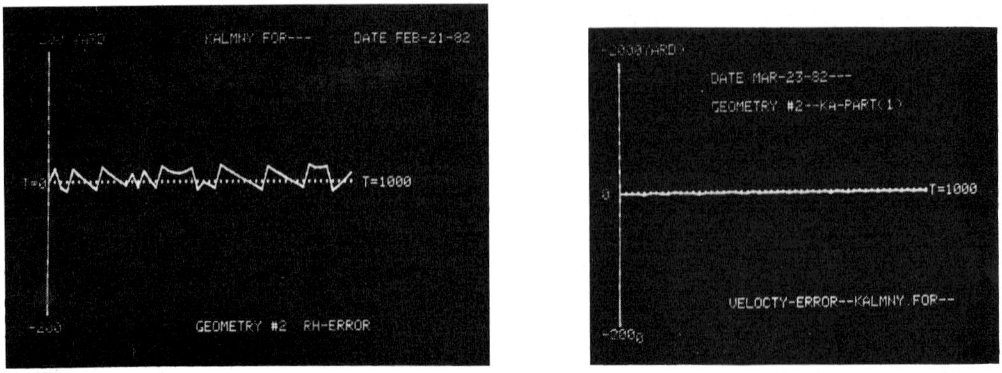

(a) Range error

(b) Velocity error

Figure 13. Results of EKF with σ_τ = 1 m sec, $\sigma_\beta = 0.5^o$ for second geometry

Table 8

Time index	ERROR of RH					ERROR of velocity				
1-5	22	- 8	-14	21	11	.0	.0	.0	-.2	.1
6-10	4	- 4	-11	24	15	.1	-.1	-.1	.2	.0
11-15	8	1	- 5	13	- 8	.0	.0	-.2	.0	.0
16-20	18	3	- 8	25	18	.0	.0	.1	.1	-.1
21-25	15	14	17	22	-12	-.1	.2	-.1	.0	.0
26-30	0	- 4	- 9	27	20	.0	.1	-.1	.1	-.1
31-35	15	9	4	- 1	- 6	.0	.1	-.1	.2	-.2
36-40	-12	25	18	13	7	.2	.0	-.1	.0	.1
41-45	2	- 3	- 8	27	22	-.1	.1	-.1	.1	-.1
46-50	22	24	-14	- 6	3	.1	-.1	.1	-.1	.1
51	14					-.1				

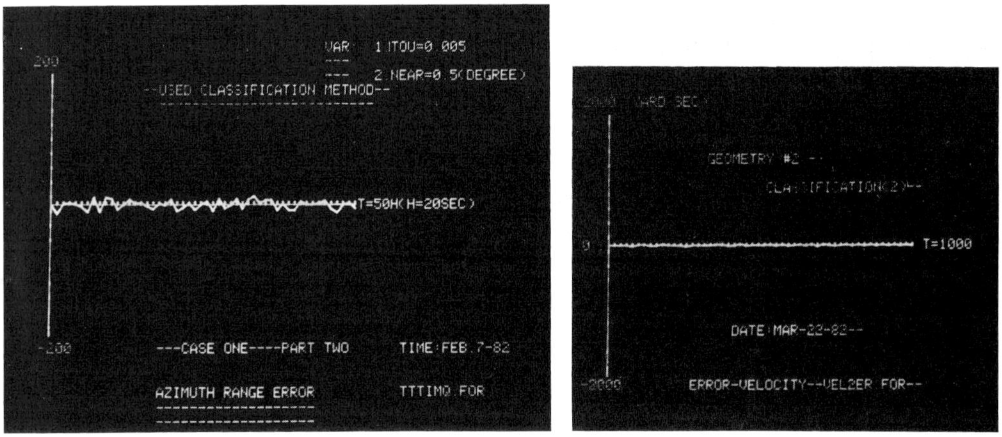

(a) Range error

(b) Velocity error

Figure 14. Results of classification method with σ_τ = 5 m sec, σ_β = 0.5^o for second geometry

Table 9

Time index	ERROR of RH					ERROR of velocity				
1-5	-15	1	3	0	-4	.0	.0	.0	-.2	.1
6-10	-12	11	-12	11	9	.1	-.1	-.1	.2	.0
11-15	- 8	0	8	2	3	.0	.0	-.2	.0	.0
16-20	- 3	- 7	2	- 1	8	.0	.0	.1	.1	-.1
21-25	-10	- 5	2	- 8	0	-.1	.2	-.1	.0	.0
26-30	5	-10	9	- 6	-1	.0	.1	-.1	.1	-.1
31-35	9	- 7	8	13	2	.0	.1	-.1	.2	-.2
36-40	7	7	- 9	1	-9	.2	.0	-.1	.0	.1
41-45	- 9	5	0	0	2	-.1	.1	-.1	.1	-.1
46-50	6	- 7	-13	- 1	3	.1	-.1	.1	-.1	.1
51	- 6					-.1				

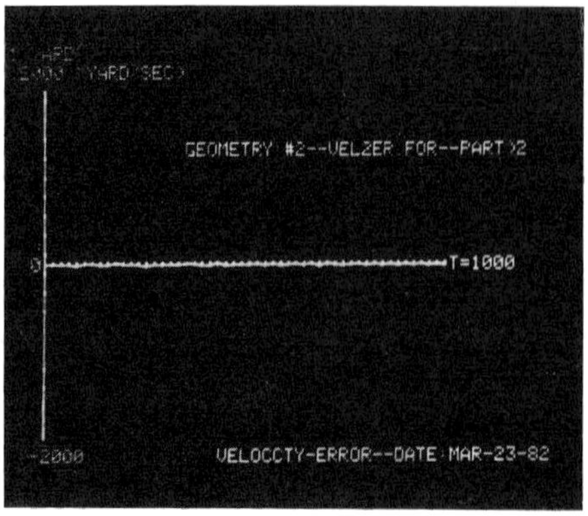

(a) Range error

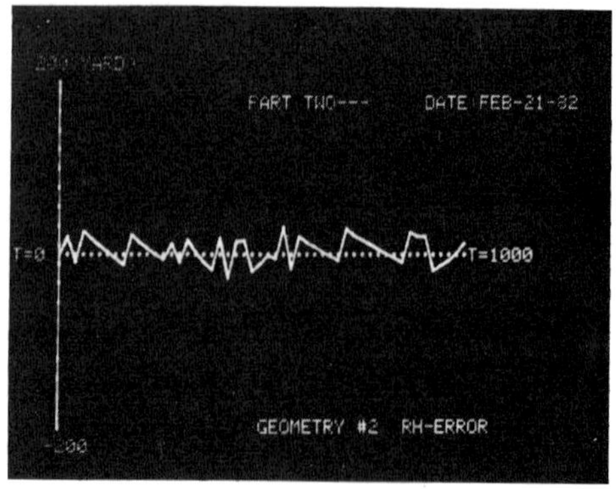

(b) Velocity error

Figure 15. Results of EKF with $\sigma_\tau = 5$ m sec, $\sigma_\beta = 0.5^o$ for second geometry

Table 10

Time index	ERROR of RH					ERROR of velocity				
1-5	22	- 8	28	19	11	.0	.0	.0	-.2	.1
6-10	4	- 4	-11	24	15	.1	-.1	-.1	.2	.0
11-15	8	1	- 5	13	- 8	.0	.0	-.2	.0	.0
16-20	18	3	- 8	-17	20	.0	.0	.1	.1	-.1
21-25	-27	16	17	-19	-10	-.1	.2	-.1	.0	.0
26-30	0	- 4	33	-16	22	.0	.1	-.1	.1	-.1
31-35	15	10	4	- 1	- 6	.0	.1	-.1	.2	-.2
36-40	31	23	18	13	7	.2	.0	-.1	.0	.1
41-45	2	- 3	- 8	27	22	-.1	.1	-.1	.1	-.1
46-50	22	-18	-12	- 6	3	.1	-.1	.1	-.1	.1
51	4					-.1				

spatial clustering method must be adopted in defining the potential target regions with the aid of a priori knowledge. This is particularly important when the multiple targets are likely to be present. The pattern recognition processing has indeed provided a new dimension to the solution of the three-dimensional TMA problems. The investigation of other effects such as the target maneuvers and of the real multipath data processing should also be considered.

It should be emphasized that the performance difference of the two methods discussed in this paper is partially due to the fact that different amount of measurements is used in processing. The results do indicate that the classification method can be very effective using all available delay measurements, and that better result in extended Kalman filtering is available with the aid of pattern recognition.

Part II
Time Delay Estimation

2.1 Introduction

Time Delay estimation from the generalized correlator without window exhibits thresholding effect as the signal-to-noise ratio decreases. By using the windows proposed by Hassab and Boucher [8], and averaging the periodograms of the data segments, the thresholding effect can be considerably reduced. Our experimental study [9] has shown that the optimum windows derived by Hassab and Boucher perform very well when the signal model is known or can be estimated for white or colored additive Gaussian noise. Let y(t) be the signal and $Z_1(t)$ and $Z_2(t)$ be the output measurements of two sensors,

$$\begin{aligned} Z_1(t) &= y(t) + n_1(t) \\ Z_2(t) &= \alpha y(t - \tau) + n_2(t) \end{aligned} \tag{1}$$

where $n_1(t)$ and $n_2(t)$ are stationary random noise processes, α is an amplitude factor assumed as 1 in this study, and τ is the time delay between the two sensors. Both α and τ in general can be assumed as constant over the observation time interval. For random signal the Hassab-Boucher windown II takes a simple form,

$$W_{II}(w) = \frac{\Phi_y(w)}{\Phi_{z_1}(w) \cdot \Phi_{z_2}(w)} \tag{2}$$

where $\Phi(w)$ is the spectral density. For signal of known spectral density the window can be determined whether the noise is white or not. For white noise but unknown signal spectrum, the Burg's maximum entropy spectral analysis method can be used to estimate $\Phi_y(w)$ by subtracting $\Phi_{z_1}(w)$ with the final prediction error (FPE). Here the signal is modeled by an autoregressive (AR) model. If noise is white then $Z_1(t)$ should be represented by an ARMA model. The AR parameter estimate for $Z_1(t)$ will be biased. Subtracting $\Phi_{z_1}(w)$ by the FPE will partially offset the effect of bias and provide a good estimate of signal spectrum. It is noted that the Burg's method essentially assumes an AR model for the measurement. The Akaike FPE criterion can be used to determine the order of the model.

2.2 Experimental Results

Some experimental results will be briefly described in the following. The random signal is generated by passing white noise through a recursive filter with the impulse response Bn exp(- Bn). The Gaussian white noises at the outputs of two sensors are independent and have been generated by the Monte Carlo method. The time delay between the sensor elements is selected as 16 or 10 units in different cases. An 128 point data segment is processed for estimating the time delay. Averaging of several segments is often necessary to improve the estimate especially at low signal-to-noise ratio. Figure 16a is the correlator output with B = 17, SNR = - 30db, time delay

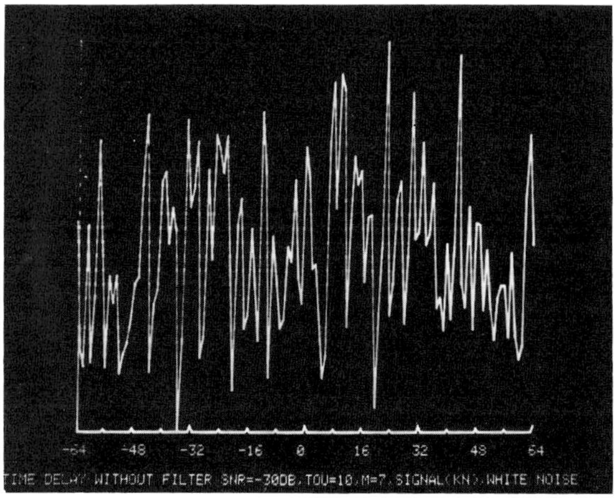

Figure 16(a)

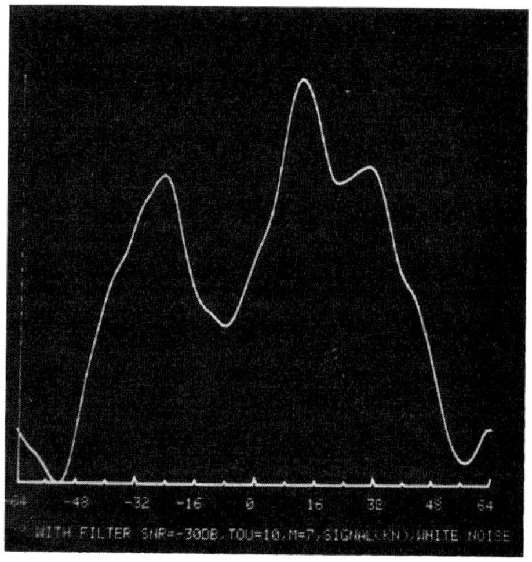

Figure 16(b)

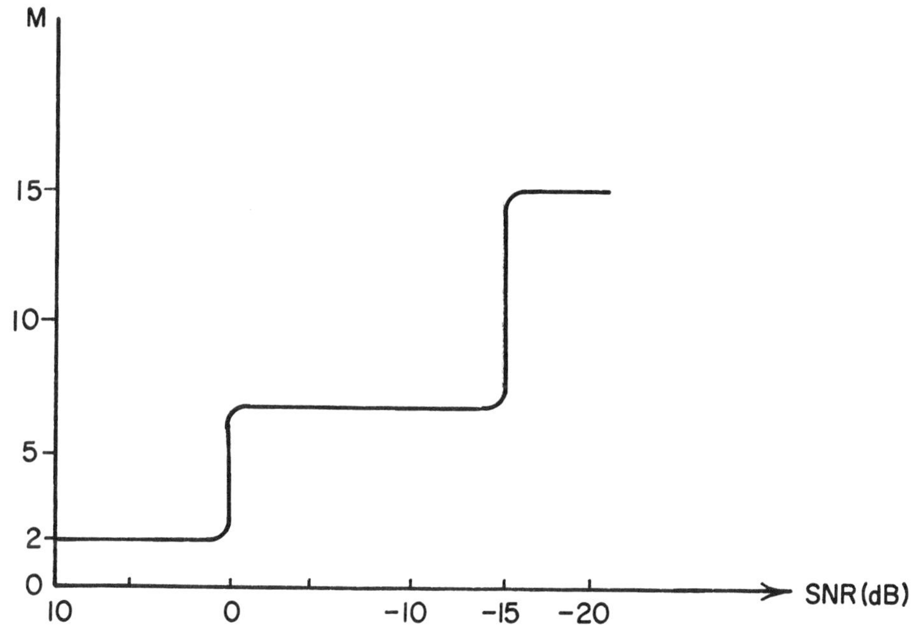

Figure 17. An empirical relationship of signal-to-noise ratio (SNR) and segment number M.

τ = 10 (units), and M = 7, where M is the number of data segments used. The estimated time delay is τ = 22. If window II is used the correct delay is estimated as shown in Fig. 16b. After extensive experimentation, an empirical relationship of signal-to-noise ratio (SNR) and optimum segment number M is determined as shown in Fig. 17. The signal is random of known form Bt exp(- Bt), as described earlier.

For unknown signal plus white noise, the signal spectrum is estimated by Burg's method from which the optimum window II is computed. Figure 18a is the correlator output with unknown signal, SNR = -10 db, and actual delay τ= 16. Also M = 1 and the true signal parameter B = 3.33. The estimated delay is τ = - 6. Figure 18b is the window II function. Figure 18c is the same as Fig. 18a except that the window function is used. The estimated delay is τ = 11. Figure 18d is the same as Fig. 18c but with M = 7 and estimated delay of τ = 11. Figure 18e is also the same as Fig. 18c but with M = 15 and estimated delay of τ = 14. For colored Gaussian noise and known signal, the correlator outputs without window are shown in Figs. 19a, 20a, and 21a corresponding to SNR = 10 db, 0dB. and -10 db respectively, and B = 0.333, M = 1, and true delay of τ = 10. The delays are all estimated correctly. If the optimum widnow II is used, the correlator outputs are sharper near the peak, with results shown in Fig. 19 b (M = 1), Fig. 20b (M = 4) and Fig. 21b (M = 6) respectively. For unknown signal and colored noise a general approach is not available. A suggested procedure is to use the recursive maximum likelihood algorithm in an adaptive noise cancelling (ANC) model to estimate the color noise and then subtract it from the noisy measurement [10].

Experimental study by Hassab and Boucher [11] indicates that the distributions represented by the histograms of the time delays from raw data are not necessarily Gaussian as the number of false detections increases with decreasing signal-to-noise ratio. This leads to a biased estimation of the time delays, and a standard deviation greater than the Cramer-Rao bound. Such bound is theoretically available when the Gaussian assumption of the time delay estimate is valid. The effect of the non-Gaussian components in addition to Gaussian signal and noise components can be measured by the increased variance of the time delay estimation. Gating and smoothing have been suggested in reducing the effect of non-Gaussian components [11][12][13]. Gating is achieved by limiting the range of time delay estimate. This obviously reduces the variance of the estimate but also decreases the probability of detection. A good a priori knowledge is also required for proper gating. Smoothing has its limit also as is evidenced by the optimum number of data segments in averaging the periodograms. A combination of both gating and smoothing is suggested for reducing the effect of non-Gaussian components in time delay estimation.

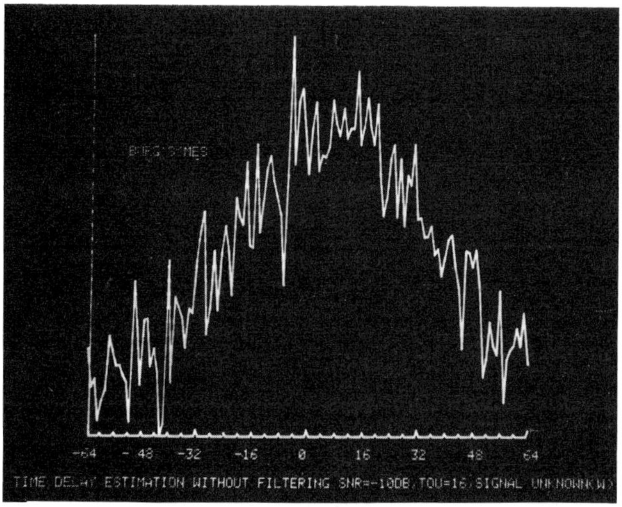

Figure 18(a)

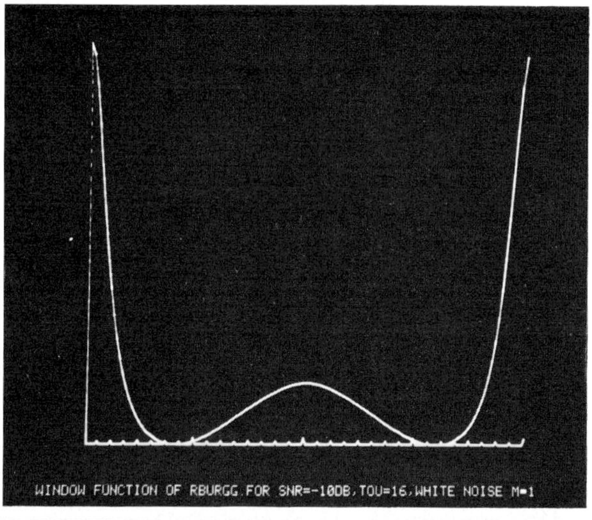

Figure 18(b)

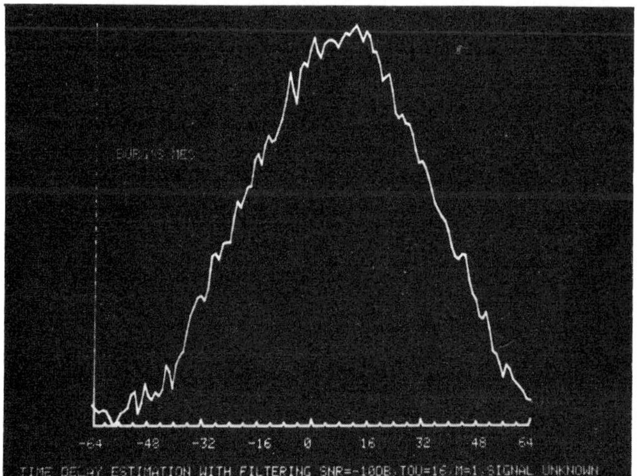

Figure 18(c)

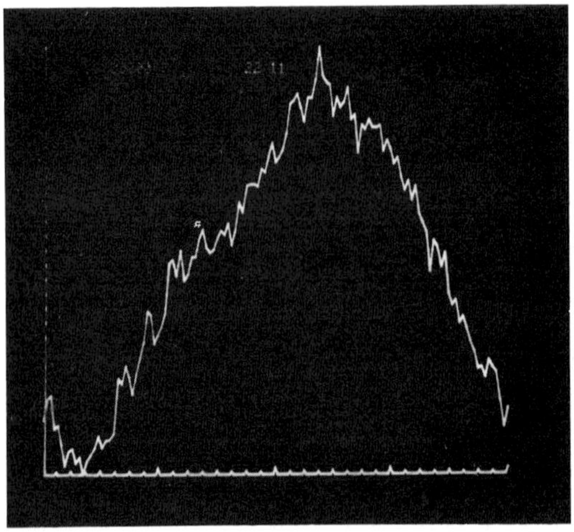

Figure 18(d)

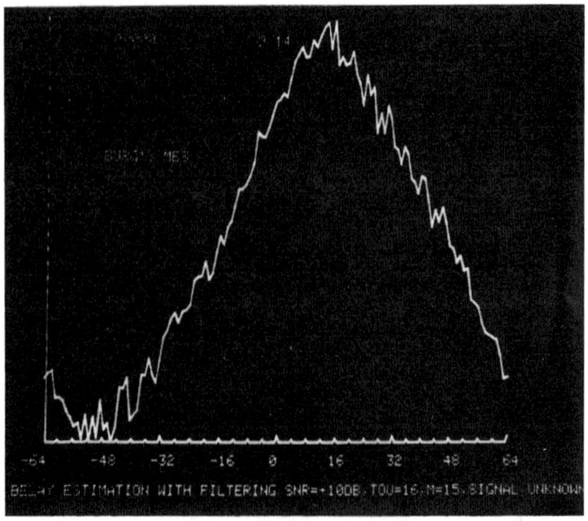

Figure 18(e)

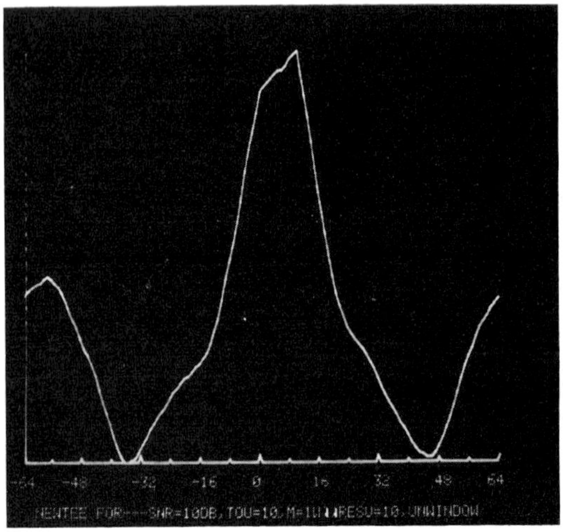

Figure 19(a)

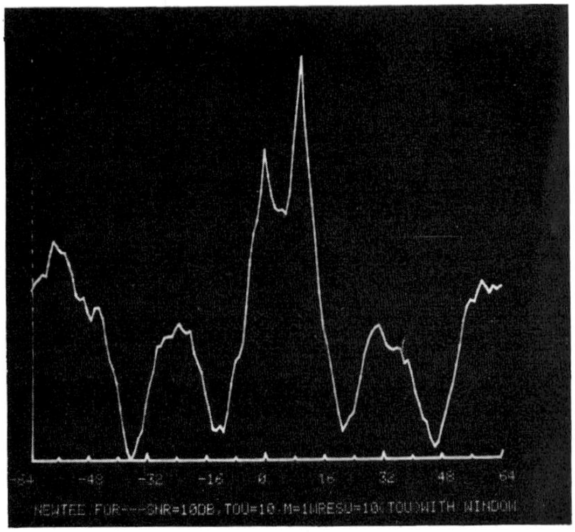

Figure 19(b)

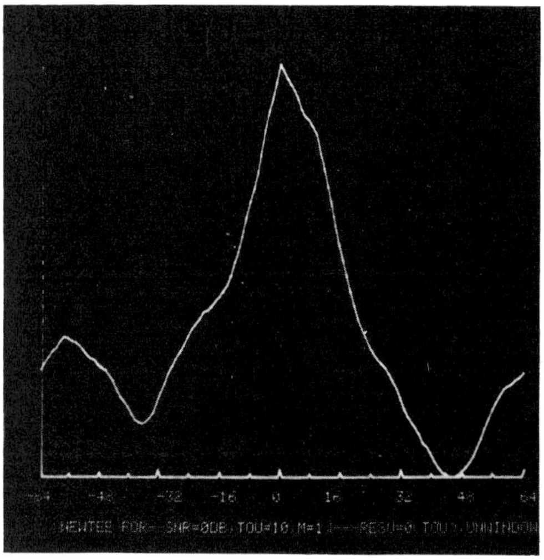

Figure 20(a)

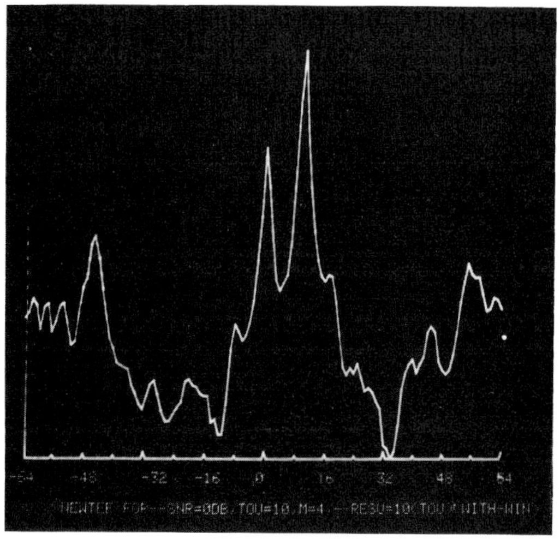

Figure 20(b)

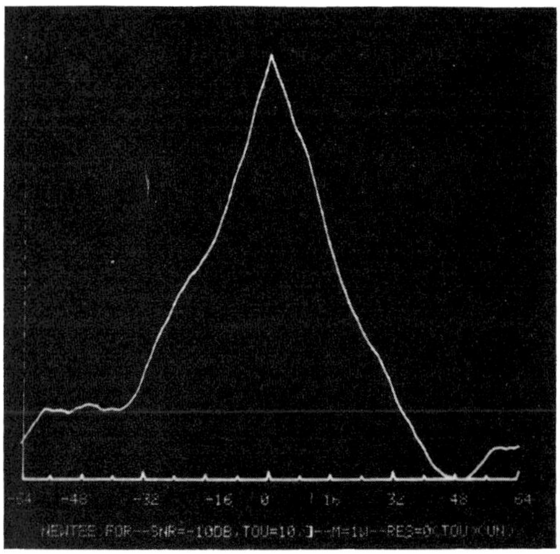

Figure 21(a)

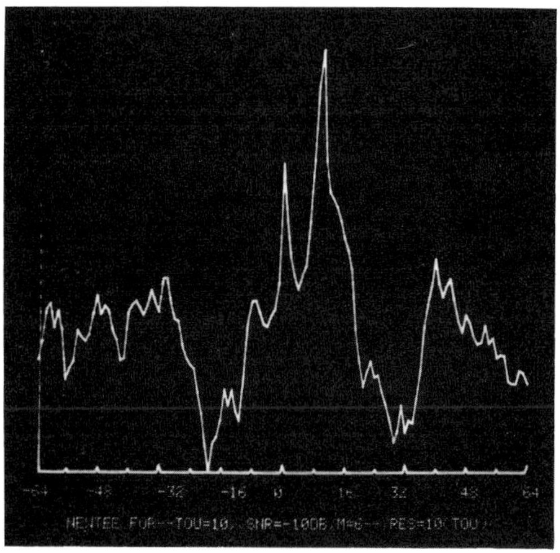

Figure 21(b)

Part III
Parameter Estimation of a Moving Source

3.1 Introduction

Now consider the basic problem of passive localization and motion estimation of a source observed from a linear array. Two time delay measurements obtained from three sensors constitute the minimum number needed to localize at each instant of time. Good time delay estimate is obviously necessary to insure small localization errors. As the source moves, the time delay measurements are taken sequentially from which the motion parameters can be estimated. The estimator structure described in Ref. 14 is to employ an expanding memory filter that uses successive time-delay observations together with an appropriate physical model of the source motion to obtain smooth source-trajectory estimates. The time delay measurements are used directly in the extended Kalman filter to obtain the localization and motion solutions.

3.2 Experimental Results

Based on the work of Ref. 14, three examples of the source parameter estimation are presented in the following. The first example employs the same source-receiver geometry as in Ref. 14 for simulation. It has an initial contact at a range of 20 kiloyards and a true bearing of 180°. The source moves on a course of 90° at a speed of 20 knots while the observer maintains a constant course of 120° with a speed of 10 knots. All angles are referenced to north. In all three examples, the array spacing is 100 ft., the sampling rate is 1 sample every 10 seconds, and the data are in the form of time delays corrupted by additive, zero mean, white Gaussian noise with a standard deviation of 5 μs. The processor used is the third one described in Ref. 14 that uses time delays directly in the extended Kalman filter to obtain the localization and motion solutions. Figure 22 shows (a) range, (b) bearing, (c) course and (d) speed along with their mean errors and rms errors as a function of time. The results match very well with those reported in Ref. 14.

The second example has two target legs at 10 minutes each and one observer (own ship) leg of 20 minutes. The initial range is 10 kiloyards and the true bearing is 180°. The target moves at a course of 45° and a speed of 10 knots in the first leg, and a course of 0°, speed of 20 knots in the second leg. The observer has a course of 120° and speed of 10 knots. Figure 23 shows (a) range, (b) bearing, (c) course and (d) speed along with their mean errors and rms errors as a function of time. Figure 23(e) shows the plant noise which is incorporated so that the system can adapt to the target manuever at the end of the first target leg. The algorithm for target manuever processing as used in the simulations is based on that developed by Davis [15] in which the selection of q, the covariance of the plant noise sequence provides the most probable sequence of residuals [7].

The third example uses a more complicated geometry. The initial range and bearing are 11700 yards and 330° respectively. There are three target legs of 10 minutes each at a course sequence of 45°, 45°, and 180°, and at a speed sequence of 10, 20, 20 knots respectively. The observer has six legs of 5 minutes each at a course sequence alternating between 90° and 30°, but the speed remains constant at 10 knots throughout the 30 minutes time interval. Figure 24 shows (a) range, (b) bearing, (c) course, (d) speed, and (e) plant noise of the estimator. It is noted that the target depth is not estimated in these simulations. The addition of plant noise for target manuever is necessary for the constant velocity model assumption. Other target manuever models have been considered. Let V(t), S(t) and C(t) be the velocity, speed and course respectively during target manuever. The target model suggested by Guimond [16] is to express

$$V_{xt}(t) = S_t(t)\sin C_t(t)$$

$$V_{yt}(t) = S_t(t)\cos C_t(t)$$

and assume $\dot{C}_t(t)$, $\dot{S}(t)$ being constant. These assumptions are very reasonable and the model appears to be suitable for a large class of target manuevers.

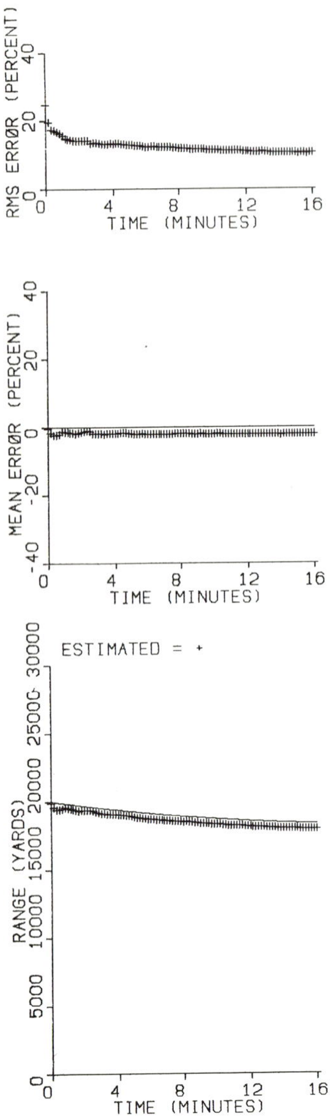

Figure 22(a)

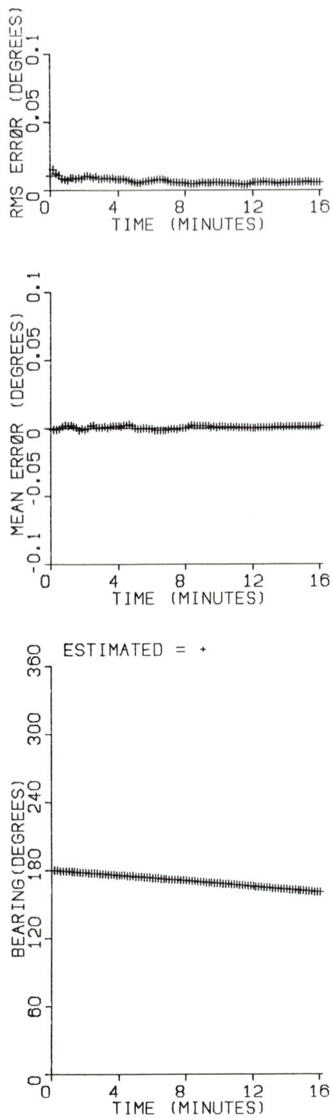

Figure 22(b)

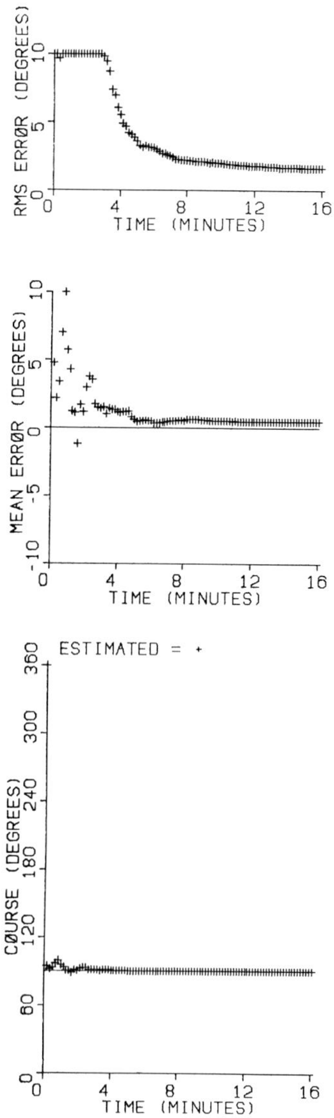

Figure 22(c)

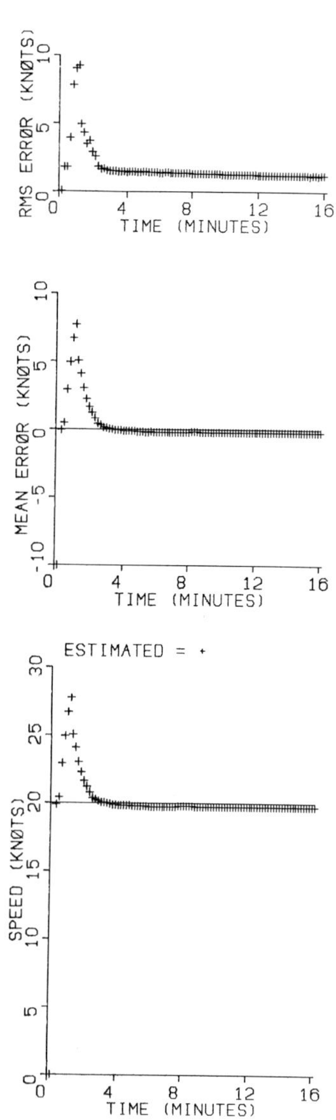

Figure 22(d)

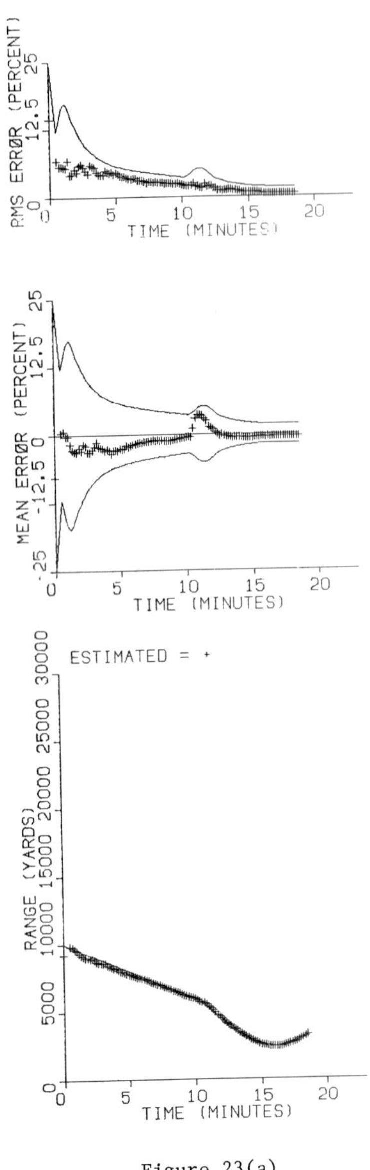

Figure 23(a)

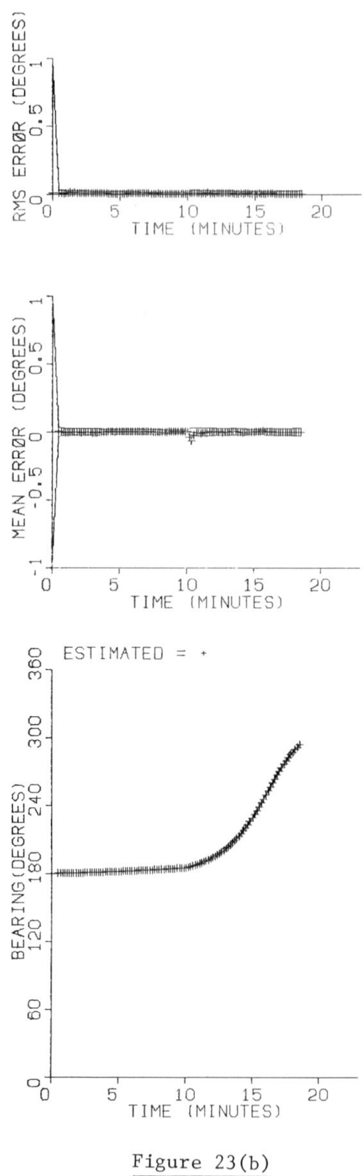

Figure 23(b)

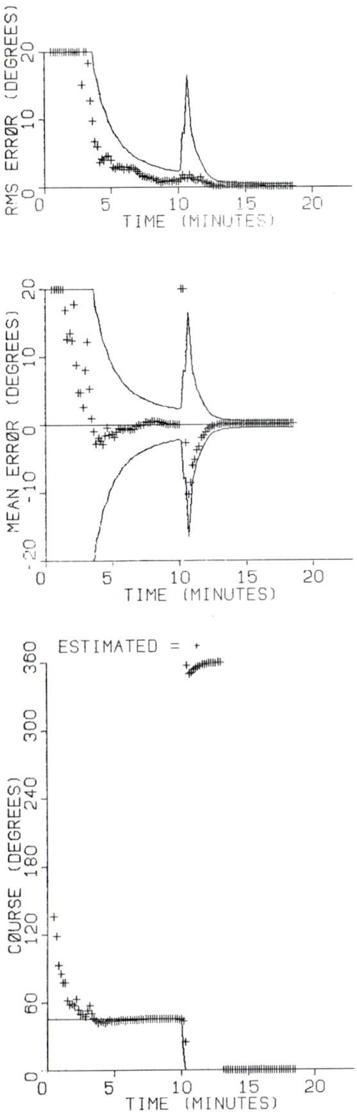

Figure 23(c)

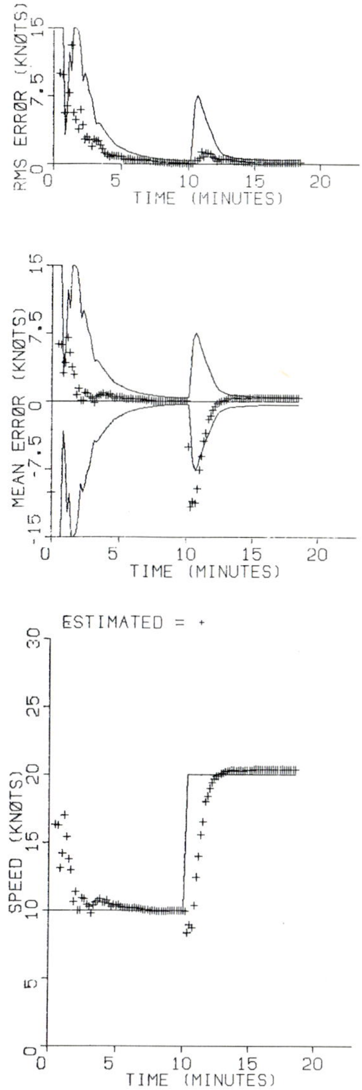

Figure 23(d)

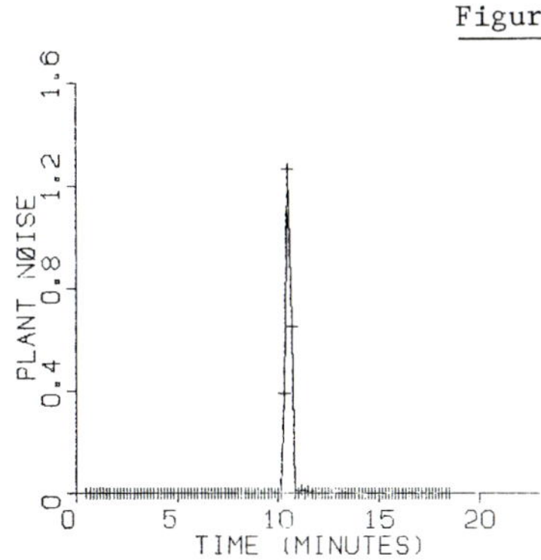

Figure 23(e)

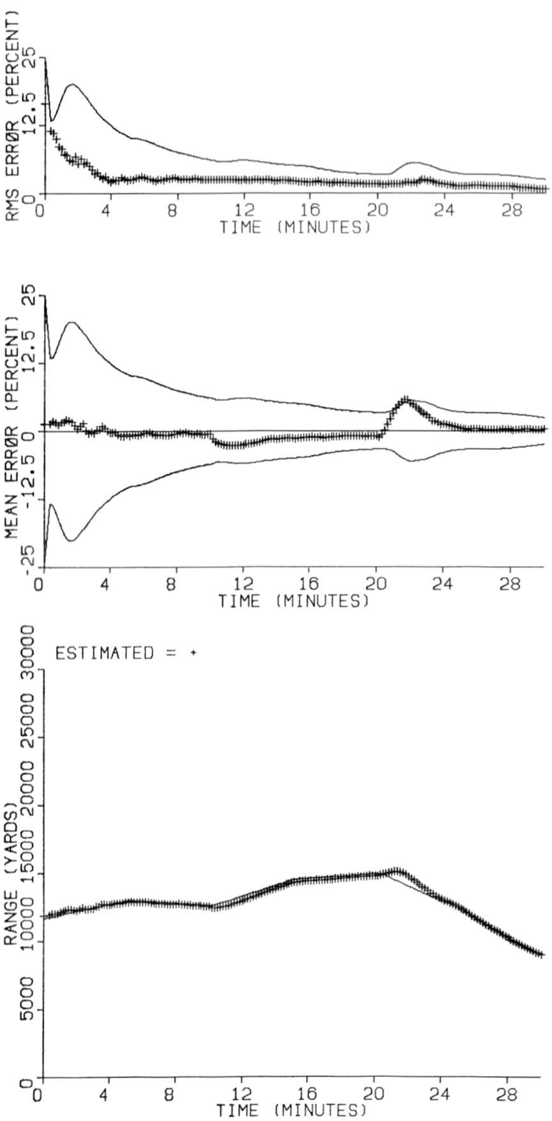
RMS ERRØR (PERCENT)
0
12.5
25
0 4 8 12 16 20 24 28
TIME (MINUTES)
MEAN ERRØR (PERCENT)
-25
-12.5
0
12.5
25
0 4 8 12 16 20 24 28
TIME (MINUTES)
ESTIMATED = +
RANGE (YARDS)
0
5000
10000
15000
20000
25000
30000
0 4 8 12 16 20 24 28
TIME (MINUTES)

Figure 24(a)

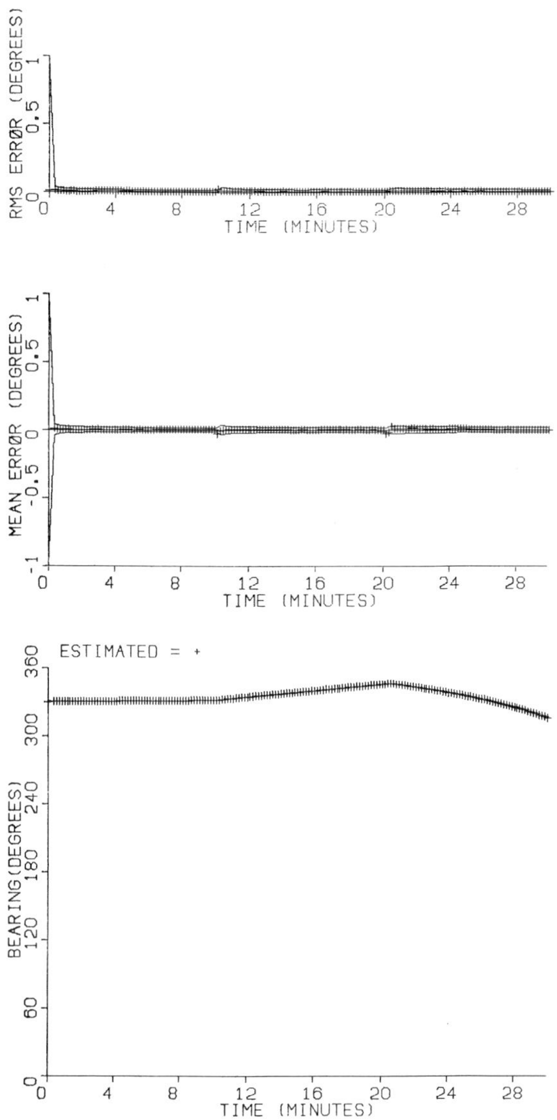
RMS ERRØR (DEGREES)
0
0.5
1
0
4
8
12
16
20
24
28
TIME (MINUTES)
MEAN ERRØR (DEGREES)
-1
-0.5
0
0.5
1
0
4
8
12
16
20
24
28
TIME (MINUTES)
ESTIMATED = +
BEARING(DEGREES)
0
60
120
180
240
300
360
0
4
8
12
16
20
24
28
TIME (MINUTES)

Figure 24(b)

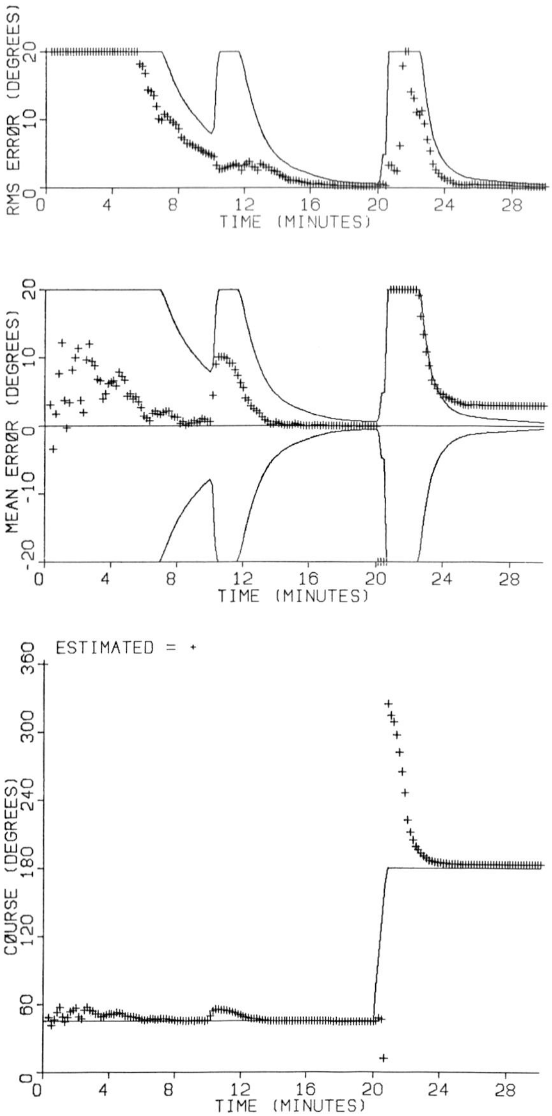
RMS ERRØR (DEGREES)
TIME (MINUTES)
MEAN ERRØR (DEGREES)
TIME (MINUTES)
ESTIMATED = +
CØURSE (DEGREES)
TIME (MINUTES)

Figure 24(c)

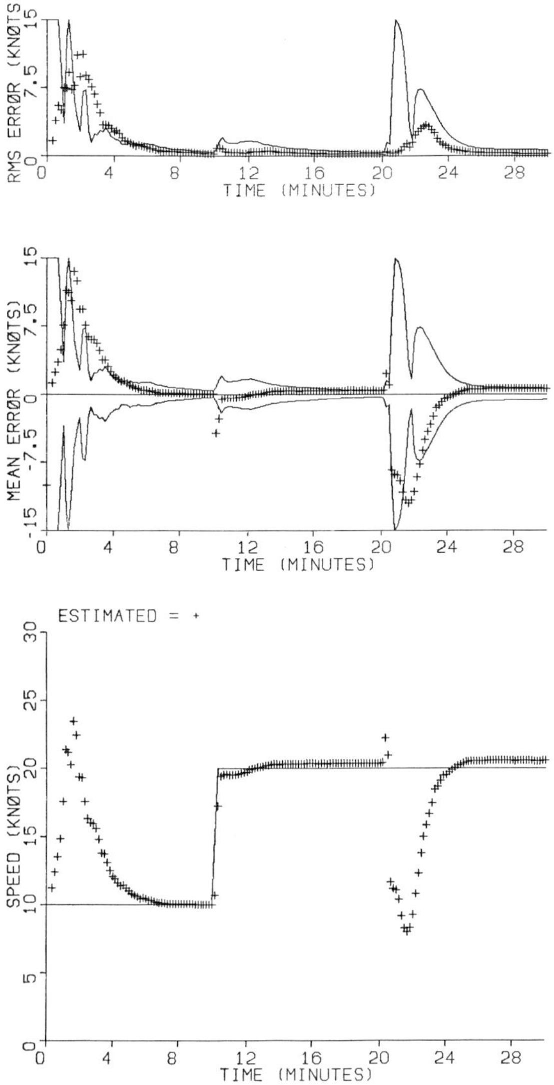

Figure 24(d)

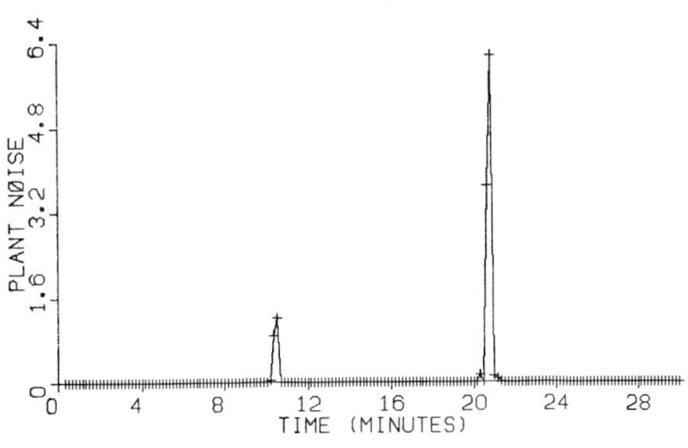

Figure 24(e)

Part IV
Multipath Ranging Pattern Recognition

It is well known that multipath ranging can provide good estimate of target motion and range and target depth solutions, as illustrated in Part I. Independent wavefront curvature and multipath range solutions can be averaged to yield a better range estimate. The multipath solutions depend critically on the correct estimate of multipath structure, i.e. the identification of paths from the cross-correlation and autocorrelation data. Multisensors certainly provide more information than the single sensor. Still there is much ambiguity in determining the multipaths. Satisfactory procedures to identify automatically the paths are still yet to be developed. This is the area that pattern recognition can help. Pattern mathcing techniques have been proposed for multipath ranging [3]. The idea is to match the correlograms generated from actual sea data and those predicted by an ocean model in a manner similar to the fingerprint identification. The predicted correlograms can be obtained from the sonar model. The difficulty here is that the number of possible multipath combinations to be matched can be very large. Both the a priori knowledge (such as own ship depth and estimated target depth) and the signal processing techniques should be used for preprocessing to enhance the correlograms and to formulate a multipath structure which best fits the data. Detailed study of this topic will be reported in the near future in classified domain.

Acknowledgement

Parts I, III and IV are supported by the NUSC under IPA Agreement with the author. Part II was supported also by NUSC under Contract No. N0014-81-M-MM54. The author would like to thank John S. Davis, Brian W. Guimond and Dr. Joseph C. Hassab for support and suggestions on this work.

References

1. S.I. Chou, "Anti-submarine warfare (ASW) passive target tracking", Final Report of Contract N00014-79-WR90084, January 1980 (Limited distribution document).
2. J.C. Hassab, "Homomorphic deconvolution in reverberant and distorional channels: an analysis," Journal of Sound and Vibration, Vol. 58, No. 2, pp. 215-231, 1978.
3. "Multipath ranging pattern recognition," (confidential), unpublished document by Rockwell International Autonetics Marine Systems Division, 1981.
4. T. McAndrew, "Large array processing systems," Technical presentation (confidential) at NUSC, January 1982.
5. W.C. Knight, R.G. Pridham and S.M. Kay, "Digital Signal processing for sonar," Proc. of the IEEE, Vol. 69, No. 11, p. 1451, November 1981.
6. K.F. Gong and J.S. Davis, "Evaluation of target motion analysis in a multipath environment,' NUSC Technical Report 4814, March 2, 1976, NUSC Newport Laboratory, Newport, R.I. (Limited distribution document.)
7. A.H. Jazwinski, "Stochastic Processes and Filtering Theory," Academic Press, New York, 1970.
8. J.C. Hassab and R.E. Boucher, "Optimum Estimation of Time Delay by a Generalized Correlator," IEEE Trans. on ASSP, Vol. 27, No. 4 pp. 373-380, August 1979.
9. C.H. Chen and S.J. Chern, "Time Delay Estimation for Known and Unknown Signals," unpublished report, December 1981.

10. B. Friedlander, "System Indentification Techniques for Adaptive Signal Processing," IEEE Trans. on ASSP, Vol. 30, No. 2, pp. 240-246, April 1982.
11. J.C. Hassab and R.E. Bouncer, "An Experimental Comparison of Optimum and Sub-Optimum Filters' Effectiveness in the Generalized Correlator," Journal of Sound and Vibration, Vol. 76, No. 117-118, 1981.
12. J.C. Hassab and R.E. Boucher, "The Influence of Signal and Data Processing Techniques in Arrays on the Estimation of Source Location," Proceedings of the Time Delay Estimation and Applications Conference, edited by J.. Hassab, Vol. I, pp. Q-1 to Q-20, July 1979.
13. J.C. Hassab and R.E. Bouncer, "Passive Ranging Estimation from an Array of Sensors," Journal of Sound and Vibration, Vol. 67, No. 2, pp. 289-292, 1979.
14. J. Hassab, B.W. Guimond and S.C. Nardone, "Estimation of Location and Motion Parameters of a Moving Source Observed from a Linear Array," Journal of the Acoustical Society of America, Vol. 70, No. 4, pp. 1054-1061, October 1981.
15. J.S. Davis, "Noisy Bearing only Measuring Target Problem," Master of Science Thesis, Dept. of Electrical Engineering, Southeastern Massachusetts University, N. Dartmouth, MA, June 1976.
16. B. Guimond, unpublished work.

ACOUSTICAL CLASSIFICATION OF THE SEAFLOOR

Jens M. Hovem
Rune M. Holt

Electronics Research Laboratory
University of Trondheim
N-7034 TRONDHEIM-NTH

NORWAY

ABSTRACT

Mainly as a result of offshore exploration there is a rapid development of measurement methods for the determination of the properties of the seafloor soils. Acoustic or seismic methods are now mainly used to obtain bottom and sub-bottom profiles. Although these methods give valuable information there is a need for improvements to obtain higher resolution and more quantitative information.

This paper describes two techniques for determination of the properties of the upper layers of the sea-floor. The wide angle reflection measurements using source and receivers close to the bottom with relative short baseline gives compressional velocities of the layers. For the determination of the shear velocity a technique based on the measurement of boundary waves can be used. The velocity of a boundary wave depends strongly on the shear velocity structure. In an inhomogeneous medium the boundary waves are dispersive, hence the measurement of velocity as function of frequency can give information about the shear modulus as a function of depth.

NATO ASI Series, Vol. F1
Issues in Acoustic Signal/Image Processing and Recognition
Edited by C.H. Chen

1. INTRODUCTION

Engineering activities on the ocean floor require a knowledge of the properties of the sea floor soils. Such activities include foundations for offshore platforms, excavations for cables and pipelines, slope stability analysis, anchoring and problems related to sediment erosion and drift and to the trafficability of the sea floor.

The objective of a complete geotechnical investigation is to delineate the various soil strata and to determine their physical characteristics and strength parameters. Index properties such as grain size, water content, density and mineralogy are used mostly to determine the general physical type of the different strata. The strength characteristics to be evaluated include the shear strength, stress-strain behavior, compressibility and dynamic load response.

Seismic or acoustic methods in geotechnical investigations are presently confined to obtaining high resolution bottom and sub-bottom profiles and for detection and mapping of surficial topography. Acoustic profiling usually forms part of any integrated investigation and can be used to identify the general sediment types. The use of acoustic methods in actually evaluating the geotechnical properties is still in its infancy and further research is required before the methods can become reliable quantitative measurements of geotechnical properties.

The objective of the acoustic measurement of the seabed could be to obtain information such as porosity, density, elastic moduli, permeability and layering information. Determination of these parameters from acoustic measurement must be based on a detailed knowlegde about sound propagation in porous and solid media and along interfaces.

The purpose of this presentation is to give, in the spirit of a workshop, an overview on some of the newer methods which are being investigated for acoustical mapping of the properties of the seabed.

2. ACOUSTICAL AND PHYSICAL PARAMETERS

A theoretical starting point is the Biot wave equations for the compressional waves in a homogeneous solid medium. [Refs 1-8]

$$\nabla^2(He-C\xi) = \frac{\partial^2}{\partial t^2}(\rho e - \rho_f \xi) \tag{1}$$

$$\nabla^2(Ce-M\xi) = \frac{\partial^2}{\partial t^2}(\rho_f e - \rho_c \xi) - \frac{\eta}{B}\frac{\partial \xi}{\partial t} \tag{2}$$

where e is the dilation of the bulk material, ξ is the dilation of the fluid relative to the solid frame.

The elastic parameters H, C, M are related to the bulk modulus K_f of the fluid, the bulk modulus K_s of the solid grains and the bulk-modulus K_b and the shear modulus μ of the frame. In principle H, C, M can be determined by performing a "jacketed test" and an "unjacted test" as described by Stoll [4]. The densities of the fluid and the solid are respectively ρ_f and ρ_s and the density of the composite medium $\rho = \phi\rho_f + (1-\phi)\rho_s$ where ϕ is the porosity. ρ_c is a mass coupling density. The last part of Eq(2) represents the loss due to the flow of a viscous fluid in the porous medium. Here B is the absolute permeability and η is the viscosity of the fluid.

The equations predict two compressional waves. Due to the viscous loss both waves are dispersive but the dispersion of the first ordinary wave is very small and its velocity is

$$v = \sqrt{\frac{K + \frac{4}{3}\mu}{\rho}} \tag{3}$$

where K is the closed system elastic modulus, The attenuation of the first wave is due to friction and viscosity and will only be important for long range transmission.

The second wave is highly dispersive and very strongly attenuated. At low frequencies the complex wavenumber of the second wave will approximately be

$$\ell_2 \approx \frac{1}{\sqrt{2}}(1-i)\sqrt{\omega \frac{\eta}{B} \frac{H}{K_b \cdot M}} \tag{4}$$

giving a velocity $v_2 = \sqrt{2 \frac{\omega B}{\eta} \frac{K_b \cdot M}{H}}$ and an attenuation $\alpha_2 = \frac{\omega}{v_2}$. Fig. 1 shows as an example the calculated velocities and attenuation coefficient for the two waves as a function of frequency where δ represents the frictional attenuation modelled as an imaginary component of K_b.

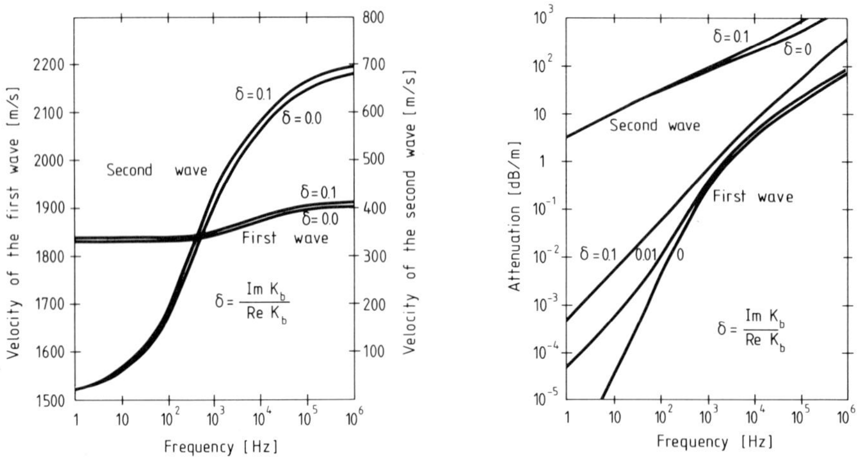

a) VELOCITIES OF THE FIRST AND SECOND WAVE AS A FUNCTION OF FREQUENCY.

b) VISCOUS ATTENUATION OF THE FIRST AND SECOND WAVE AS A FUNCTION OF FREQUENCY.

Fig. 1

In addition to the two compressional waves there will also exist shear waves. The viscosity of the fluid may also cause the shear wave to be dispersive but the dispersion will be very small. The shear wave velocity will therefore be

$$v_s = \sqrt{\frac{\mu}{\rho}} \tag{5}$$

The physical and acoustical properties of marine sediments have been the subject for many studies. The majority of these are experimental giving semi-empirical relationships between the acoustical and physical properties in various geographical areas. [Refs 9-13] Reference is in particular made to the many papers by Hamilton which represent much of the present knowledge.

Acoustic measurements can give values for the following parameters: compressional wave velocity v_p, shear wave velocity v_s and acoustic impedance Z. Under special conditions it may also be possible to obtain values for the attenuation coefficients but such measurements and their interpretations are considerably more difficult than for those listed above.

From the measurements of the acoustical parameters listed above it is possible to calculate the following physical parameters

Bulk modulus (system) $$K = \rho[v_p^2 - \frac{4}{3} v_s^2] \tag{6}$$

Shear modulus $$\mu = \rho v_s^2 \tag{7}$$

Density $$\rho = Z/v_p \tag{8}$$

These physical parameters are, however, not independent but can be mutually related to other parameters, and one of the most important is the porosity. The effect of porosity on the density is obvious but also other parameters as K_b and M are related to the porosity as shown by Hamilton [9].

3. REFLECTION MEASUREMENTS

The reflected signal from the bottom using a normal echosounder will depend on the bottom roughness and the impedance contrast between the water and the bottom. If the bottom appears flat at the frequency which is used the reflection coefficient at normal incidence for the water/bottom interface is given by

$$R = \frac{Z_b - Z_w}{Z_b + Z_w} \tag{9}$$

where Z_b and Z_w are, respectively, the acoustic impedance of the bottom and the water. Under conditions where the roughness and layering of the bottom is not important, measurement of the first reflected signal is essentially a measurement of impedance ratio. The arrival time is used in the determination of the bottom topography.
For high porosity media the frame bulk modulus becomes quite small and K_b and μ will have a rather limited effect on the compressional wave velocity which then approaches

$$v_p = \left\{\left[\frac{\phi}{K_f} + \frac{1-\phi}{K_s}\right]\left[\phi\rho_f + (1-\phi)\rho_s\right]\right\}^{-\frac{1}{2}} \tag{10}$$

The impedance ratio will then approach

$$\left(Z_b/Z_w\right) = \left\{\left[\phi + (1-\phi)\rho_s/\rho_f\right]\left[\frac{1}{\phi+(1-\phi)\cdot K_f/K_s}\right]\right\}^{\frac{1}{2}} \tag{11}$$

In practice K_s, K_f, ρ_s and ρ_f are relatively constant, and the reflection coefficient then becomes very strongly dependent on the porosity.

Dependent on source level, frequency, bandwidth and beamwidth the bottom penetration sounders and light seismic equipment based on boomer, sparker and airguns will also yield layering information and reflection coefficients for the interfaces. These techniques are used extensively in surveys and have proved their usefullness.

A basic limitation with the normal incidence methods is that they can not give velocity information. In order to determine both layer thickness and velocities a bistatic measurement with separate transmitting/ receiving station may be used. In principle both refraction and reflection techniques can be applied. The refraction method, however, can only be used when the wave velocity is an increasing function of depth and the presence of a low velocity layer can not be detected by this method. Loose deposit sediments can have a sound velocity lower than the velocity of water in which case the refraction method can not be used. In addition, this method requires a rather long transducer-receiver distance. The reflection measurement with a wide variation in the source-receiver separation does not have these limitations.

The wide angle reflection technique originally used sonobuoys and applied to deep waters using a large source-receiver separation. More recently the technique has been developed to give detailed information about the shallow bottom structure. [Refs 14-17] Both stationary and towed experiments have been conducted and the technique is now approaching the practical stage for engineering purposes.

In order to illustrate the method and the results a simple experiment conducted by the authors will be described. The main objective with the experiment was to gain experience with the operational aspects and assess the accuracy and resolution of the method. Other and similar experiments have been described in the litterature.

The experiment was conducted in the Tromsø area in August 1981. The research vessel F/F "Johan Ruud" was used with the assistance from the workboat (sjark) "Hyas". The transmitter was suspended from F/F "Johan Ruud" and the receiver from "Hyas" with direct cable connection to F/F "Johan Ruud" (Fig. 2). Two sites were investigated, both with waterdepths about 100-120 meters and then the depth of transmitter and receiver was chosen to about 70 meters. This choice was dictated from the necessity of having all the sub-bottom reflections to arrive before the surface reflected signal. The baseline, i.e. the distance between the transmitter and receiver, varied from about 20 meters to about 200 meters with straight line propagation. This gives a range of incidence angles from approximately 18^0 to 73^0.

The transmitter was a prestressed piezoelectric transducer ring or cylinder with a diameter of 40 cm, height 16 cm and a weight in air of about 25 kg.

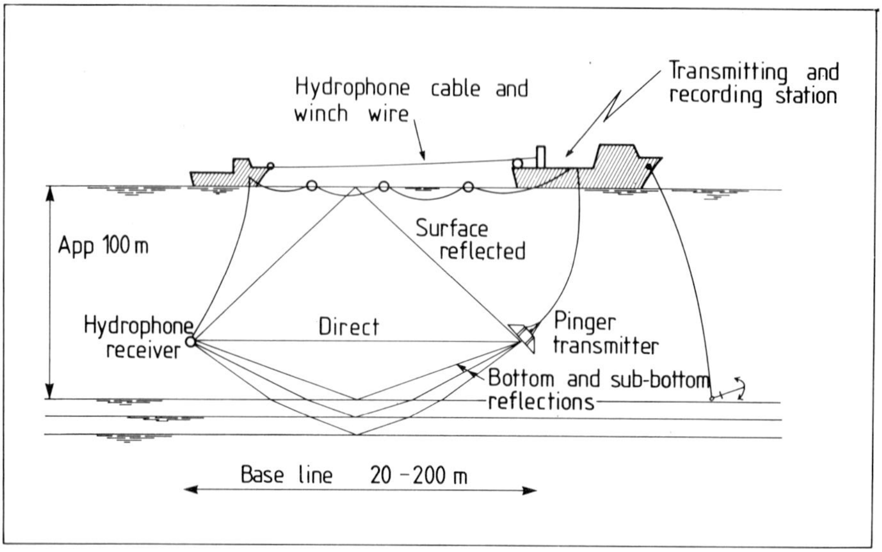

Fig. 2. Variable angle reflection experiment

In the experiment the frequencies used were 2, 4 and 6.3 kHz and source level at those frequencies was about 200 dB rel 1 μ Pascal. The signal received by the hydrophone was preamplified and transmitted to F/F "Johan Ruud" by cable via the assisting boat "Hyas". After amplification and band filtering in 1/3 octave filters the received signals were recorded on a high resolution paper recorder. A particular run started with "Hyas" bringing the hydrophone and cable out to about 200 meters in the down wind direction. With suspended hydrophone and its engine shut off, "Hyas" was then slowly pulled back to F/F "Johan Ruud". This took about 30 minutes which is therefore the duration of a complete recording.

The results are recorded signals as functions of time for 3 different frequencies. Fig. 3 shows one example of the recordings. The most significant arrivals are indicated by continuous lines and these arrivals have also been analysed. From the recordings one can deduce the source-receiver separations and therefore plot arrival times as a function of distance for the bottom and the sub-bottom reflections. Fig. 4 is the time-distance plots for the three frequencies for this location.

The results from the different frequencies were analysed separately by use of the ray parameter method. [Ref 14] The time-distance data

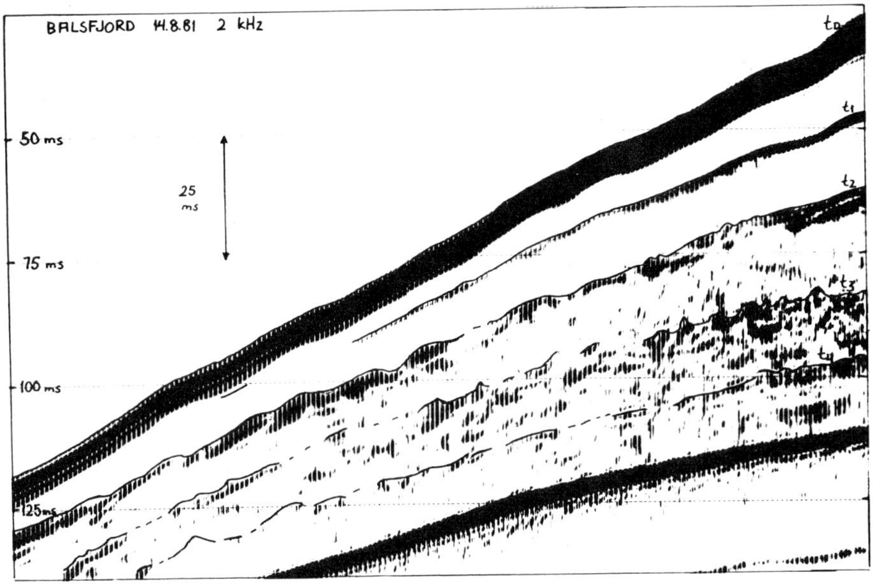

Fig. 3. Recorded signals in Balsfjord, f = 2 kHz

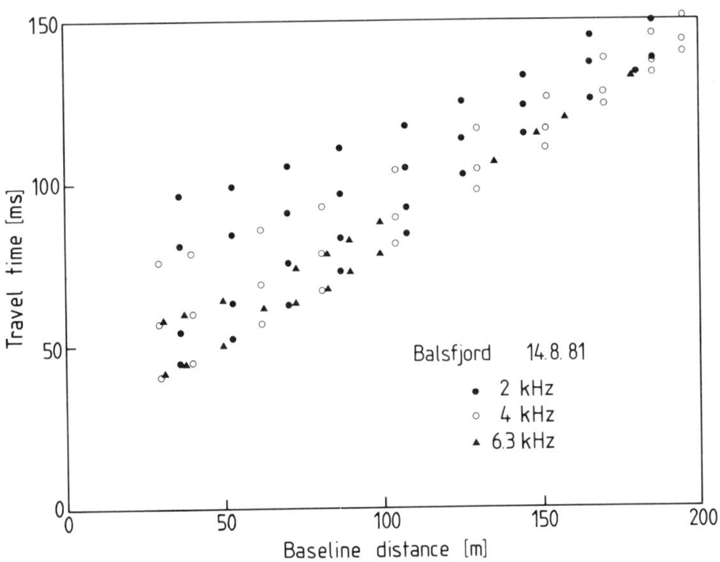

Fig. 4. Time-distance plot for three frequencies (Balsfjord)

was first fitted to a polynominal $t^2(x) = A + Bx^2$ by using a minimum square error fitting in x^2, t^2. The calculated wave velocities and layer thicknesses are given in Table I.

TABLE I: Balsfjorden

Layer no	Frequency [kHz]	Sound Velocity [m/s]	Thickness [m]
0	2	1423	26.6
	4	1444	26.2
	6.3	1450	27.4
1	2	1512	12.8
	4	1513	13.7
	6.3	1520	12.9
2	2	1501	16.3
	4	1491	14.9
	6.3	-	-
3	2	1773	13.3
	4	-	-
	6.3	-	-

Layer 0 is the water layer below the transducer and the hydrophone. Since the amplitudes of the received signals vary with the baseline length and penetration depth, while the detection level of the recorder is constant, there is a tendency to underestimate the sound velocities. In particular, this is the case at 2 kHz, where the rise time of the pulse is $\sim$ 3 ms. A deviation of 1.5 ms at 100 m baseline length is enough to explain the difference between the sound velocity found from CTD-data and the velocity calculated from the recorded traces. This problem can be solved by taking the available information on signal amplitudes into account. This also provides the possibility of including acoustic absorption in the classification procedure.

The experiment has demonstrated the possibility of obtaining detailed layering information by using a pinger source and relatively short baselines. Layer thickness down to 1.5 meters have been resolved and deep layers down to appriximatively 40 meters have been identified and analysed. The resolution is better than obtained with a sparker and the penetration is about the same. Several improvements in the operation, the data collection and the analysis are possible and should be implemented in future experiments.

4. SHEAR MODULUS DETERMINATION USING INTERFACE WAVES

Neither of the techniques discussed so far can give any valuable information of the shear modulus of sea-floor sediments. Knowledge of this modulus (modulus of rigidity) provides, however, a basic element in the judgement of the mechanical strength properties of a sediment or a porous rock. The shear modulus μ can be measured through the velocity v_s of acoustic shear waves ($v_s = \sqrt{\mu/\rho}$, ρ: mass density). Remote sensing of shear waves is, however, difficult, since liquids do not transmit such modes of propagation. In addition, shear wave excitation is rather inefficient with present methods, and the attenuation (particularly in soft sediments) is fairly large.

There exists, however, indirect means by which shear wave properties can be extracted from acoustic measurements. The dominating parts of seismic signals originating from earthquakes are identified as boundary waves propagating near the earth's surface with an exponential decay inwards. These modes are Rayleigh-type waves and Love-waves [18-20].

Already a century ago, Rayleigh [18] proved the existence of a surface wave along the free surface of a semiinfinite solid. This mode of propagation is called a Rayleigh-wave. It is characterized by a displacement amplitude decreasing exponentially inwards in the solid, such that the acoustic energy is confined roughly to the upper wavelength of the medium. The polarization is elliptical and lies in the sagittal plane (i.e. the plane of the wavevector and the normal to the surface). The Rayleigh-wave velocity v_R is deduced from the dynamic elasticity equation and the boundary conditions (stress free surface). For an isotropic solid

$$v_R = \frac{0.87 + 1.12\,\sigma}{1 + \sigma}\, v_s \tag{12}$$

where σ is Poisson's ratio. v_R is hence a proper measure of the shear modulus.

If the solid is in contact with a liquid, the wave solutions change because of the change in boundary conditions. For a typical water-soft sediment interface, the boundary wave velocity can be expressed empirically as

$$v = (0.87 + 0.25\,\sigma - 0.18\,\rho_1/\rho_2)v_s \tag{13}$$

Here ρ_1/ρ_2 is the ratio between the densities of the water and the bottom material. This wave mode is termed a Scholte-wave (or often in literature cited as a Stoneley-wave) [22-24]. Again, the boundary wave velocity is a good measure of v_s.

Normally the solid medium is not perfectly isotropic. Both shear modulus and density increase with increasing depth, either in a continuous manner or forming a layered structure. Since the boundary waves sense mainly the properties of the upper wavelength, this means that the wave propagation becomes dispersive. Relating the frequency dependence of the interface wave velocity to the nonhomogeneity in the elastic parameters of the sea-floor is a rather difficult task [see for instance Refs 25,26]. Numerical methods exist [27] which may be of great importance in connection with experimental work.

A simple model of a nonhomogeneous structure is to assume a soft homogeneous layer above a harder homogeneous substrate. Love [19] showed that horizontally polarized shearwaves may propagate mainly inside the layer with an exponential decrease in amplitude in the substrate. These surface waves (Love-waves) are also dispersive and may be used in modelling the shear modulus structure of the sea-floor along with Scholte-waves. In that case it may be convenient to assume a continuously increasing shear modulus and solve the shear wave equation. The solution is reminent of surface waves.

Field measurements of boundary wave propagation along the ocean bottom have been performed to some minor extent since the beginning of the 1960's. Bucker et al. [28] found that the attenuation of the Scholte-wave in a layer of sand was 0.2 - 0.3 dB/m for f $\sim$ 20-25 Hz. Both Davies [29] and Rauch [30] found the normal dispersion for boundary waves on a layered bottom. Rauch was able to fit his results to a detailed model for the stratified structure of the upper 10 m of the bottom. Similar experiments were performed in the North Sea by Schirmer [31].

In most of the prior experiments known by the authors, explosive charges fired in the water column or near the bottom were used for excitation of the waves. In the work which at the present is conducted by the authors, a vibration exciter is used as signal source. Similar techniques have been used by Stoll [32] and in geotechnical prospecting on land by Prange [33]. The signal may be pulsed or CW, and the frequency is continuously tunable above $\sim$ 10 Hz. This makes it particularly well suited for studying dispersive wave propagation. The boundary waves are detected by seismometers which may be oriented in the sagittal plane (Scholte/Rayleigh-waves) or normal to this plane (Love-waves). Detailed studies of seismometer configurations in boundary wave experiments and deepsea noise are reported in [34]. The experimental set-up is shown in Fig. 5.

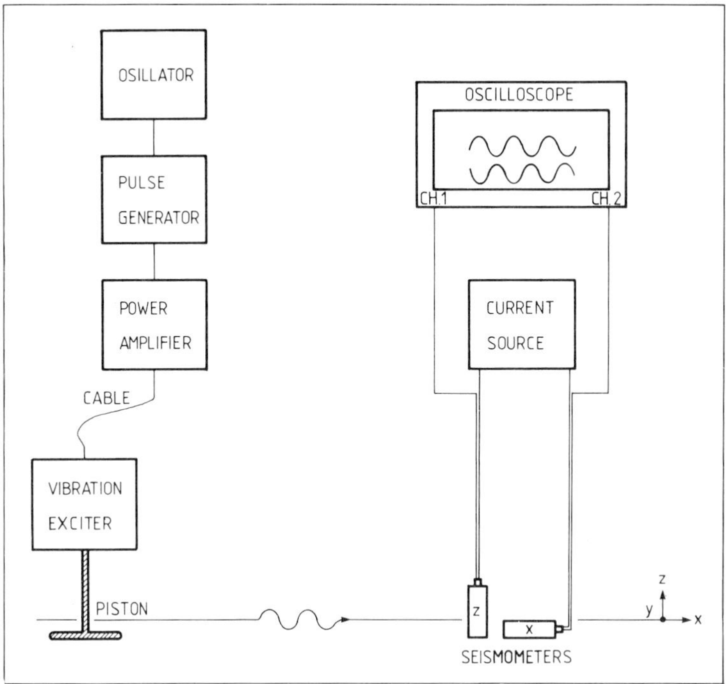

Fig. 5. Experimental setp-up for Rayleigh wave measurements

Some preliminary measurements were performed on land at Øysand beach near Trondheim. Both the elliptical polarization characterizing Rayleigh-waves and the horizontal displacement of Love-type waves were detected. By burying the seismometers into the ground, the exponential decay of the signals was mapped. Wave velocities were calculated from measured phase differences between two equally oriented seismometers at a fixed spacing. The frequency dependence of the Rayleigh-wave velocity shown in Fig. 6 proves the potential of the interface wave technique in determining the shear modulus profile of the upper few meters of the sea-floor.

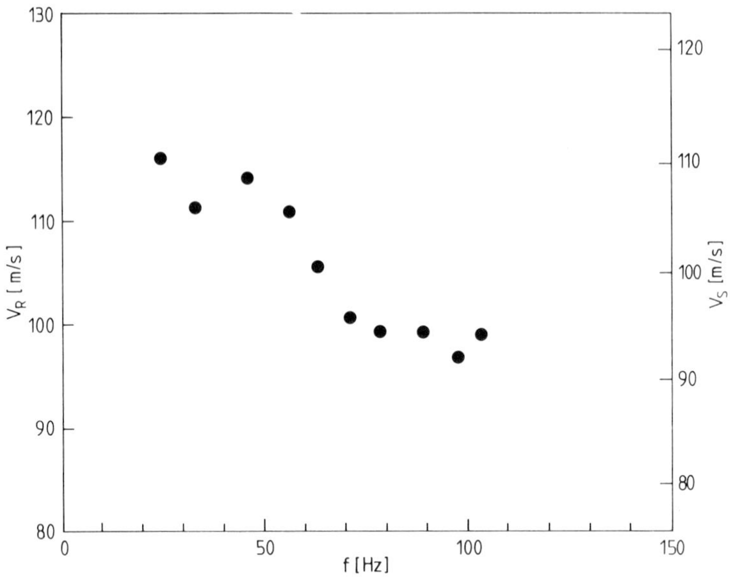

Fig. 6. Measured velocity v_R (left scale) vs frequency. Calculated shear velocity v_S is shown by the right hand scale. f = 20 Hz corresponds to $\lambda \approx 6$ m, f = 100 Hz to $\lambda \approx 1$ m. As a rule of the thumb, v_S at a given frequency corresponds to v_S at a depth of $\lambda/2$.

In conclusion, more research is needed, both theoretically and experimentally, to develop Scholte- and possibly Love-type measurements into a field method, but the results obtained so far are encouraging.

DISCUSSION

This paper has mainly described two new techniques for the determination of the compressional and shear wave velocities of the upper layers of the ocean bottom. The experiments reported here as well as other similar experiments reported in the literature have demonstrated that such velocity structures can be mapped. Further developments are, however, required if the methods are to be used in practical investigations. Suitable signal sources and geometry of source/receiver have to be found and the operational procedure must be improved. The actual determination of the shear and compressional moduli requires also the solution of an inverse problem. In particular for the boundary wave technique an inverse method must be found to convert dispersive velocity data into shear modulus as a function of depth.

REFERENCES

1. M.A. Biot: "Theory of propagation of elastic waves in a fluid-saturated porous solid, I. Low-frequency range". J. Acoust. Soc. Am. 28, 168-178 (1956).
2. M.A. Biot: "Theory of propagation of elastic waves in a fluid-saturated porous solid, II. Higher frequency range". J. Acoust. Soc. Am. 28, 179-191 (1956).
3. R.D. Stoll and G.M. Bryan: "Wave attenuation in saturated sediments". J. Acoust. Soc. Am. 47, 1440-1447 (1969).
4. R.D. Stoll: "Acoustic waves in saturated sediments" in Physics of Sound in Marine Sediments, edited by L.D. Hampton (Plenum Press, New York, 1974).
5. R.D. Stoll: "Acoustic waves in ocean sediments", Geophys. 42, 715-725 (1977).
6. T. Gertsma and D.C. Smit: "Some aspects of elastic wave propagation in fluid-saturated porous solids". Geophys. 26, 169-181 (1961).
7. J.M. Hovem and G.D. Ingram: "Viscous attenuation of sound in saturated sand". J. Acoust. Soc. Am. 66, 1807-1812 (1979).
8. J.E. White: "Seismic Waves: Radiation, Transmission and Attenuation". McGraw Hill 1965.
9. E.L. Hamilton: "Elastic properties of marine sediments". J. Geophys. Res. 76, 579-604 (1971).
10. E.L. Hamilton: "Prediction of Deep-Sea Sediment Properties, State-of-the-art", in A.L. Inderbitzen (Ed) Deep-Sea-Sediments, Physical and Mechanical Properties (Plenum Press 1974).
11. E.L. Hamilton: "Geoacoustic Models of the Seafloor" in Physics of Sound in Marine Sediments, edited by L.D. Hampton (Plenum Press, New York 1974).
12. E.L. Hamilton: "Geoacoustic Modeling of the Seafloor", J. Acoust. Soc. Am. 68, 1313-1340 (1980).
13. S. Buchan, D.M. McCann and D. Taylor-Smith: "Relationship between the acoustic and geotechnical properties of marine sediments". Q. Jl. Engng. Geol. Vol. 5, 1972, pp 265-284.
14. G.M. Bryan: "The hydrophone-pinger experiment". J. Acoust. Soc. Am. 68 (5), 1403-1408 (1980).
15. G.M. Bryan: "Sonobuoy measurements in thin layers", in Loyd Hampton (Ed) "Physics of Sound in Marine Sediments", p 119-130 (Plenum Press 1974).
16. K. Wrolstad, "Interval velocity and attenuation measurements in sediments from marine seismic reflection data", J. Acoust. Soc. Am. 68, 1415-1435 (1980).
17. A.G. McKay and P.M. McKay: "Compressional-wave velocity measurement in the seabed materials by use of equipment deployed near, but above the bottom", J. Acoust. Soc. Am. 71 (4), 871-877, April 1980.
18. Lord Rayleigh: "On waves propagated along the plane surface of an elastic solid", Proc. London Math. Soc. 17, 4-11 (1885).
19. A.E.H. Love: "Some problems of Geodynamics", Cambridge University Press (1911).

20. H. Überall: "Surface waves in acoustics" in W.P. Mason and R.N. Thurston (Ed) "Physical Acoustics, Vol. X", p 1-60 (1973).

21. I.A. Viktorov: "Types of acoustic surface waves in solids (review)", Sov. Phys. Acoust. 25, 1-9 (1979).

22. R. Stoneley: "The elastic waves at the interface of separation of two solids", Proc. Royal Soc. London A 106, 416 (1924).

23. J.G. Scholte: "The range of existence of Rayleigh and Stoneley waves", Mon. Not. R. Astron. Soc. Geophy. Suppl. 5, 120-6 (1947).

24. E. Strick and A.S. Ginzbarg: "Stoneley-wave velocities for a fluid-solid interface", Bull. Seismol. Soc. America 46, 281-92 (1956).

25. F.S. Grant and G.F. West: "Interpretation Theory in Applied Geophysics", McGraw-Hill (1965).

26. W.M. Ewing, W.S. Jardetzky, and F. Press: "Elastic Waves in Layered Media", McGraw-Hill (1957).

27. F.R. Di Napoli and R.L. Davenport: "Numerical models of underwater acoustic propagation", in "Ocean Acoustics", Ed. by J.A. DeSanto, Springer (1979).

28. H.P. Bucker, J.A. Whitney, and D.L. Keir: "Use of Stoneley waves to determine the shear velocity in ocean sediments", J. Acoust. Soc. Am. 36, 1595-6 (1964).

29. D. Davies: "Dispersed Stoneley waves on the ocean bottom", Bull. Seismol. Soc. America 55, 903-18 (1965).

30. D. Rauch: "Experimental and theoretical studies of seismic interface waves in coastal waters" in "Bottom-interacting Ocean Acoustics" ed. by W.A. Kuperman and F.B. Jensen, Plenum Press (1980).

31. F. Schirmer: "Experimental determination of properties of the Scholte wave in the bottom of the North Sea" in "Bottom-interacting Ocean Acoustics" ed. by W.A. Kuperman and F.B. Jensen, Plenum Press (1980).

32. R.D. Stoll: "Theoretical aspects of sound transmission in sediments", J. Acoust. Soc. Am. 68, 1341-9 (1980).

33. B. Prange: "Primary and secondary interferences in wavefields", Proc. of "Dynamical Methods in Soil and Rock Mechanics", Karlsruhe 1977, publ. Rotterdam (1978).

34. "Marine Geophysical Researches", Special issue of "The Lopez Island OBS intercomparison Experiment" ed. by B.D. Loncarevic (1981).

MACHINE ANALYSIS OF ACOUSTICAL SIGNALS

by

Joseph N. Maksym, Anthony J. Bonner,
C. Ann Dent, and Gavin L. Hemphill

Defence Research Establishment Atlantic
Box 1012
Dartmouth, Nova Scotia, Canada.
B2Y 3Z7

Abstract

Research toward machine analysis of acoustical signals is described. The approach is that of the **expert system**, where knowledge about the physical world that produced the signals is used by the system in the interpretation process. Expert systems have already matched human capabilities in certain applications such as chemical structure analysis and certain areas of medical diagnosis, suggesting that acoustical signal analysis will meet with similar success. The basic approach is described in the paper, using examples from an experimental system being developed for the interpretation of acoustical signals. The paper also identifies some of the more important unresolved research problems.

1 INTRODUCTION

Expert systems have been around for about a decade, and currently represent a substantial portion of experimental Artificial Intelligence (AI) research. In general, expert systems are computer programs that perform complex problem-solving tasks, normally associated with an area of human expertise. Some of the earlier examples are **MYCIN**, a diagnostic expert system in medicine **[Shortliffe, 1976]**, **DENDRAL**, an interpretation system for mass spectrogram data **[Feigenbaum, 1977a]**, and **HEARSAY**, an interpretation *(understanding)* system for speech in a limited task domain **[Lesser, 1977]**. The science of expert systems, known as knowledge engineering, is still relatively young and in a stage of rapid development. Consequently there is, as yet, little agreement among AI researchers on the best general design principles, with each new system acting as a test bed for research ideas suggested by the application area.

In this paper we describe the expert system approach to the interpretation of signals from underwater acoustic sensors. The analysis of signal data requires not only the kind of numerical computation normal in conventional signal processing, but also the problem-solving techniques employed by human beings. The analysis of acoustical signals depends upon two things; an appropriate display of the signal information, and extensive knowledge about the signal domain. The human analyst must know how to associate features in the signal data with physical processes in the ocean. For example, drive mechanisms of ships are known to manifest themselves as harmonically-related sets of spectral lines characteristic of the different drive-system components (engines, pumps, turbines, and so forth). In addition, the properties of sensor systems and sound propagation will affect the way in which the noise mechanisms appear in the processed signal data.

These issues will be illustrated with examples from an experimental expert system, called **INTERSENSOR**, that is being developed at DREA. The first step in the development of **INTERSENSOR** is an examination of the way that human experts think about the signal interpretation process.

NATO ASI Series, Vol. F1
Issues in Acoustic Signal/Image Processing and Recognition
Edited by C. H. Chen

2 INTERPRETATION OF ACOUSTICAL SIGNALS BY HUMAN EXPERTS

Figure 1 shows a display of underwater acoustical signals in the form of a spectrogram. The figure represents twenty minutes of signal data recorded at sea on one beam of an experimental sensor array. The horizontal axis is frequency, increasing to the right. The vertical axis represents successive spectral records or "looks", each "look" representing several seconds of signal data.

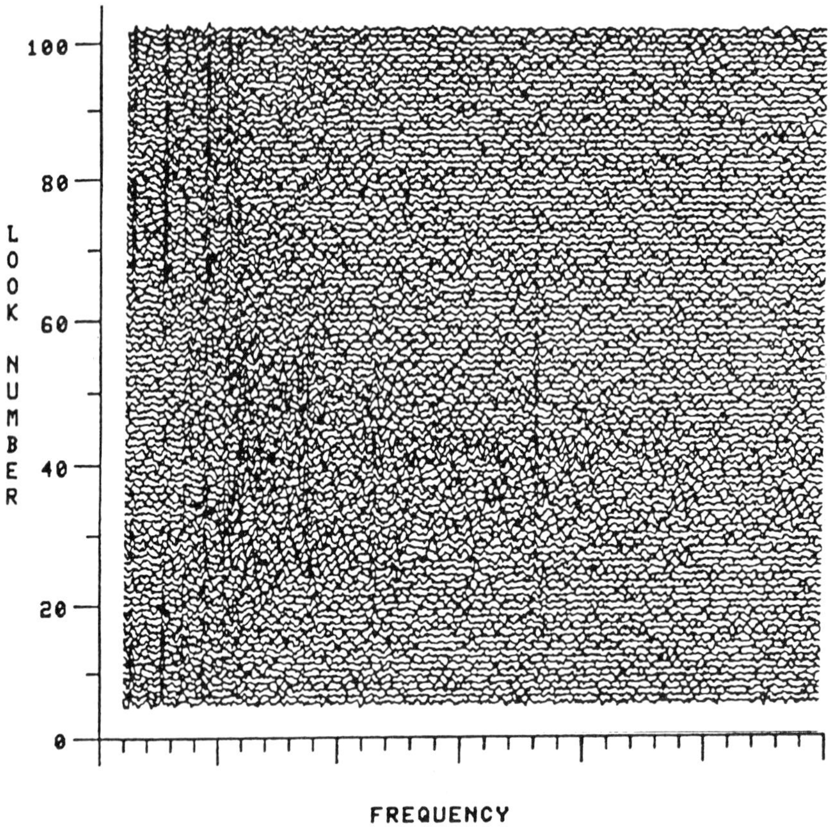

Figure 1. **Spectrogram Display**

The spectrogram displays patterns that relate to specific generators of acoustic signals received by an underwater sensor. Sets of harmonically-related spectral lines, for example, are typical manefestations of rotating machines. There is evidence of these in Figure 1, where the lines are very likely produced by the propellers of merchant ships in the vicinity of the acoustic sensing system.

Consideration of the expert acoustic analyst's interpretation process suggests a multi-level representation of the data as in Figure 2. Although it is not clear what happens at the lowest levels of this data hierarchy, the human expert's interpretation process appears to be carried out by the manipulation of symbolic labels. The labels (track samples, lines, harmonic sets, etc) refer to complex conceptual objects in the mind of the expert analyst. A consistent and correct interpretation is equivalent to the formation of a node-link graph similar to Figure 2.

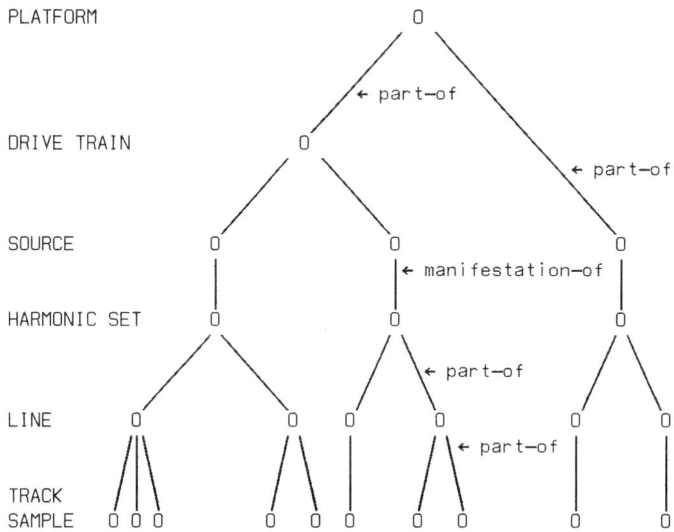

Figure 2. Multi-level Data Representation

As an example, consider the recognition of a sound source that emits harmonically-related spectral lines. The first step is the recognition of lines. An expert analyst looks at Figure 1 and matches the picture elements *(pixels)* in the spectrogram against his concept of a line. The line concept not only defines the properties of pixels that comprise it, but introduces new properties associated with a line, such as frequency, linewidth, stability, start time, end time, and so forth. It then becomes possible to consider more complex objects such as harmonic sets in terms of syntactic and semantic properties of lines, rather than the simpler properties of the original acoustic spectrogram data. In similar fashion the presence of acoustic sources is deduced by comparing the properties of harmonic sets (number, strength, stability, etc. of harmonics) against expectations generated by each potential source hypothesis.

This kind of multi-level data representation is central to the expert system approach to machine analysis and interpretation of acoustical signals.

3 A MODEL EXPERT SYSTEM

Figure 3 shows a model expert system (adapted from **[Shortliffe, 1976]**). The system is composed of a number of separate components. In the acoustic data analysis context, the ***ANALYSIS SYSTEM*** takes in a variety of ***ACOUSTIC DATA*** and effects the actual ***INTERPRETATION***, using a store of analysis information (***KNOWLEDGE BASE***). The ***EXPLANATION SYSTEM*** responds to queries from the user about the way in which the analysis is being done. The ***KNOWLEDGE ACQUISITION SYSTEM*** is used by experts (who are not necessarily computer experts) to update the ***KNOWLEDGE BASE*** as required.

One of the characteristic features of such systems is careful separation of the specialized knowledge of the problem domain (***KNOWLEDGE BASE***) from the code (sequence of program instructions) that applies the knowledge. In the ***KNOWLEDGE BASE***, information is represented *explicitly*. If this knowledge were incorporated directly into the code, it would be difficult for acoustic analysis experts to verify its authenticity. Also, it would not be clear what knowledge is used and where. Consequently, changing the knowledge upon which the system is based would be very difficult (i.e., the system would be hard to modify and extend).

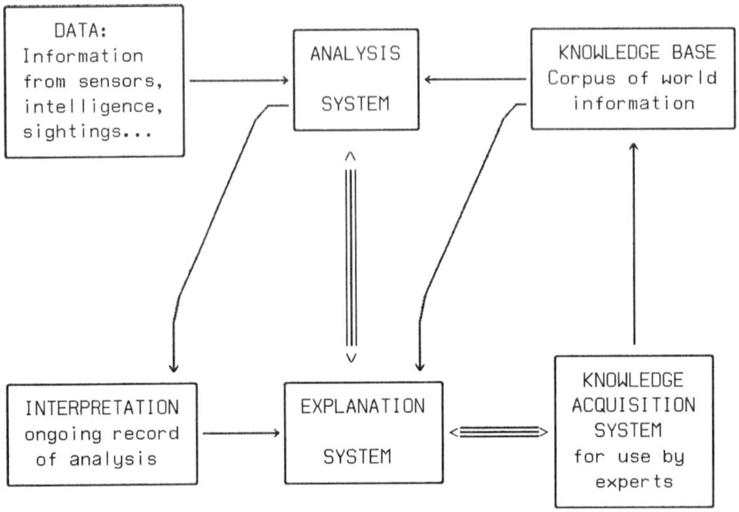

Figure 3. Expert Analysis System

It has been found that high performance in these programs depends critically on domain-specific knowledge. We therefore focus our discussion on the ***KNOWLEDGE BASE***--specifically on how acoustic analysis knowledge is represented and used in forming an interpretation.

4 REPRESENTATION OF KNOWLEDGE

In AI, as in numerical analysis, the *data structures* are of fundamental importance. For numerical analysis, matrices of coefficients provide convenient data structures, and the concepts of matrix algebra provide a powerful abstraction for thinking about matrix manipulations. In AI, the data structures are often generalized networks of symbols which may be thought of as node-link graphs with properties associated with the nodes and links **[Feigenbaum, 1977a]**.

Node-link graphs can be used to represent the structure of objects in the acoustic signal world (e.g., platforms, engines, harmonic sets, lines). A node-link graph similar to that in Figure 2 can form the basis for data interpretation in a computer program. The nodes in the graph represent levels of data abstraction, and the links represent relations between these nodes.

The lowest level in the graph is the ***TRACK SAMPLE*** level. Each node at this level represents a point in the acoustic data. The second level is the ***LINE*** level. A line is made up of track samples. So, each track sample node is related to a line node by a ***part-of*** relation. The third level is the ***HARMONIC SET*** level.

At the next level in the hierarchy, each harmonic set is attributed to a possible ***SOURCE*** (such as a rotating propeller blade). A harmonic set is a ***manifestation-of*** a source in the acoustic data. Some sources are combined into ***DRIVE TRAINS*** using knowledge about drive-train structure and how it affects lines in the acoustic data. Finally, drive trains and other sources are associated with ***PLATFORMS*** using knowledge about the drive trains and sources present on particular classes of platforms.

In **INTERSENSOR**, the experimental expert system for acoustical signal interpretation at DREA, many node-link graphs such as the one in Figure 2 are stored. They form models of the acoustic signal world. The system uses these models as templates for building node-link graphs to represent interpretations of data.

The node-link graph is one of two kinds of knowledge-encoding in the **INTERSENSOR** knowledge base. The other is the ***production rule.*** Rules *control* the analysis--they create nodes and link them together. Production rules are an integral part of many expert systems, including **INTERSENSOR**. Discussions of production systems as a programming methodology can be found in the literature **[Davis and King,1976]**, and are outside the scope of this paper. We give an example rule from **INTERSENSOR** in Section 5.

Node-link graphs are implemented in **INTERSENSOR** as networks of units. A unit is a data structure with a set of slots for storing information about the unit itself and about its links to other units. Each node in Figure 2 is implemented as a unit; each link is implemented as a slot.

Units are the fundamental entities of the **Unit Package [Stefik, 1980]**, **[Smith, 1980]**, a knowledge representation and acquisition system incorporated into **INTERSENSOR**. The **Unit Package** is representative of the newer knowledge representation languages. Such tools were not available to the designers of earlier expert systems for machine interpretation of signals, such as **HASP [Nii, 1978]** or **SIAP [Drazovich, 1978]**.

The **Unit Package** (developed for the planning of experiments in molecular genetics **[Friedland, 1979]**) has been modified and extended to meet the specific requirements of acoustic analysis. The modified version, named **Unit* [Smith et al., in preparation]** to distinguish it from the original, is used in **INTERSENSOR**.

One of the major advantages of such packages is that they incorporate sophisticated editors that assist an analysis expert to understand and modify a knowledge base. There is evidence that *direct* interaction (i.e., without the presence of a computer scientist intermediary) eases construction of larger, more accurate knowledge bases **[Friedland, 1979]**.

4.1 THE KNOWLEDGE BASE

4.1.1 PROTOTYPES

Each unit in an **INTERSENSOR** interpretation is constructed from a ***prototype*** unit in the **INTERSENSOR** knowledge base (Figure 3) and is called an ***instance*** of its prototype. A line unit, for example, is an instance of a line prototype. A prototype specifies the slots its instances will have and the range of data values that can fill them. This is called **property inheritance**.

Figure 4 shows the TYPE-X-TRAWLER prototype. Every instance of TYPE-X-TRAWLER will have this set of slots. Most of the slots in Figure 4 are filled with ***value-restrictions***. The SPEED slot, for example, contains the restriction *between 0.0 and 15.0 knots*. The value of the SPEED slot in an instance of TYPE-X-TRAWLER must satisfy this restriction. As long as it satisfies the restriction of its prototype, a slot in an instance may itself hold a restriction. For example, the SPEED slot in an instance of TYPE-X-TRAWLER could hold the restriction *between 6.0 and 8.0 knots.*

Not all restrictions are numerical ranges. The DRIVE-TRAINS slot of the TYPE-X-TRAWLER unit, for example, states that every TYPE-X-TRAWLER has one drive train made up of a single eight-cylinder, two-stroke diesel engine directly coupled (DD stands for Diesel Direct) to a three-bladed propeller.

4.1.2 THE GENERALIZATION HIERARCHY

An **INTERSENSOR** interpretation (Figure 3) is made up of instances of prototype units. As the interpretation is developed, slots in these instances are filled with increasingly restrictive values. This is one way in which **INTERSENSOR** makes an interpretation more specific. Another way is by giving an instance a more specific prototype. This is possible because units are organized in a ***generalization hierarchy***.

Figure 5 displays part of the generalization hierarchy in the **INTERSENSOR** knowledge base.

```
Unit: TYPE-X-TRAWLER
  SIGNALS:          [A SIGNAL]
  BEARING:          [0 360]
  DRIVE-TRAINS:     [A DRIVE-TRAIN with
                      ENGINES: [An ENGINE with
                                 #-CYLINDERS: 8
                                 STROKE: 2]
                      PROPELLERS: [A PROPELLER with
                                    #-BLADES: 3]
                      MODE: [DD]
  RANGE:
  SPEED:            [0.0 15.0]
  LOAD:             [EMPTY FULL]
```

Figure 4. TYPE-X-TRAWLER Prototype

```
OBJECT
    MACHINE
        PLATFORM
            *PLATFORM1*
            SURFACE-VESSEL
                MERCHANT
                    FISHING-TRAWLER
                        TYPE-X-TRAWLER
                            *TYPE-X-TRAWLER1*

        DRIVE-TRAIN
        SOURCE
            ENGINE
            PROPELLER
            SHAFT

    SIGNAL
        HARMONIC SET
        LINE
        TRACK-SAMPLE
```

Figure 5. Units Generalization Hierarchy

("*" before and after a unit name means that the unit represents an instance and is part of an interpretation.) The unit immediately above and to the left of another unit in the hierarchy is called its ***generalization***.

Each unit inherits the slots of its generalization. Figure 6 shows the MERCHANT unit printed in a mode that shows inheritance information. MERCHANT inherits the slots of the PLATFORM unit. These include the SIGNALS slot, which is itself inherited by PLATFORM from the MACHINE unit. In addition MERCHANT can have slots of its own such as the LOAD slot (acoustic emissions can vary with loading of a merchant vessel). The LOAD slot is marked "*Top*" because its definition is first encountered in MERCHANT. Moving down the generalization hierarchy, units become more specific in two ways: (i) by further restricting the available ranges of slot values; and (ii), by adding more slots.

5 A SAMPLE INTERPRETATION BY INTERSENSOR

In this section we illustrate the formation of the node-link graph, as shown in Figure 2, during a sample interpretation of acoustic data. For this example we have assumed that the data is a

```
Unit: MERCHANT
Generalization: PLATFORM
  SIGNALS:          from MACHINE     [A SIGNAL]
  BEARING:          from PLATFORM    [0 360]
  DRIVE-TRAINS:     from PLATFORM    [A DRIVE-TRAIN]
  RANGE:            from PLATFORM
  SPEED:            from PLATFORM
  LOAD:             *Top*
```

Figure 6. Property Inheritance by a Unit from its Generalization

spectrogram showing the typical features of a fishing trawler that we have named *type-x-trawler*. The interpretation starts at the track sample level of Figure 2. As the system analyzes data, it constructs a similar node-link graph, working from bottom-to-top. A complete graph is an outline of the final interpretation. **INTERSENSOR** then proceeds to make this interpretation more specific--to fill in the details and determine exactly which class of platform is manifest in the data.

To simplify the initial coding, we have implemented **INTERSENSOR**'s low-level signal-processing as a separate front-end module. The designs of the signal-processing and higher-level interpretation modules have been carried out concurrently and interactively. The front-end processor at this stage of our experimental design of **INTERSENSOR** is a sequential likelihood ratio tracker, described later in this paper. Its main function is to pass a list of detected track samples to the rest of the system as each "look" of acoustic signal data is processed.

Each track sample passed to the rest of analysis system includes a frequency, smoothed frequency, level, linewidth, cumulative likelihood and track number (a label identifying track samples that represent the same line at successive looks). **INTERSENSOR** constructs a track sample unit from each track sample and stores the associated information in its slots (Figure 7). These track sample units form the nodes in the lowest level of Figure 2. This is the start of the **INTERSENSOR** interpretation of the data.

```
Unit: TRACK-SAMPLE3
  LEVEL:                 5.3516
  BANDWIDTH:             1.1623
  FREQUENCY:             30.112
  LIKELIHOOD:            9.2859
  LOOK-NUMBER:           18
  SMOOTHED-FREQUENCY:    30.315
  TRACK-NUMBER:          18003
```

Figure 7. TRACK SAMPLE Unit

Using the track number, **INTERSENSOR** associates track sample units from successive looks with single line units (These are the associations found by the front-end.). Figure 8 illustrates a line unit. The links in Figure 2 between a line node and its track sample nodes are represented by the slot called ***TRACK SAMPLES***.

INTERSENSOR also uses information other than track number to associate track samples into lines. Because lines can be several frequency bins wide, the front-end often detects more then one track sample within each line. Using linewidth information, **INTERSENSOR** associates track sample units from the *same* look with single line units.

Having grouped the track samples into lines, **INTERSENSOR** now groups the lines into harmonic sets. This operation is not straightforward in general, because of the combinatorics of the problem. A *generate-test-prune* approach is used to first generate potential harmonic sets, and then to prune out unlikely ones. For this example, eight harmonic sets were generated, and all but one were pruned (Figure 9).

```
Unit: LINE1
  LEVEL:            5.7623
  BANDWIDTH:        1.0347
  END-TIME:         68
  FREQUENCY:        30.315
  TRACK-SAMPLES:    [TRACK-SAMPLE3, TRACK-SAMPLE7,
                            TRACK-SAMPLE12, ...]
  START-TIME:       18
```

Figure 8. **LINE Unit**

```
Unit: HARMONIC-SET1
  END-TIME:         68
  HARMONICS:        [LINE1, LINE2, LINE3, ...]
  SPACING:          10.105
  START-TIME:       18
```

Figure 9. **HARMONIC SET Unit**

To generate harmonic sets, **INTERSENSOR** first compares each line against every other line. When a pair of lines is found such that the frequency of either one is an integer multiple of their frequency difference (within some tolerance), then we have the beginnings of a harmonic set. **INTERSENSOR** then tries to fit the other lines into this harmonic set. A line fits into a harmonic set if its frequency is an integer multiple of the harmonic set's spacing (within some tolerance.) If a line fits, then it is added to the harmonic set, and a new estimate of the spacing is determined (using a least squares fit).

Up to this point **INTERSENSOR** has been using only *signal processing knowledge*. This has allowed **INTERSENSOR** to construct the bottom three levels of Figure 2. (This is processing that a human analyst does with his eye.) In order to interpret the data further, different knowledge is required--knowledge of platforms (their capabilities and noise sources) and how noise sources manifest themselves in spectral data. This additional knowledge allows the formation of an interpretation up to the platform level. As the rest of our example interpretation will show, **INTERSENSOR** can also use this knowledge to construct all the upper levels of Figure 2.

After the harmonic sets have been generated and pruned, **INTERSENSOR** attempts to deduce the sources of the remaining harmonic sets. We examine, in particular, the instance unit in Figure 9 and illustrate the action of specific knowledge to infer that there is evidence for a particular acoustic source.

The structure in Figure 10 is constructed by knowledge sources that use information stored in the Harmonic-Set1 unit in Figure 9. Knowledge sources are those collections of rules of analysis obtained from human experts that effect the interpretation. The knowledge sources include a rule that says:

If 1) Harmonics are broad, and

2) Every n^{th} harmonic is dominant.

Then there is suggestive evidence (CF=0.8) that the acoustic source is a propeller shaft with n blades.

The action of the above rule on the information in ***HARMONIC-SET1*** results in the formation of the instance units, ***SHAFT1*** and ***PROPELLER1***, in Figure 10. The parameter *CF* in the rule is a

```
Unit: PLATFORM1
  SIGNALS:          [HARMONIC SET1]
  BEARING:
  DRIVE-TRAINS:     [DRIVE-TRAIN1]
  RANGE:
  SPEED:

Unit: DRIVE-TRAIN1
  SIGNALS:          [HARMONIC SET1]
  RATE:             10.105
  RPM:              606.30
  ENGINES:
  GENERATORS:
  MOTORS:
  PROPELLERS:       [PROPELLER1]
  SHAFTS:           [SHAFT1]
  TURBINES:
  MODE:

Unit: PROPELLER1
  SIGNALS:          [BLADE-LINES1]
  RATE:             30.315
  RPM:              606.30
  #-BLADES:         3

Unit: SHAFT1
  SIGNALS:          [SHAFT-LINES1]
  RATE:             10.105
  RPM:              606.30
```

Figure 10. Units in a Skeleton Interpretation

measure of the inferential strength of the rule, a property described further in Section 7. Next, the system invokes the following fragment of knowledge: a propeller is driven by a shaft, a shaft and a propeller are parts of a drive train, and a drive train is part of a platform. This knowledge is used to build shaft, drive train and platform units, and to fill in appropriate values in slots of the skeleton interpretation in Figure 10. This is analogous to the skeletal interpretation initially formed by a human analyst.

To make its interpretation more specific, the system tries to invoke another fragment of knowledge: if the propeller has three blades and the eighth shaft line is dominant (ignoring blade lines), then the platform is probably a *type-x-trawler*. **INTERSENSOR** concludes that in the track sample data, the eighth shaft line is dominant. It has already concluded that the propeller has three blades. So, this knowledge is invoked, and the platform unit is classed as a *type-x-trawler*. This is the final interpretation of the data.

It is important to note that the machine interpretation gives the most specific interpretation that is consistent with the data. If insufficient data is available to support a trawler hypothesis, there can still exist a more general hypothesis such as surface vessel with three-bladed propeller, for example.

6 ARCHITECTURE OF THE INTERSENSOR SYSTEM

6.1 SIMILARITIES BETWEEN INTERSENSOR AND HEARSAY

HEARSAY II [Lesser, 1977] is an expert system for speech understanding that introduced a number of generally-useful concepts for constructing expert systems. There are a number of analogues between **INTERSENSOR** and the **HEARSAY II** speech understanding system. These analogues are:

1) A *BLACKBOARD*. The blackboard is the place where the system records the results of its reasoning. The blackboard has a number of different levels. Each level represents an interpretation of the raw data. (In **INTERSENSOR**, the blackboard levels are: Platform, Noise-Source, Harmonic-Set, Line and Line-Segment.)

2) *KNOWLEDGE-SOURCES*. A knowledge-source (KS) uses information at one level to infer information at another level. Each knowledge-source has available to it all of the results on the blackboard. The various knowledge sources are independent. That is, each one knows nothing about the others. This makes programming easier and understandable since the problem is now broken down into independent subproblems. (In **INTERSENSOR**, the ***Deduce-Trawler*** knowledge-source uses information at the ***Noise-Source*** level to infer information at the ***Platform*** level, but the ***Deduce-Trawler*** KS knows nothing about the ***Make-Harmonic-Set*** KS). The **HEARSAY** architecture is well suited to problems which can be broken down in this way. However, when such a breakdown is not possible, programming in this architecture is rather clumsy **[Davis and King,1976]**.

3) A *SCHEDULER*. When a change occurs at one level which some KS can use to infer a change at another level, then that KS is scheduled for execution. At any given time, there may be many KSs awaiting execution. The Scheduler decides in which order they will be executed. (In **INTERSENSOR**, the scheduler is currently designed to make the system data-driven; that is, the low-level KSs are executed first. Much work needs to be done on the **INTERSENSOR** scheduler so that it can make more informed decisions).

4) A *BLACKBOARD-HANDLER*. Each knowledge-source records its conclusions by using the blackboard-handler. The blackboard-handler is a collection of routines used for changing the blackboard. (In **INTERSENSOR**, they include Unit-Access functions, such as PUTVALUE, PUTFIELD etc.).

When a KS is created, the blackboard handler must be informed of the kind of changes that the KS will respond to. Then during run-time, the blackboard-handler schedules a KS for execution when those changes occur. For example, when the blackboard-handler adds new ***Line***-units to the blackboard, it also schedules the ***Make-Harmonic-Set*** KS for execution.

6.2 STRUCTURE OF INTERSENSOR

In the preceeding section we discussed analogues between **INTERSENSOR** and the **HEARSAY II** system. Here we describe the structural features of the ***INTERSENSOR SYSTEM*** itself. in ***INTERSENSOR*** the KSs are independent and can in principle be implemented in different ways with different techniques. A KS may be a piece of SAIL or LISP code, a signal processing algorithm, a production system, or a KS may have a **HEARSAY** architecture within itself (with its own local blackboard and KSs.)

To date, most **INTERSENSOR** KSs have been **INTERLISP** functions. This approach was taken for initial simplicity. And, as long as each KS did not contain too much knowledge, this approach was feasible. However, as the quantity and diversity of knowledge within each KS has increased, the **INTERLISP** functions have become unwieldy. Where possible, the KSs have been broken down into a number of smaller, independent KSs. However, independence has not always been possible to achieve. For these KSs, we are experimenting with another approach that appears promising. This is described in the following paragraphs.

Each KS is implemented as a combination of **INTERLISP** function and an associated unit. The **INTERLISP** function represents *procedural* knowledge, and the unit represents *declarative* knowledge **[Winograd, 1975]**. Procedural knowledge is a description of *how* to do something (algorithms are procedural). Declarative knowledge is a description of *what* something is (the value of a global variable, for example, is declarative).

We have separated the declarative knowledge that an **INTERLISP** function uses from the function definition (ie., its LISP code.) The declarative knowledge is now stored in the unit associated with the function.

This declarative knowledge includes such things as threshold settings, ratios, frequency ranges, heuristic rules, etc. For example, a slot called ***THRESHOLD*** in a unit called ***TRACKER*** might contain the detection-threshold for a line-tracking function. Similarly, a slot called ***TEST*** in a unit called ***MAKE-HARMONIC-SET*** might contain a list of heuristics-rules. The function associated with this unit could use these heuristics for testing the "goodness" of a harmonic-set.

By taking this information out of the function definition and placing it in a unit, the function definition is made simpler. The resulting LISP code is cleaner, and easier to understand and modify. By storing the information in a unit, all of the declarative knowledge used by the function is clearly identified and gathered together in one location. Furthermore, the declarative knowledge can be edited through the **UNITS EDITOR** by someone who has no knowledge of **INTERLISP**.

These advantages are typical when facilities such as the **UNIT PACKAGE** are exploited. We have found the unit package invaluable for organizing declarative knowledge and making it possible to maintain clean dividing lines between the many components of the **INTERSENSOR** system. This has made it easier to keep track of the knowledge that **INTERSENSOR** is based upon, which is often detailed and easy to forget.

7 INFERENCE MECHANISMS

In the preceding sections we described the formation of an interpretation of acoustical data in the form of a node-link graph similar to that in Figure 2. There are two problems with this:

1) Acoustic signals are corrupted by noise so that we are never absolutely certain about conclusions drawn from them.

2) Knowledge that implements the interpretation process is often uncertain, so that conclusions, even from perfectly reliable evidence, are uncertain.

In this section we examine inference mechanisms, methods of quantifying the certainty of conclusions made from uncertain observations and knowledge. Our work with **INTERSENSOR** implements two methods of quantifying inferences: a probabilistic approach that, so far, is used only at the lowest level of the data hierarchy, and a pseudo-probability approach that uses subjective measures of belief similar to those in the **MYCIN** system **[Shortliffe, 1976]**.

7.1 PSEUDO-PROBABILITY METHODS

The judgmental rules that are used by acoustic experts are very similar in form to the diagnostic rules in the **MYCIN** system. Typically they are approximate implications such as:

$$E_1 \text{ suggests } H_1 \tag{1}$$

and,

$$E_2 \text{ AND } E_3 \text{ suggests NOT } H_1 \,. \tag{2}$$

In general, each rule has a left-hand-side (premise) that involves a logical predicate of pieces of evidence: E_1, E_2, E_3, etc. In contrast to predicate calculus, the outcome of judgmental rules is not simply true or false, but is quantified by a measure of belief (say, a number in the range 0 to 1) according to the expert's judgment of the inferential strength of the rule. The measure of belief does not correspond to our ordinary notion of probability. For example, in the approximate implication in equation 1, it is not always true that:

$$MB(H_1,E_1) = 1 - MB(NOT\ H_1,E_1)\ . \tag{3}$$

This appears to be characteristic of judgmental reasoning. An expert acoustic signal analyst may admit, for example, that the probability of a source is, say 0.8, when a certain harmonic set is present, but may be unwilling on the same evidence to say that the probability of the source being absent is 0.2, as required by probability theory.

For this reason, evidence for a hypothesis is collected separately from evidence against the hypothesis. The total measure of belief, based upon both kinds of evidence, is contained in a single factor, the confidence factor (CF) given by the difference:

$$CF(H,E) = MB(H,E) - MB(NOT\ H,E)\ , \tag{4}$$

where, E is the total of the evidence processed to this point. (In the example given above, which involves only two rules, the total evidence E would consist of E_1, E_2, and E_3.)

The confidence factor (CF) for the outcome of a rule in the **MYCIN** system is a number in the range -1.0 to +1.0, arrived at as follows: Each piece of evidence has an associated measure of certainty (CF) between -1.0 and +1.0. In evaluating the logical predicates in the condition part of a rule, the CF associated with AND is the minimum CF over the arguments, whereas the CF associated with OR is the maximum CF over the arguments. For a rule to succeed in **MYCIN**, the CF for the logical predicate that comprises the condition part must exceed a threshold value equal to 0.2. If this happens the CF for the conclusion is calculated as the product of the CF of the premise and the CF of the rule itself. In general, the conclusion of a rule (with associated CF value) provides a new piece of evidence that can be acted upon by subsequent rules.

In our experimental work with **INTERSENSOR** thus far we have used similar methods for judgmental reasoning (i.e. those rules that take track samples and form the interpretation graph). In **INTERSENSOR** the parameter CF is an integer between -1000 and +1000 for convenience in processing by **INTERLISP**. The **INTERSENSOR** system allows the selection of a different threshold for each rule instead of the constant 0.2 used by **MYCIN**. Since the process of interpretation is, to a large extent, data driven there can be a large number of alternative interpretations. Setting an appropriately high threshold is one way to limit these to a reasonable number.

Further discussion of pseudo-probability methods can be found in the literature, such as **[Davis and King,1976]**, or **[Shortliffe, 1976]**.

7.2 PROBABILISTIC METHODS

Where noise and uncertainty in observations or in knowledge can be described by probabilistic models, the well-known results of statistical decision theory provide not only the means of quantifying belief in hypotheses, but also the mathematical tools with which to measure performance.

In a typical probabilistic model we define the possible states of the world as a set of exhaustive, and mutually exclusive hypotheses, H_1, H_2,...,H_n. Occurrence of states is described by the a priori probabilities $\{P(H_i) : i=1,n\}$. The true state is assumed to be observed by an uncertain or noisy mechanism resulting in pieces of evidence, say, E_1, E_2,...,E_n. The noise and uncertainty are described by a set of conditional probabilities $\{P(E_j|H_i) : j=1,m\ ;\ i=1,n\}$.

When a probabilistic model such as the above can be defined, straightforward application of Bayes' rule gives the a posteriori probability on each hypothesis H_i as follows:

$$P(H_i|E) = \frac{P(E|H_i)P(H_i)}{\sum_{k=1}^{n} P(E|H_k)P(H_k)} , \tag{5}$$

where, $E = E_1, E_2, \ldots, E_m$.

Expert systems usually do not process all of the evidence at once. Instead, pieces of evidence are considered sequentially as knowledge sources are applied. Thus, the sequential form of equation 5 is more appropriate. If it can be assumed that the pieces of evidence E_j are conditionally independent on each hypothesis H_i, i.e. if:

$$P(E_1, E_2, \ldots, E_m|H_i) = \prod_{j=1}^{m} P(E_j|H_i) , \tag{6}$$

and a new piece of evidence E_{m+1} also satisfying the conditional independence assumption is processed, the probabilities on the hypotheses are updated according to the expression:

$$P(H_i|E,E_{m+1}) = \frac{P(E_{m+1}|H_i)P(H_i|E)}{\sum_{k=1}^{n} P(E_{m+1}|H_k,E)P(H_k|E)} . \tag{7}$$

There has been some controversy about the use of Bayesian methods to update prior probabilities on multiple hypotheses **[Pednault, 1981]**, **[Szolovits and Pauker,1978]** that stems from the assumptions of conditional independence and exhaustiveness of the hypotheses. These discussions are outside the scope of the present paper. In practice, however we have found the methods to work as long as the hypotheses are organized into an exhaustive set, and prior probabilities on all hypothesis are updated by the processing of evidence as required by equation 7.

7.2.1 LOW-LEVEL PROCESSING IN SPECTROGRAMS

In this section we examine Bayesian techniques for interpretation at the lowest levels in the data hierarchy in Figure 2, the inference of track samples from the spectrogram data. We look first at a simple application of equation 7 to find track samples that do not change frequency. We assume that there are two states: H_1 denoting the presence of a track in a given frequency bin, and H_2 denoting the absence of a track. It is assumed that each look of data presents a new piece of evidence, namely, the spectrogram value x(f) in the frequency bin centered at frequency f.

The probabilistic model is described by the two probability density functions $p(x|H_1)$ and $p(x|H_2)$. The detection of narrowband signals in noise has been studied extensively, and theoretical models for the above density functions can be found in the literature (see for example, **[Woinsky, 1972]**). From equation 7 we obtain a sequential form for the a posteriori probability of H_1 in a given frequency bin:

$$P(H_1|x,E) = \frac{L(x)P(H_1|E)}{1 - P(H_1|E) + L(x)P(H_1|E)}, \tag{8}$$

where, L(x) is the likelihood ratio:

$$L(x) = \frac{p(x|H_1)}{p(x|H_2)}. \tag{9}$$

and, E is the evidence (spectrogram values) already processed.

Before applying equation 8 to the spectrogram in Figure 1 the data values are normalized. Normalization is an attempt to remove background level variations. We have found a split-window two-pass procedure to be suitable. Both the split window (boxcar with central gap) and multiple passes reduce the effect of large signal spikes on nearby normalized levels. The first pass forms a rough estimate of average background level. Spikes larger than, say, twice this are removed prior to a second, more accurate background level computation. Normalization is accomplished by dividing the original data by this estimate of the average background level. Stability of the background estimate is important in preserving the shape of the probability distribution of the data. Thus, if the variance and bias of the background estimate are small relative to the mean background level, the probability distribution of the resulting normalized data (original divided by background estimate) will tend to be a scaled version of the original distribution since we are dividing the original data by what is essentially a constant.

Figure 11 illustrates sequential updating of the probability of the signal hypothesis in each frequency bin. To obtain this display the spectrogram in Figure 1 was normalized. Since the a priori probability of a signal in a frequency bin is unknown, we arbitrarily selected a small value, $P(H_1)=0.1$, for each bin at look 1. The probabilities on the hypotheses were then updated at subsequent looks according to equation 8. As is evident in Figure 11, frequency bins containing signal on subsequent looks yield an a posteriori probability that approaches unity, while in bins containing only noise this probability approaches zero. These probabilities may be used as measures of belief in constructing the rest of the interpretation graph as described earlier.

7.2.2 SEQUENTIAL LIKELIHOOD RATIO TRACKING

In the real world spectral lines vary in frequency in response to physical mechanisms such as changes in rotation rate of an engine. The inference process for track samples must, therefore, contain some mechanism for following such changes. In **INTERSENSOR** we use a sequential likelihood ratio tracking algorithm (SLRT) to follow frequency changes while maintaining a cumulative likelihood ratio. (Updating the cumulative likelihood ratio is related to updating the probabilities on the hypotheses). The algorithm uses the sequential likelihood ratio test introduced by Wald **[Wald, 1947]**, with the addition of a procedure for choosing the next data sample to be included in the test. This involves using track parameters to predict an association window for the next look of data and extending the track to the data point in this window that results in the highest cumulative likelihood.

The criterion for deciding when an association of data points represent a detected track compares the cumulative likelihood (cl) with a lower threshold (T_l) and an upper threshold (T_u) as follows: 1) If $cl<T_l$ then decide no track. 2) If $cl>T_u$ then decide track. 3) If in between, defer decision and consider more data.

Data points that do not associate with old tracks start new tracks if the likelihood is above the lower threshold.

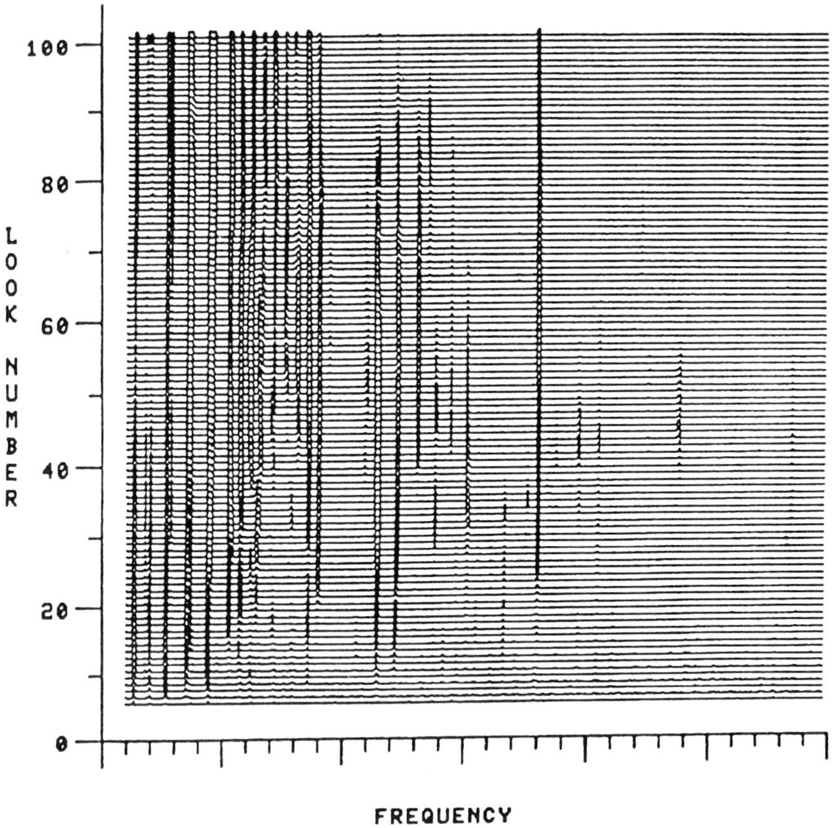

Figure 11. A Posteriori Probability in Frequency Bins

7.2.3 PRUNING OF SPURIOUS TRACKS

The tracking procedure described thus far generates many tentative tracks which are either duplicates or can readily be shown to be artifacts of the algorithm. The tracking algorithm includes sorting and pruning procedures which delete most of the spurious tracks according to a set of locally applied logical tests. For example, if two tracks follow the same frequency path for more than one look it is reasonable to delete the one created most recently. We have observed, for example, that large data values on a strong track tend to associate with any nearby tentative tracks caused by noise. Keeping the oldest track differentiates between the true track and these spurious ones.

7.2.4 ESTIMATION OF TRACK PARAMETERS

Interpretation of track samples as parts of lines involves knowledge of parameters such as frequency. Parameter estimation makes sense only for tracks resulting from real sources, as these will conform to real dynamical constraints. We, therefore, restrict the parameter estimation process to those tracks which have been detected (those with $cl>T_u$).

The estimation process for frequency and width of lines is based upon sample moments of signal levels in the association window. Stable estimates of these moments are obtained by considering only the excess of signal over the mean noise level (which is unity because of the normalization). This is considered to give a linewidth estimate similar to human perception of linewidth.

The moment estimates are used as inputs to smoothing algorithms yielding smoothed estimates of track frequency, linewidth, and so forth.

Figure 12 is a display of the tracks found in the spectrogram in Figure 1 by the SLRT algorithm. The original spectrogram was normalized before the tracking algorithm was applied.

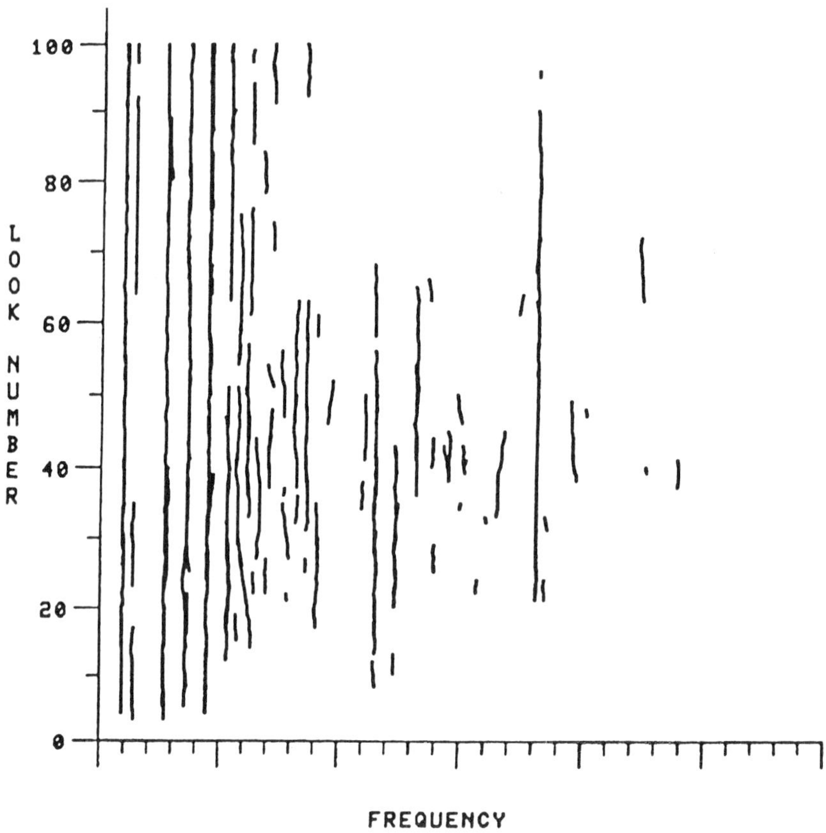

Figure 12. Tracks Generated by SLRT Algorithm

8 CONCLUSIONS AND DISCUSSION OF RESEARCH ISSUES

We have described contributions from the field of Artificial Intelligence to the analysis and interpretation of acoustical signals. These contributions come from the applied side of AI research, an area that is known as knowledge engineering. The vehicle for this research is the expert system, a collection of computer programs that can acquire and use human expertise in a specific field to emulate intelligent human problem-solving behavior.

There are two aspects of expert systems that are emphasized in this paper. The first is very much related to computer science and concerns programming methodology. Major research issues here are refinements of the production system programming formalism, and the question of procedural versus declarative knowledge representation. The other aspect is more familiar to those working in areas such as pattern recognition and signal analysis. It involves the modelling of the signal analysis world and research on problem-solving and inference-forming mechanisms.

An expert system is a programming structure that is initially incomplete, and may in fact never reach a state of final completion in the conventional programming sense. The programming methodology is usually such that additions to the system in terms of declarative knowledge (what something is), and in terms of procedural knowledge (how to do something) are readily made at any time, and with little knowledge of the workings of the rest of the system. This contrasts with more conventional programming techniques, where the knowledge tends to be embedded into the code of the programs. **UNITS** and **UNIT*** are representative of the trend toward general representation packages that make this kind of interaction easy. They represent the current stage of development, but are probably still very primitive.

The problems in constructing a new expert system are still very great. A large body of knowledge needs to be encoded. Some of this is knowledge in the form of procedures, and therefore suffers from the previously stated difficulty of knowledge embedded in code. The technique, mentioned in the paper, of associating a unit data structure with a procedure appears to separate the embedded declarative knowledge from the procedure itself, thus allowing the procedure to be written in more general form.

We have examined inference mechanisms for coping with noise and uncertainty in data entering into the interpretation process, as well as the uncertainty associated with judgmental rules obtained from human experts. Our work has tended to use pseudo-probability methods similar to those in the **MYCIN** system for the interpretation at levels in the data hierarchy above the track sample level. This is consistent with the acquisition of judgmental rules of analysis. A problem with the pseudo-probability approach is that performance is difficult to predict theoretically. It is generally necessary to investigate operation on a case-by-case basis, as is done in medical systems like **MYCIN**. This means that system development is a lengthy process of rule addition and subsequent assessment of performance. This is the way that humans learn, but it is not clear whether this is the best method for expert systems.

The paper describes some investigations into probabilistic models. Thus far these models have been applied at the lowest level of the interpretation graph. Well-studied Bayesian inference techniques appear to be a useful model. It is interesting to speculate whether the probabilistic models can be extended to other levels in the interpretation. One reason for wanting to do this is to derive theoretical measures of performance.

References

[Davis, 1975]
R. Davis, B. G. Buchanan, and E. H. Shortliffe, *Production Rules as a Representation for a Knowledge-Based Consultation Program.* STAN-CS-75-519 (AIM-266), Dept. of Computer Science, Stanford University, October 1975.

[Davis and King,1976]
R. Davis, and J. King,An Overview of Production Systems, **Machine Intelligence**, vol. 8,1976.

[Drazovich, 1978]
R. J. Drazovich and S. Brooks, Surveillance Integration Automation Project. *Proceedings of the ARPA Distributed Sensor Net Symposium*, December 1978, pp. 119-123.

[Feigenbaum, 1977a]
E. A. Feigenbaum, R. S. Engelmore, and C. K. Johnson, A Correlation Between Crystallographic Computing and Artificial Intelligence Research. *Acta Cryst.* (1977). A33, pp. 13-18.

[Feigenbaum, 1977b]
E. A. Feigenbaum, The Art Of Artificial Intelligence: I. Themes And Case Studies Of Knowledge Engineering. *Proc. 5th Int.Joint Conf. Artificial Intelligence*, Cambridge Massachussetts, 1977, pp. 1014-1029.

[Friedland, 1979]
P. E. Friedland, *Knowledge-Based Experiment Design In Molecular Genetics*. STAN-CS-79-771 (HPP-79-29), Dept. of Computer Science, Stanford University, October 1979.

[Lesser, 1977]
V. R. Lesser, and L. D. Erman, A Retrospective View of the Hearsay-II Architecture, *Proc. 5th Int. Joint Conf. Artificial Intelligence*, Cambridge Massachussetts, 1977, pp. 790-800.

[Nii, 1978]
H. P. Nii and E. A. Feigenbaum, Rule-Based Understanding Of Signals. In D. A. Waterman and F. Hayes-Roth (Eds.), *Pattern-Directed Inference Systems*. New York: Academic Press, 1978. Pp. 483-501.

[Pednault, 1981]
E. P. D. Pednault, S. W. Zucker, and L. V. Muresan, On The Independence Assumption Underlying Subjective Bayesian Updating, *Artificial Intelligence* vol. 16, pp. 213-222, 1981.

[Shortliffe, 1976]
E. H. Shortliffe, **MYCIN: Computer-Based Medical Consultations**. New York: American-Elsevier, 1976.

[Shortliffe, 1979]
E. H. Shortliffe, B. G. Buchanan, E. A. Feigenbaum, Knowledge Engineering For Medical Decision-Making: A Review Of Computer-Based Clinical Decision Aids. *Proc. IEEE*, Vol. 67, No. 9, September 1979, pp. 1207-1224.

[Smith, 1980]
R. G. Smith and P. Friedland. *Unit Package User's Guide*. DREA TM 80/L, Defence Research Establishment Atlantic, Dartmouth, N.S., Canada, December 1980. (Also published as Stanford Heuristic Programming Project Memo HPP-80-28, Dept. of Computer Science, Stanford University.)

[Smith et al., in preparation]
R. G. Smith and C. A. Dent, *A Guide To UNIT**. DREA, Technical Memo (in preparation), Defence Research Establishment Atlantic, Dartmouth, N.S., Canada.

[Stefik, 1980]
M. J. Stefik, *Planning With Constraints*. STAN-CS-80-784 (HPP-80-2), Dept. of Computer Science, Stanford University, January 1980.

[Stefik et. al., 1980]
M. J. Stefik, J. Aikins, R. Balzer, J. Benoit, L. Birnbaum, F. Hayes-Roth, E. Sacerdoti, The Organization of Expert Systems, A Tutorial. *Artificial Intelligence* vol. 18, pp. 135-173, 1982.

[Szolovits and Pauker,1978]
P. Szolovits, and S G. Pauker, Categorical and Probabilistic Reasoning in Medical Diagnosis. *Artificial Intelligence* vol. 11, pp. 115-144, 1978.

[Wald, 1947]
A. Wald, **Sequential Analysis**, New York, Wiley, 1947.

[Winograd, 1975]
T. Winograd, Frame Representations and the Procedure/Declarative Controversy, **Representation and Understanding**, Bobrow and Collins, eds., Academic Press.

[Woinsky, 1972]
M. N. Woinsky, Nonparametric Detection Using Spectral Data, *IEEE Transactions on Information Theory*, Vol. IT-18, No. 1, January 1972.

PROBLEMS OF SPATIAL SIGNAL PROCESSING IN SHALLOW WATER

R. Klemm
FGAN-FFM-EL
D-5307 Wachtberg-Werthoven
F.R. Germany

Abstract

This paper gives an introduction into the field of spatial array processing for underwater applications. The field is subdivided into two parts, namely interference rejection (mainly for active systems) and estimation of source parameters, such as bearing, range and depth. Special emphasis has been laid on applications in a random dispersive channel such as the shallow water waveguide, which has influence on the techniques to be applied. After a brief discussion of the methods and principles of spatial array processing a statistical model of signals and interference is given. The model is based on a normal mode field description. The normal mode setup is extended by a simple intermode phase fluctuation model. By applying the processing techniques to the signal model a number of results on interference rejection and source position estimation are obtained. Although this paper gives an overview over several earlier publications of the author a number of unpublished results are presented. In conclusion, the dispersive channel offers additional possibilities of passive position finding, if the channel parameters are known and the fluctuations are small. Otherwise, the dispersive field leads to gain reductions and bearing errors.

1. INTRODUCTION

The use of sensor arrays has been treated in a large number of publications in the last 10 to 20 years. Considerable interest came up after Bryn /1/ and Mermoz /2/ published their first papers on adaptive arrays. A number of important optimization problems have been presented by Lo et. al. /3/. In the meantime several textbooks on array design have been published, such as /4, 5, 6/. A great variety of applications has been published in the literature, for instance, problems of high resolution of sources, position estimation, detection in presence of interference, superdirective arrays a.s.o. Most of the work is based on rather simple assumptions on the received wave field, namely single plane waves as response of a point shaped source.

NATO ASI Series, Vol. F1
Issues in Acoustic Signal/Image Processing and Recognition
Edited by C. H. Chen

Although the theory of wave propagation in arbitrary media is well established there are not so many publications available, that take the special conditions of the propagation medium into account. It seems that there is not very much communication between wave propagation experts and signal processing people. Only a few papers, e.g. /7, 8, 9/, are influenced by the existing knowledge on wave propagation.
The objective of this paper and several other papers of the author /10, 11, 12, 13/ is to connect the existing knowledge on array processing and signal processing methods with the special conditions of acoustic wave propagation in shallow water. The shallow water sound propagation channel is a waveguide with two different boundaries. The far field of an acoustic source can be expressed as a sum of eigenfunctions of the channel, the well known normal modes. Each of the modes travels at its individual modal velocity determined by the modal eigenvalue of the channel (modal wavenumber). Thus the field appears to be spatially dispersed. If the boundaries are random as in case of surface fluctuations or roughness of the bottom phase fluctuations among modes may occur which generate random wavefronts. Therefore, the response of a sensor array to an acoustic source can be described in general by a positive definite spatial covariance matrix. This has considerable consequences on the design and performance of spatial signal processors.

There are two main objectives of spatial signal processing: detection in presence of interference and localisation of sources. In the first case the task is to increase the power of an expected signal as compared with interference and noise. This is usually done in two steps, namely rejection of the interfering field components and matching to the signal to increase the signal-to-noise ratio. Localisation of sources can be done by comparing a measurement in some way with a set of replicas of all possible signals (for instance, a set of beamformer weights), so that a power maximum or minimum indicates, which of the replicas is closest to the measured information.
In the following a number of processors for both interference suppression and parameter estimation are discussed. The influence of sound propagation in shallow water on the design and the performance of spatial processors is emphasized. After some general comments on the nature of array processors a statistical model for signal and interference is derived. Applying the processors to the signal model yields a number of useful results which are presented in the subsequent sections.

2. METHODS FOR SPATIAL SIGNAL PROCESSING

PRELIMINARY REMARKS

In this section a brief overview over the processing schemes used further below is given. At first an important property of the inverse of a matrix is pointed out.

Any hermitian positive definite matrix can be written in the following way

$$\underline{R} = \underline{M}\,\underline{M}^* + \lambda_{min}\,\underline{I} \qquad (1)$$

If $\underline{R}$ is a covariance matrix then (1) means splitting the matrix up into a deterministic part $\underline{M}\,\underline{M}^*$ and a white noise term $\lambda_{min}\,\underline{I}$, where λ_{min} is the minimum eigenvalue of $\underline{R}$. Using the following matrix identity /14/

$$(A - B\,\underline{D}^{-1}\,\underline{C})^{-1} = \underline{A}^{-1} + \underline{A}^{-1}\,\underline{B}\,(\underline{D} - \underline{C}\,\underline{A}^{-1}\,\underline{B})^{-1}\,\underline{C}\,\underline{A}^{-1}$$

one obtains

$$\underline{R}^{-1} = (\underline{M}\,\underline{M}^* + \lambda_{min}\,\underline{I})^{-1} = \lambda_{min}\,(\underline{I} - \underline{M}\,(\lambda_{min}\,\underline{I} + \underline{M}^*\,\underline{M})^{-1}\underline{M}^*) \qquad (2)$$

Normalising to λ_{min} and considering the noise-free case, i.e. $\lambda_{min} \rightarrow 0$, yields

$$\underline{P} = \lambda_{min}\,\underline{R}^{-1} = \underline{I} - \underline{M}\,(\underline{M}^*\,\underline{M})^{-1}\,\underline{M}^*, \qquad (3)$$

which is an orthogonal projection matrix that projects any vector on a vector subspace orthogonal to $\underline{M}$. In particular,

$$\underline{P}\,\underline{M} = (\underline{I} - \underline{M}\,(\underline{M}^*\,\underline{M})^{-1}\,\underline{M}^*)\,\underline{M} = \underline{O}.$$

In other words, the inverse of a covariance matrix can be interpreted as an approximate orthogonal projection on the subspace given by $\underline{M}$.

DETECTION IN PRESENCE OF INTERFERENCE

The optimum quadratic detector

Let

$$\underline{x} = \begin{pmatrix} x_1 \\ \vdots \\ x_N \end{pmatrix}$$

be the vector of spatial field samples as received by a sensor array. Basically one should write $\underline{x}$ (t) or $\underline{x}$ (ω); however, we assume that the input is sufficiently narrowband. Therefore, one can drop the time dimension and confine to the purely spatial case. $\underline{x}$ may either be a set of noise samples or contain signal and noise. If signal and noise are both random the optimum processor is quadratic:

$$y = \underline{x}^*\,\underline{K}\,\underline{x} \qquad (4)$$

where $\underline{K} = \underline{Q}^{-1}\ \underline{S}\ \underline{Q}^{-1}$ is the processor matrix and $\underline{Q} = E\{\underline{n}\ \underline{n}^*\}$ and $\underline{S} = E\{\underline{s}\ \underline{s}^*\}$ are the covariance matrices of interference + noise and signal respectively. (4) is optimum in the sense of the detection index which has been introduced for optimization instead of the SNR in order to avoid fourth order moments /15/. $\underline{Q}$ is positive definite and can be factorised: $\underline{Q} = \underline{F}\ \underline{F}^*$. The role of $\underline{F}$ in (4) can be interpreted as a spatial whitening filter, because the covariance matrix of the transformed noise vector $\underline{y} = \underline{F}^{-1}\ \underline{x}$ is $E\{\underline{y}\ \underline{y}^*\} = E\{\underline{F}^{-1}\ \underline{x}\ \underline{x}^*\ \underline{F}^{-1*}\} = \underline{F}^{-1}\ \underline{F}\ \underline{F}^*\ \underline{F}^{-1*} = \underline{I}$. Another interpretation is that by means of the transform $\underline{y} = \underline{Q}^{-1}\ \underline{x}$ the received signal vector becomes approximately orthogonal to the inference subspace of $\underline{Q}$ (see equ.(2)). If there is no noise the projected vector $\underline{y}$ is precisely orthogonal to the interference which causes the signal-to-noise and interference ratio (SNIR) to become infinite. If there is white noise the approximate orthogonal projection matrix $\underline{Q}^{-1}$ provides an optimum compromise between interference rejection and signal reduction. The covariance matrix $\underline{S}$ is a generalised matched filter (or beamformer). In practical applications $\underline{Q}^{-1}$ has to be replaced by an adaptive spatial filter, e.g. $\hat{\underline{Q}}^{-1} = [\frac{1}{T}\sum_{t=1}^{T} \underline{n}_t\ \underline{n}_t^*]^{-1}$, and instead of $\underline{S}$ a set of beamformer matrices $\underline{S}\ (\Theta)$ depending on a set of parameters Θ (e.g. bearing, range, depth) is used.

The performance of the quadratic processor can be judged by calculating the gain in SNIR /15/

$$G = \frac{\mathrm{tr}\ (\underline{K}\ \underline{S})P_Q}{\mathrm{tr}\ (\underline{Q}\ \underline{K})P_S} \tag{5}$$

with $P_Q = \mathrm{tr}\ \underline{Q}$ and $P_S = \mathrm{tr}\ \underline{S}$.

For the optimum processor $\underline{K} = \underline{Q}^{-1}\ \underline{S}\ \underline{Q}^{-1}$ (5) becomes

$$G_{opt} = \mathrm{tr}\ (\underline{S}\ \underline{Q}^{-1}) \tag{6}$$

Linear processors

Linear processors are simply described by a scalar product

$$y = \underline{x}^*\ \underline{h} \tag{7}$$

with $\underline{h}$ being the vector of processor weights. In the random signal case the optimum linear processor is obtained by maximizing the SNIR

$$\max_{\underline{h}} \frac{\underline{h}^*\ \underline{S}\ \underline{h}}{\underline{h}^*\ \underline{Q}\ \underline{h}}$$

The solution is found to be the eigenvector belonging to the maximum eigenvalue of the generalised eigenproblem

$$(\underline{S} - \lambda\, \underline{Q})\ \underline{h} = \underline{o}.$$

If the signal is deterministic we get as optimum processor

$$\underline{h} = \underline{Q}^{-1}\ \underline{s}\ . \tag{8}$$

where $\underline{Q}^{-1}$ is again an interference suppression filter and $\underline{s}$ a vector matched to the expected signal.

The gain in SNIR of any linear processor $\underline{h}$ is

$$G = \frac{\underline{h}^*\ \underline{S}\ \underline{h}\ P_Q}{\underline{h}^*\ \underline{Q}\ \underline{h}\ P_S} \tag{9}$$

which, in the optimum case, simplifies to

$$G_{opt} = \underline{h}^*\ \underline{Q}^{-1}\ \underline{h}\,\frac{P_Q}{P_S} \tag{10}$$

Suboptimum approximations

The implementation of the optimum processors as adaptive spatial filter for suppression of interference in real-time is cumbersome and expensive. Therefore, one frequently looks for suboptimum replacements of the crucial matrix inversion $\underline{Q}^{-1}$. A logical approach is to proceed from (3), that means, to find a filter that projects the received signal vector onto an interference-free vector subspace. Instead of the processor matrix $\underline{K} = \underline{Q}^{-1}\ \underline{S}\ \underline{Q}^{-1}$ in (4) one uses $\underline{K} = \underline{P}\ \underline{S}\ \underline{P}$ with

$$\underline{P} = \underline{I} - \underline{M}\ (\underline{M}^*\ \underline{M})^{-1}\ \underline{M}^* \tag{11}$$

The corresponding linear processor is

$$\underline{h} = \underline{P}\ \underline{s} \tag{12}$$

The approximation of $\underline{Q}^{-1}$ by $\underline{P}$ is the better the higher the interference-to-noise ratio is. The matrix $\underline{M}$ contains the information about the interference. The problem is how to choose $\underline{M}$. The mathematical way is to make an eigenvector factorisation of $\underline{Q}$ and separate the interference eigenvalues from the noise eigenvalues. Then the matrix $\underline{M}$ has to be filled with those eigenvectors corresponding to the interference eigenvalues. Then $\underline{P}$ is orthogonal to the interference. This procedure is adaptive in that it is based on an estimate of $\underline{Q}$. No a priori knowledge about the interference is required. However, the computational expense is even higher than the one required for the matrix inversion. In some simple cases, in particular, if the interfering field is known, $\underline{M}$ can be designed in a simple way (for example, nulling of an interfering plane wave with known direction).

Another possibility of cancelling one plane wave is the design of difference sensor patterns which is done by a projection matrix of the form

$$\underline{P} = \begin{pmatrix} 1 & & & & 0 \\ -a & 1 & & & \\ & \ddots & \ddots & & \\ & & -a & 1 & \\ 0 & & & -a & 0 \end{pmatrix} \qquad (13)$$

where a = exp ( ) is the complex phase factor between two subsequent sensors of an equidistant array caused by a plane wave, given by a vector

$$\underline{a} = \begin{pmatrix} 1 \\ a \\ a^2 \\ \vdots \end{pmatrix}$$

Another class of suboptimum approaches are the well known shading methods. Instead of $\underline{Q}^{-1}$ or $\underline{P}$ a diagonal matrix $\underline{D}$ containing the shading coefficients is used for interference suppression:

$$y = \underline{x}^* \, \underline{D} \, \underline{s} \, .$$

There are many kinds of shading functions, such as cosine, $\cos^2$, binominal, Dolph-Tschebycheff and others. They all reduce the sidelobe level of a beamformer, but broaden the main beam. Their advantages are easy computation and validity for any constellation of interfering sources.

Finally the so called pre-transform methods should be mentioned. The received signal vector $\underline{x}$ is subjected to a suitable linear pre-transform (L x N matrix) in order to reduce the order of the system:

$$\underline{y} = \underline{T} \, \underline{x} \, , \qquad (14)$$

and the usual adaptive processing according to (4) or (8) has now to be applied to the transformed variable $\underline{y}$. The corresponding covariance matrices are $S_T = E\{T\, ss^*\, T^*\} = \underline{T}\,\underline{S}\,\underline{T}^*$ and $\underline{Q}_T = E\{\underline{T}\,\underline{n}\,\underline{n}^*\,\underline{T}^*\} = \underline{T}\,\underline{Q}\,\underline{T}^*$. The pre-transform should be chosen so that

- $L << N$ in order to spare arithmetic operations
- L is larger than the number of sources
- no signal energy is lost.

The third requirement can be met by steering a beam in the hypothetical target direction. Typical pre-transforms are

$$\underline{T} \equiv \begin{pmatrix} 1 & & & \underline{b}^* \\ & 1 & & \\ & & \ddots & \\ & & & 1 & 0 \end{pmatrix} \qquad \text{(auxiliary sensor transform)}$$

or

$$\underline{T} \equiv \begin{pmatrix} \underline{b}_1^* \\ \underline{b}_2^* \\ \vdots \\ \underline{b}_L^* \end{pmatrix} \qquad \text{(auxiliary beam transform).} \qquad (15)$$

The multiple beam transform (15) has been shown to be very effective when applied to linear equally spaced arrays /16/.

PARAMETER ESTIMATION METHODS

In the following we are talking briefly about a family of methods which are widely used to estimate parameters such as frequency (spectral analysis), arrival time (matched filter), bearing (beamformer) and others. The idea of these methods is to compare the observed data with a set of steering matrices $\underline{H}(\theta)$ in such a way that a power peak indicates which value of θ is characteristic. In the subsequent discussion θ stands for bearing, range or source depth. There are basically two types of processors

The first is

$$P(\theta) = \operatorname{tr}(\underline{R}\,\underline{H}(\theta)) \qquad (16)$$

where $\underline{R}$ is the spatial covariance matrix of the observed field and $P(\theta)$ the processor output power. $\underline{R} = \underline{S}(\theta_0) + P_W \underline{I}$ with $\underline{S}(\theta_0)$ being the field covariance matrix and P_W the power of additional white noise. It follows from the generalised Schwartz-inequality

$$/\operatorname{tr}(\underline{R}\,\underline{H}(\theta))/^2 \leq \operatorname{tr}(RR)\operatorname{tr}(\underline{H}\,\underline{H})$$

that $P(\theta)$ in (16) becomes maximum for $\underline{H}(\theta) = \underline{S}(\theta_0)$. The power estimator (16) covers all kind of matching techniques such as spectral analysis, beamforming, field matching an so on. In particular, it optimises the SNR. If the signal $\underline{s}$ (θ) is deterministic, i.e. $E\{\underline{s}\ \underline{s}^*\}$ is a dyadic, (16) can be written as

$$P(\theta) = \underline{h}^*(\theta)\ \underline{R}\ \underline{h}(\theta) \tag{17}$$

with $\underline{h}$ (θ) being a steering vector.

The second type of power estimator is of the form

$$P(\theta) = \frac{1}{tr\,(\underline{B}\ \underline{H}\,(\theta))} \tag{18}$$

where $\underline{B}$ is usually some modification of the covariance matrix $\underline{R}$, or, in other words, of the measured data. More precisely, if $\underline{B}$ is orthogonal to $\underline{S}$ (θ_o) then P (θ) goes to infinity if $\underline{H}$ (θ) approaches $\underline{S}$ (θ_o). This is the principle which a large class of high resolution power estimators is based on. Notice that the inversion $\frac{1}{\cdot}$ in (18) is just a nonlinear scaling which sharpens peaks in the power response but has no influence on the resolution capability. The resolution capability of (18) is given by the phenomenon that the squared scalar product between two vectors change more rapidly in the vicinity of an orthogonality than in the vicinity of equality. A well known example is the power response of an array. The nulls in the array pattern (orthogonality between steering vector and field) are always narrower than the main beam (equality). This property is well known and has been widely applied, for instance, for high resolution angle estimation such as monopulse (radar) or split beam technique (sonar). If the signal is deterministic, a special form of (18) is

$$P(\theta) = \frac{1}{\underline{h}^*(\theta)\ \underline{B}\ \underline{h}(\theta)} \tag{19}$$

with $\underline{h}$ (θ) being now the steering vector (e.g., beamformer). A few examples should be mentioned:

1. $\underline{B} = \underline{R}^{-1}$ means an approximate orthogonality between $\underline{B}$ and the field as shown at the beginning of this section. The estimator

$$P(\theta) = \frac{1}{\underline{h}^*(\theta)\ \underline{R}^{-1}\ \underline{h}(\theta)} \tag{20}$$

is well known as maximum likelihood method (MLM) or minimum variance estimator. If satisfies the condition

$$\min_{\underline{g}}\ \underline{g}^*\ \underline{R}\ \underline{g} \quad \text{with } \underline{g}^*\ \underline{h}(\theta) = 1,$$

that means, a vector $\underline{g}$ is determined so that the response is minimum under the constraint that the response in the steering direction is unity.

2. Let $\underline{B} = \underline{r}\,\underline{r}^*$ with $\underline{r} \equiv$ 1. column of $\underline{R}^{-1}$.

The one gets

$$P(\theta) = \frac{1}{|\underline{h}^*(\theta)\,\underline{r}|^2} \tag{21}$$

If $\underline{R}$ is Toeplitz (stationary process) (21) is the well known maximum entropy method. It follows from equation (2) that $\underline{r}$ is approximately orthogonal to the data vector. In the subsequent discussion this method is referred to as MEM.

3. $\underline{B} = \underline{P} = \underline{I} - \underline{M}(\underline{M}^*\underline{M})^{-1}\underline{M}^*$. $\underline{M}$ has to be chosen to satisfy $\underline{S}(\theta_0) = \underline{M}\,\underline{M}^*$, which can be accomplished by eigenvector factorisation of $\underline{R}$ and inserting the s i g n a l eigenvectors into $\underline{M}$. This is a consequent extension of (20) and yields higher resolution, because the limiting white noise part of $\underline{R}$ is eliminated in $\underline{P}$.

4. $\underline{B} = \underline{e}\,\underline{e}^*$, with $\underline{e}$ being the eigenvector belonging to λ_{min} (i.e., to the noise). As all the eigenvectors of $\underline{R}$ are orthogonal $\underline{e}$ is orthogonal to the signal space. The power estimater $P(\theta) = 1/|e^*\,h(\theta)|^2$ can be considered a noise-free version of (21). This method is used several times in the discussion and is abbreviated as EVM (eigenvector method).

In case of random signals the steering vector dyadic $\underline{h}(\theta)\,\underline{h}^*(\theta)$ has to be replaced by a set of positive definite steering matrices $\underline{H}(\theta)$, s. equ. (23). In this case perfect orthogonality and hence an infinite power response is obtained if the white noise part is removed from the steering matrices

$$\underline{H}(\theta) = \underline{H}(\theta) - \lambda_{min}\,\underline{e}\,\underline{e}^*$$

where $\underline{e}$ is again the eigenvector of $\underline{H}(\theta)$ corresponding to λ_{min}.

Whenever methods of the type (16) or (18) are applied in the sequel they are called "generalized methods" because the signal is generalised to the random case, and will be abbreviated as GMF or GBF, GMLM, GMEM and GEVM.

3 THE SIGNAL MODEL

In order to investigate the use of the processing schemes described above in a shallow water environment the spatial characteristics of signals or interference have to be described in terms of their spatial covariance matrix. Assuming that the shallow water sound channel is homogeneous acoustic wave propagation can be described in terms of the eigenfunctions of the channel, i.e. as a sum of normal modes:

$$P(\zeta, z, r, t) = e^{-j(\omega t - \pi/4)} \sum_{m=1}^{M} A_m e^{jk_m r} \tag{22}$$

The modal amplitudes are given by

$$A_m = P_o \frac{\omega \varrho^2}{H} \sqrt{\frac{1}{8\pi r}} \frac{u_m(\zeta) u_m(z)}{\sqrt{k_m}} e^{-\alpha_m r} \tag{23}$$

Where

k_n	modal wavenumber
P_o	source strength
ϱ	water density
H	water depth
r	range
z	receiver depth
ζ	source depth
α_m	modal attenuation coefficient
$u_m(.)$	normal mode function
M	number of modes

The number of modes depends on several parameters, such as frequency, water depth and sound velocity in the bottom. The modal attenuation coefficients include two portions, one due to bottom loss, the other one due to scattering at the rough surface /17, 18, 19/. So far the model is deterministic. It is extended now by introducing a random modal phase φ_m in order to model phase fluctuations due to the time-varying surface:

$$p(\zeta, z, r, t) = p_o^{-j(\omega t - \pi/4)} \sum_{m=1}^{M} A_m e^{j(k_m r + \varphi_m)} \tag{24}$$

Then the covariance between two receiver positions i,k becomes

$$c_{ik} = E\{ p(\zeta, z_i, r_i, t)\, p^*(\zeta, z_k, r_k, t) \} =$$

$$= E \sum_{n=1}^{M} A_n(\zeta, z_i, r_i) e^{j(k_n r_i + \varphi_n)} \sum_{m=1}^{M} A_m(\zeta, z_k, r_k) e^{-j(k_m r_k + \varphi_m)}$$

which can be separated in a power term and a mode interference term:

$$c_{ik} = \sum_{n=1}^{M} A_n(i) A_n(k) e^{jk_n(r_i - r_k)} + \sum_{\substack{n,m=1 \\ n \neq m}}^{M} A_n(i) A_m(k) e^{j(k_n r_i - k_m r_k)} E\{e^{j(\varphi_n - \varphi_m)}\} \tag{25}$$

If the surface fluctuations are small compared with the water depth the phase fluctuation term can be approximately described by a gaussian correlation function /18/:

$$c(n,m) = E\{e^{j(\varphi_n - \varphi_m)}\} = e^{-(k_n-k_m)^2 \; r^2 \sigma^2 \cdot 2H^2)} \tag{26}$$

where σ = is the RMS - wave height of the surface. It is obvious that for large fluctuations and ranges the modes tend to become uncorrelated so that (25) becomes

$$c_{ik} = \sum_{n=1}^{M} A_n\,(i)\; A_n\,(k)\; e^{\,j\,k_n\,(r_i - r_k)} \tag{27}$$

For a horizontal array the modal amplitudes are equal for different sensors. Hence (27) becomes

$$c_{ik} = \sum_{n=1}^{M} A_n \; e^{\,j\,k_n\,(r_i - r_k)} \tag{28}$$

The effect of mode propagation on a horizontal linear array is illustrated in Fig. 1. Each of the modes corresponds to a plane wave arriving at

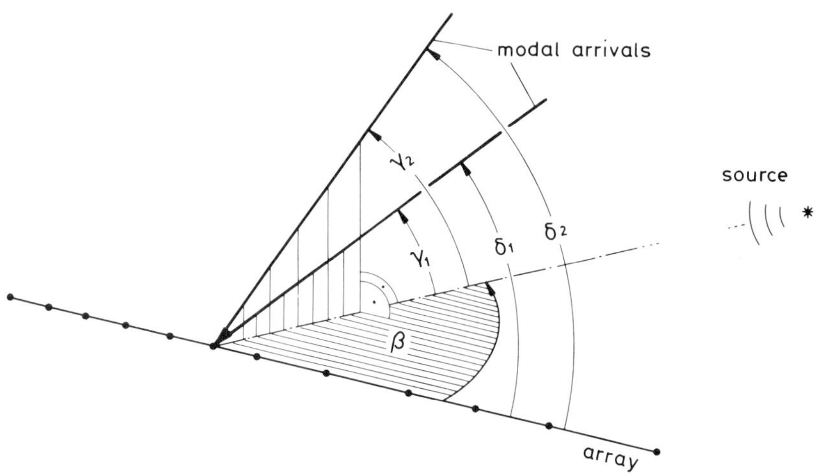

Fig. 1. Geometry of modal arrivals

vertical incident angles γ_n with $\cos\gamma_n = k_n/k_o$ ($k_o = \omega/c(z)$, $c(z)$ sound velocity at depth z). Therefore, the array perceives several arrivals with bearing angles instead of one arrival at bearing β . In other words, the signal energy appears to be smeared towards broadside direction. The covariance matrix is positive definite for all angles other than broadside (90°), where it is just a sum of equal and hence linearly dependent dyadics. That means, a source appears to be spatially coherent at broadside and spatially random everywhere else.

The term $e^{j(k_n r_i - k_m r_k)}$ in the mode interference part of (25) is range dependent which can be clarified as follows. Let $r_i = r$ and $r_k = R + d$. Then the range dependent phase term becomes $\exp(j(k_n r - k_m r - k_m d)$ which depends on r as far as $k_n \neq k_m$. If the mode interference term becomes zero due to fluctuations (25) becomes range independent. Therefore, range estimation can be done only if there is sufficient phase stability among the modes.

For a vertical array one gets $r_i = r_k = r$. Hence the spatial correlation value between two sensors becomes

$$c_{ik} \sim \sum_{m=1}^{M} \sum_{n=1}^{M} u_n(\zeta)\, u_n(z_i) u_m(\zeta) u_m(z_i)\, e^{jr(k_n - k_m)} E\{e^{j(\varphi_n - \varphi_m)}\} \quad (29)$$

which in case of uncorrelated modes reduces to

$$c_{ik} \sim \sum_{m=1}^{M} \sum_{n=1}^{M} u_n(\zeta)\, u_n(z_i) u_m(\zeta) u_m(z_i), \quad (30)$$

i.e. all phase information is lost. The covariance matrix is real, no range or depth estimation is possible.

In addition to the field covariance matrix as pointed out above white noise is assumed in order to simulate various kinds of nondirective noise such as ambient or sensor noise. The interference-to-noise level or signal-to-noise ratio is 20 dB in most of the examples.

4. SOME RESULTS ON INTERFERENCE REJECTION

In the following some results on signal detection in presence of interference are presented. They are based on the assumption that signal and interference can be described by covariance matrices as pointed out before. In particular we make the assumption that the modes are uncorrelated which is a worst case assumption as far gain is concerned; that means, the elements of the signal and noise covariance matrix are given by (28). The mode functions u(z) and wavenumbers k_n have been calculated by the normal mode propagation model SNAP /17/. Two typical results of SNAP are shown in Fig. 2 and 3. If the source is close to the boundary higher order modes are emphasised, if it is in the middle lower order modes are dominant. Dominating higher order modes occur particularly in case of summer sound velocity profiles (SVP's), s. /20/, pp. 104. In the following several results are given that illuminate the area of interference rejection in shallow water. We are more interested in the principles than in a prediction on the performance under special conditions. Therefore, comprehensive listing of environmental input parameters is omitted.

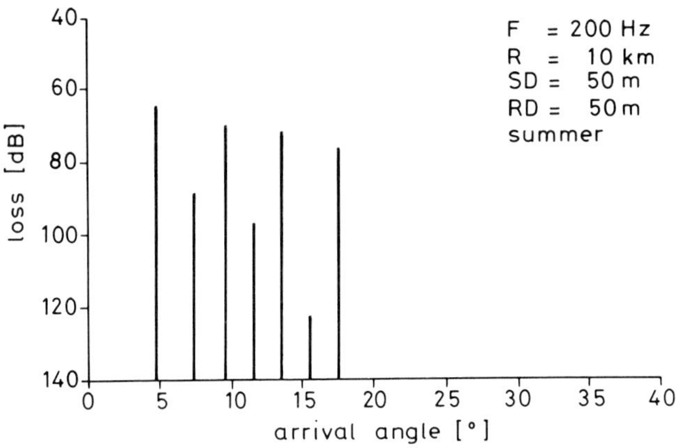

Fig. 2. Modal arrivals (source in the middle of the channel)

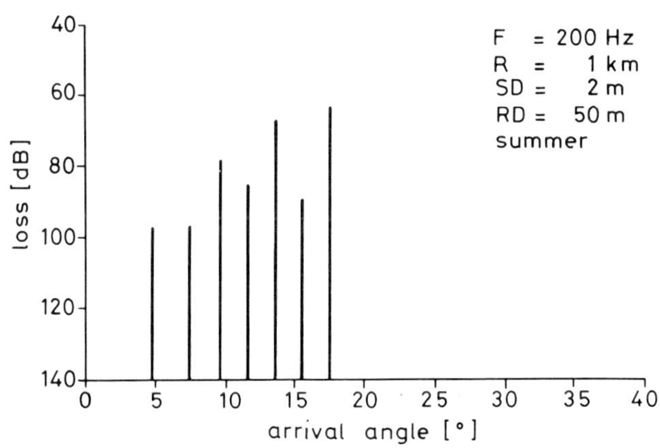

Fig. 3. Modal arrivals (source close to the surface)

The optimum processor versus nullsteering methods

Fig. 4 and 5 show some gain curves for detection of signals in presence of one interfering source. The gain in signal-to-interference ratio is plotted versus the signal direction. The interfering source is located at $\beta = 0°$ (endfire). A short (5 sensors) and a long array (160 sensors) have been used. At $\beta = 0°$ the spatial spread due to modal arrival angles is maximum which results in a rather broad gain minimum (ca. 20°) at $\beta = 0$. The optimum quadratic processor (OQP,s. (4)) is compared with two linear nonadaptive processors according to (12) with $\underline{P}$ given by equ. 13 (DIP) and equ. 11 (NUL). For the nullformer (11) $\underline{M}$ was chosen a beamformer vector

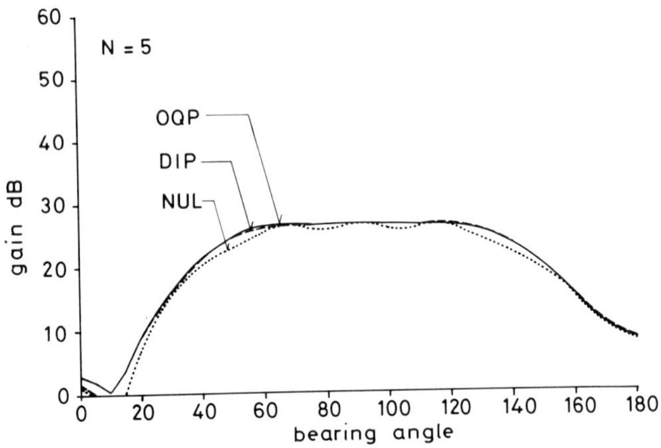

Fig. 4. Comparison of OQP and nullsteering methods

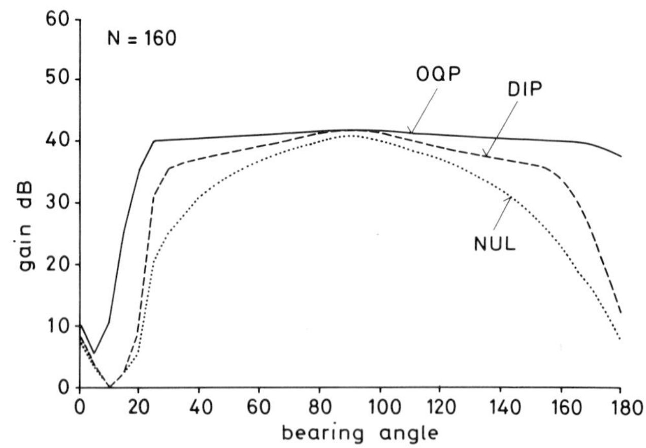

Fig. 5. Comparison of OQP and nullsteering methods

pointing in the direction of the interfering source, thus forming a single null in the array pattern at $\beta = 0°$. If the array is small (Fig. 4) all three methods are equivalent. For large arrays (Fig. 5) considerable differences in gain can be noticed. There are several reasons. Obviously the rather broad null of the dipole diagram former (equ. 13, DIP) fits very well to the width of the spatial spread of the interference caused by mode propagation. Even the projection nullformer (NUL,equ.11) gives sufficient interference cancellation although it is based on the assumption that the interference is a single plane wave. In case of a large array the null formed by the projection nullformer is narrower than the width of the interference spread, thus achieving insufficient interference rejection. A second effect is that the DIP- and NUL- processor are linear, that means, integration of signal energy is done by a linear beamformer, s. for instance (12). At broadside direction the signal appears to be spatially coherent which means that the beamformer $\underline{b}$ in (12) is matched. For all other directions there is a mismatch between $\underline{b}$ and the signal; in other words, the spread signal wave is broader than the beamwidth. Thus signal energy is lost. It is obvious from the DIP - curve, that interference cancellation works quite well because at 90° the maximum gain is reached; however, the optimum gain is obtained only at $\beta = 90°$ because of the perfect match between beamformer and signal waveform. Something else can be observed. Outside the interference region ($\beta = 0...20°$)the gain curves are white noise limited (interference-to-noise ratio = 20 dB) and, therefore, are supposed to be perfectly flat. However, even the OQP-curve in Fig. 5 shows a distinct maximum at $\beta = 90°$ and some decay to both sides. This is because the signal wave becomes random for angles other than broadside. The signal power obtained at broadside is N^2; for all other angles it is less because the signal wave integration becomes more and more incoherent.

Cos - shading and signal beam processing

In the next two figures (Fig. 6 and 7) a comparison between the OQP and two other suboptimum approaches is made, namely simple shading with coefficients $a_n = 0.5 - 0.5 \cos(2\pi(n - 0.5)/N)$, $n = 1...N$, even. This is an arbitrary choice; any other shading could be used as well. The SIB processor is quadratic in the sense of equ. (4), but includes a pre-transform of the received data according to (15) where the $\underline{b}_i$ form a bundle of 7 orthogonally overlapping beams. Five interfering sources are assumed. As can be seen the quadratic SIB-processor is almost as good as the OQP. Cos-shading in connection with linear beamforming gives a reasonable approximation for large arrays (Fig. 7). Basically shading broadens the beam which leads to loss in signal gain if the spatial signal response is narrow. If the signal is broader than the beamwidth broadening of the beam by shading can be useful. Notice that the cos-processor used in Fig. 8 is linear and, therefore, very simple. About 3 dB loss in signal-to-interference ratio has to be taken into account. For small arrays (Fig. 6, N = 20) the main beam is rather broad which results in considerable areas of mainbeam interference.

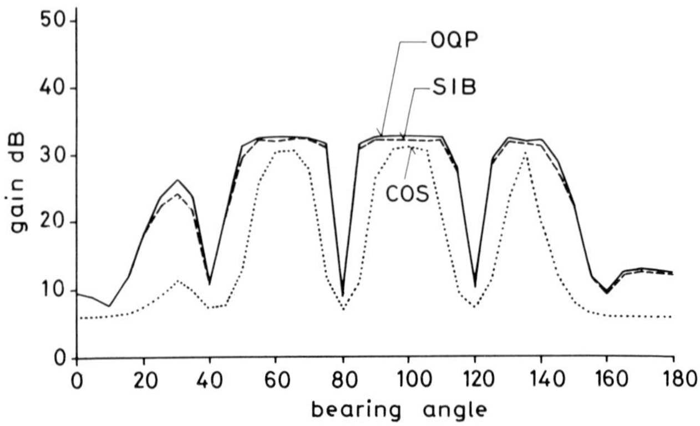

Fig. 6. Signal beam processor and cos-shading (N = 20)

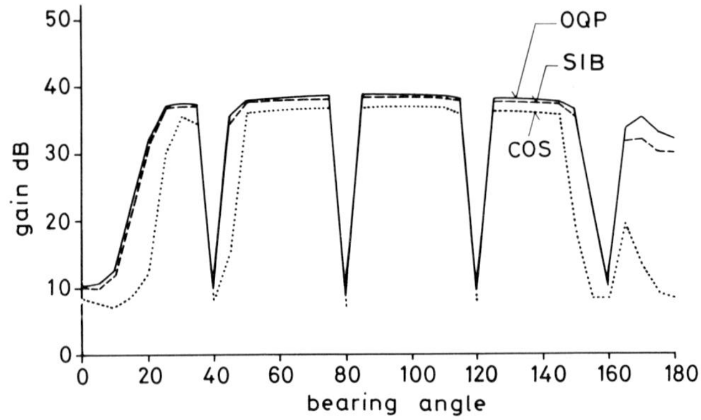

Fig. 7. Signal beam processor and cos-shading (N = 80)

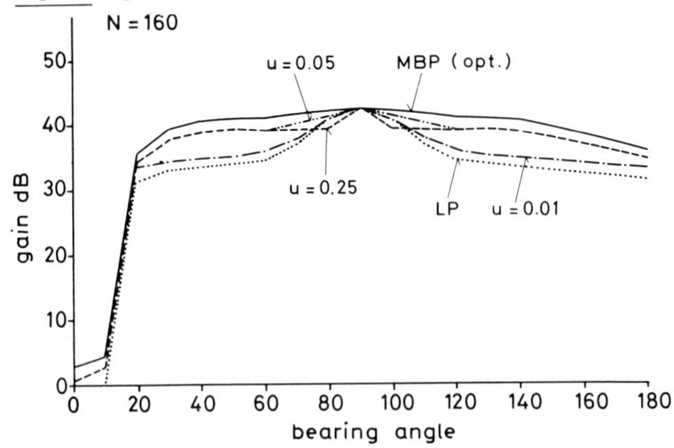

Fig. 8. Influence of beamwidth of the quadratic multi-beam processor

Suboptimum quadratic multi-beam processors

In order to obtain the gain in interference-to-noise ratio as indicated by the SIB-curves in Fig. 6 and 7 the spatial signal covariance matrix must be known for all possible directions. This, however, is difficult obtain because it depends on a variety of environmental parameters which may change with time. Instead, one might use a suboptimum signal covariance matrix which approximates the spatial characteristics of the signal. It is known that the field is spatially dispersed over a ortain angle interval, but not, how the energy is distributed in space. One might assume, therefore, that in general the signal energy is uniformly distributed in space which leads to the following steering matrix

$$\underline{S}\,(\beta) = \left(\frac{\sin\,(\,a\,(i-1))}{a\,(i-1)} \; e^{j\,b\,(i-1)} \right) \tag{31}$$

where

$a = d \cdot k \cdot \frac{u}{2} \cos\beta$, $b = d \cdot k \cos\beta$, d = spacing, β = steering direction, $k = 2\pi/\lambda$, $u = u_1 - u_2$ = difference between two directional cosines. u controls the "beamwidth" of the quadratic processor. The final steering matrix is then given by $\underline{S}_T\,(\beta) = \underline{T}\,\underline{S}\,(\beta)\,\underline{T}^*$ with $\underline{T}$ given by (15). Fig. 8 shows a comparison of the optimum multibeam processor (exact steering matrices) with the case of steering matrices (31) for different values of u. Obviously u = 0.05 is the most favourable choice. If corresponds to an angle interval of about 20° which is the typical signal spread observed at endfire direction (s. Fig. 2 and 3). The linear multibeam processor LP (1 search beam and L-1 auxiliary beams) is the worst choice because the beam is narrower than the spatial signal spread which causes loss in gain (except the region around 90° where the angular spread is within the beamwidth).

Adaptive passive detection

In an active system a "noise -alone-estimate" of the covariance matrix $\underline{Q}$ can be obtained by estimating before transmission. In a passive system interference and signal are always present. There is no way of distinction between them. Therefore, signals are suppressed the same way as interference unless the beamforming element of the processor is perfectly matched to the signal waveform /22/.Typical examples are shown in Fig. 9. The OQP- gain response is the same as before because the beamformer matrix in the OQP is perfectly matched everywhere. The other three curves show the performance of the optimum linear processor (8) for different target ranges. It can be seen how the transmission loss effects the gain reduction due to inclusion of the signal in the adaption. Only at broadside direction the beamformer of (8) is perfectly matched, thus obtaining the maximum gain.

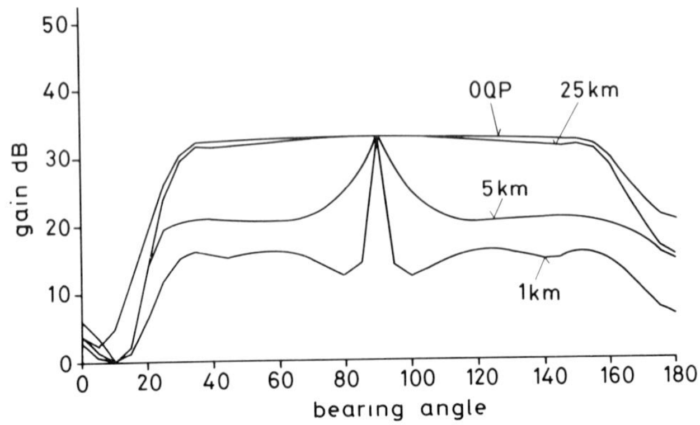

Fig. 9. Inclusion of signal in adaption

5. BEARING ESTIMATION IN A MODAL FIELD

The effect of modal propagation on bearing estimation is illustrated in Fig. 10. The arrow denotes the true position of the target whereas the line array perceives a number of modal arrivals. They are, in this case, on the right of the target position. The vertical lines denote the levels of modal amplitudes. Any power estimator based on a beamformer such as (21) will try to resolve the mode structure or look for the point of gravity of the energy distribution. In this example the conventional MEM yields two peaks at wrong positions as response to the modal spread. Using a generalised method of the form (18) one obtains one maximum at the right bearing (GMEM).

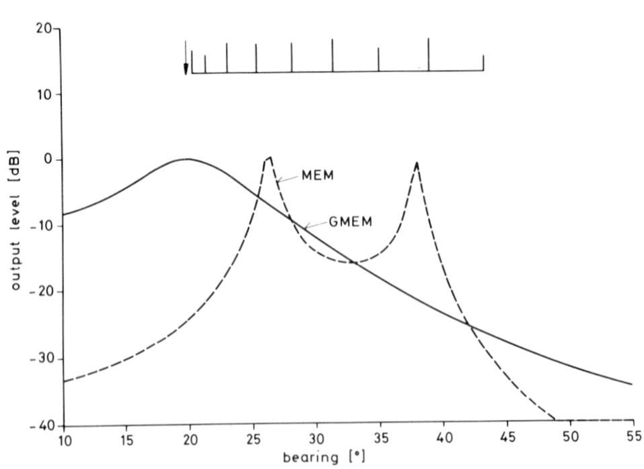

Fig. 10. Bearing estimation in a modal field

The use of generalised power estimators, such as the GMEM, for parameter estimation requires precise a priori knowledge of the statistics of the expected field, i.e. the set of steering matrices $\underline{H}$ (β) has to be known. This is a serious constraint on the efficiency of such methods because propagation models such as the normal mode solution are based on idealised assumptions which are hardly met by the conditions of a real ocean. The effects of processor mismatch on bearing estimation have been discussed in /10/.

In /21/ a method for reducing the bearing error caused by mode propagation has been described. It is based on the idea to replace the exact signal covariance matrices $\underline{H}$ (β) by substitutes of the form (31), i.e., it is assumed that the energy inside the interval of modal spread is uniformly distributed. The procedure runs as follows:

Step 1: Have a test source at known bearing.

Step 2: Estimate the first row of the covariance matrix due to the test source response.

Step 3: Approximate the covariance values by a sin ax/ax function by variation of a (estimate the angular signal spread)

Step 4: Insert the sin ax/ax - beamformer matrix in the GMEM (equ. 18 with $\underline{B} = \underline{r}\,\underline{r}^*$) and steer it over the test source.

Step 5: Observe the bearing error and use it for correction of the beamformer matrix (For more details s. /21/).

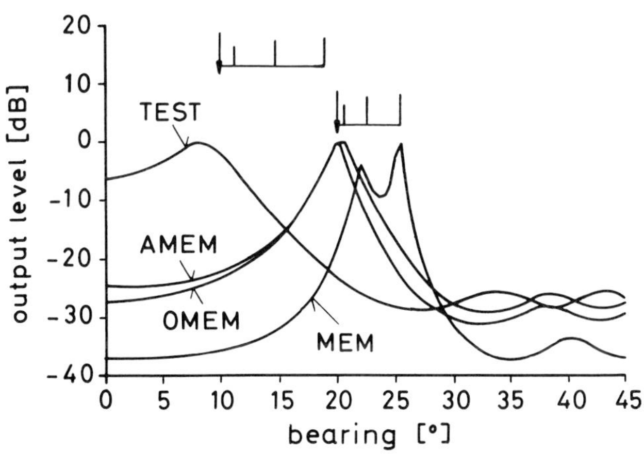

Fig. 11. The test source method

Fig. 11 shows an example. The left upper subfigure shows the test source position and the response of the channel. The right subfigure denotes the target. The test curve shows the test source response to the approximated GMEM. A slight bearing offset can be seen. This offset is used for correcting the beamformer matrices. After correction we get the AMEM-curve as power response due to the target. As can be seen the AMEM-curve is very similar to the optimum one (OMEM). It is remarkably better than the response of the conventional MEM.

6. DEPTH AND RANGE ESTIMATION BY MODE INTERFERENCE

As pointed out earlier mode interference can basically be utilised for passive range and depth estimation by applying generalised power estimators to the output data of a sensor array. All kinds of arrays may be used /13/. We confine ourselves in the following to a few examples, namely depth estimation by a vertical array. In Fig. 12 three power estimators (GBF, s. equ. 16, GMLM, s. equ. 18 case 1, GMEM, equ. 18 case 2) are used to estimate the depth of an acoustic source in a shallow water waveguide with 50 m water depth. The source is located at 25 m depth. As can be seen all three methods show different peak-to-sidelobe levels in their power response. Similar results are obtained by using a horizontal array. However, as has been shown in /13/ horizontal arrays are more sensitive to noise and fluctuations. Best performance for range and depth estimation is obtained in endfire direction because here the mode spread is maximum and, therefore, the mode structure is projected optimally on the horizontal array. At broadside direction all modal arrivals are in phase; no distinction between modes and, therefore, no evaluation of mode interference can be done.

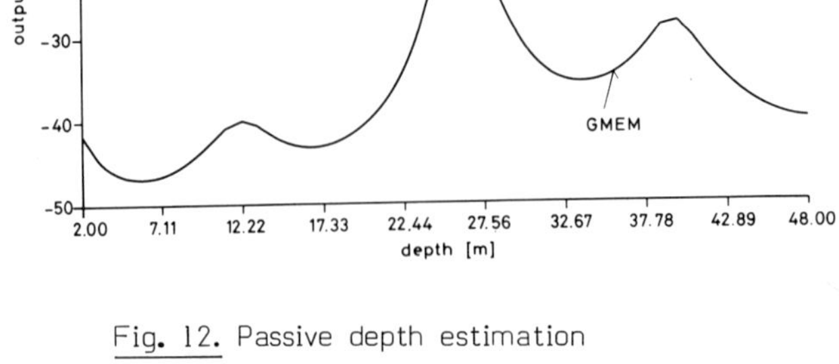

Fig. 12. Passive depth estimation

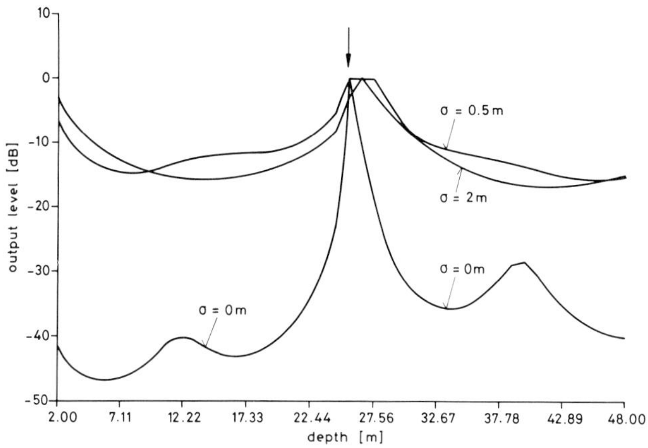

Fig. 13. Influence of fluctuations

In Fig. 13 the influence of surface fluctuations according to (26) is shown. The curves show the power responses of the GMEM for three values of σ (RMS-surface wave height) As one can see the influence of fluctuations on depth estimation is serious. One can expect that for $\sigma \approx$ 5% of the water depth no depth estimation is possible. The same is valid for range estimation.

It was supposed in Fig. 14 that the intermodal correlation due to surface fluctuations is known. Fig. 14 shows the effect of mismatch between the actual fluctuations and the assumptions on the processor side. For the processor σ_P = 1 m was assumed. If the assumption is true, i.e. the actual fluctuations are 1 m high (RMS) the source can be recognised. If there is mismatch between processor and field any arbitrary

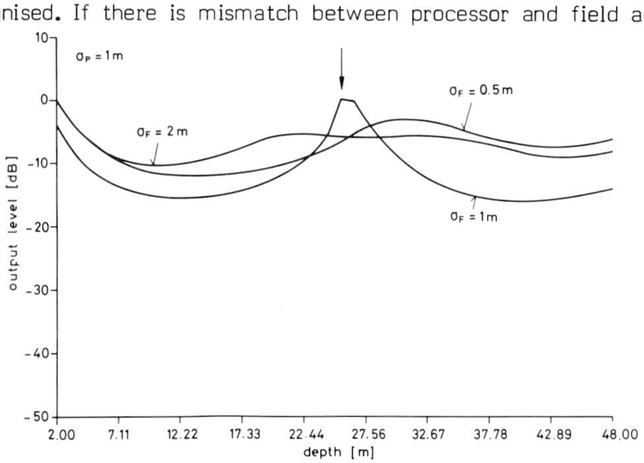

Fig. 14. Fluctuations mismatch

response can be obtained. This example shows very clearly the problems of evaluating the mode interference for parameter estimation. There are a lot of other parameters which are usually unknown. In particular bottom parameters, such as layer depth, density, sound velocity, attenuation and inhomogeneities are usually unknown which makes parameter estimation very difficult /10/.

7. RANGE ESTIMATION BY CURVATURE MATCHING

Range estimation by evaluating the curvature of a circular wavefront by horizontal arrays is well known. It is a sensitive procedure as far as the range to be estimated is large compared to the aperture of the array. In this case the estimates can be corrupted by several effects, such as unknown distortions of the array and incoherence of the field. Let us consider briefly the effect of mode propagation on this type of range estimation. If the modes are uncorrelated the horizontal array perceives a set of modal wavenumbers. Consequently, any method based on a match to one circular wavefront may obtain errorneous results. An example is shown in Fig. 15. The conventional matched filter type of processor (MF) and the MLM suffer from poor resolution. A certain range error can be noticed. If, instead, the generalised methods are used no error occurs (Fig. 16).

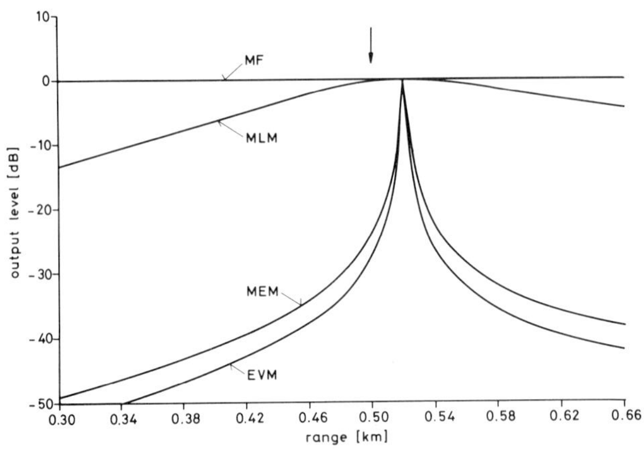

Fig. 15. Curvature matching

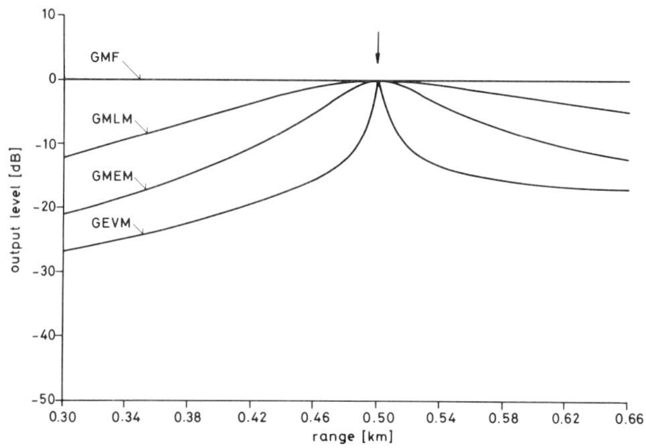

Fig. 16. Curvature matching using generalised methods

8. CONCLUSIONS

A variety of examples showing the effect of mode propagation in shallow water on detection and parameter estimation of acoustic sources has been presented. It has been shown that in principle mode propagation can be utilised for passive range and depth estimation, if the propagation medium is perfectly known. The a priori knowledge has to be inserted in the steering matrices of generalised power estimators. In a practical situation, however, the knowledge of the steering matrices may be difficult to obtain. In the case of bearing estimation an adaptive approach using a test source could be applied.

The influence of mode propagation on target detection in presence of interference has been demonstrated. Basically, the random dispersive field leads to quadratic and hence, more complicated receiver structures than in case of single plane waves, as in case of deep water or radar. It was shown that quadratic adaptive multibeam processors can approximate the performance of the optimum quadratic processor very well. For large arrays certain shading can be used to maximise the signal-to-interference ratio.

REFERENCES

/1/ F. Bryn, "Optimum signal processing of three-dimensional arrays operating on Gaussian signals and noise", J. Acoust. Soc. Amer., Vol. 34 No. 3, 1962, pp. 289-297.

/2/ H. Mermoz, "Filtrage adaptè et utilisation optimal d'une antenne", NATO Advanced Study Institute on Signal Processing with Emphasis on Underwater Acoustics, Grenoble, France, 1964, pp. 160-299

/3/ Lo, Y., T., Lee, S.W., Lee, Q. H., "Optimisation of Directivity and Signal-to-Noise Ratio of an Arbitrary Antenna Array", Proc. IEEE, Vol. 54, No 8, 1966, pp. 1033-1045.

/4/ Steinberg, B.D., "Principles of Aperture and Array System Design", John Wiley & Sons, New York.

/5/ Ma, M.T., "Theory and Application of Antenna Arrays", John Wiley & Sons, New York.

/6/ Hudson, J.E., Adaptive Array Principles. Peter Peregrinus LTD, Stevenage, UK.

/7/ Clay, C.S., "Waveguides, arrays and filters", Geophysics, 1966, 31, pp. 501-505

/8/ Bucker, H.P., "Use of calculated wave fields and matched field detection to locate sound sources in shallow water", J. Appl. Syst. Anal., 1976, 59 pp. 368-373.

/9/ Melvin J. Hinich, "Maximum-likelihood signal processing for a vertical array". J. Acoust. Soc. Amer., Vol 54, No 2, 1973, pp. 499-503.

/10/ R. Klemm, The influence of unknown bottom parameters on bearing estimation in shallow water", Proc. of Bottom Interaction Ocean Acoustics, June 9-13, 1980, SACLANT ASW Research Centre, La Spezia, Italy.

/11/ R. Klemm, "High resolution analysis of non-stationary data ensembles", Proc. of the EUSIPCO-80, Lausanne, Switzerland, Sept. 16-19, 1980 (North-Holland, Amsterdam, 1980).

/12/ R. Klemm, "Use of generalized resolution methods to locate sources in random dispersive media", IEE Proc. Vol. 127, No 1, 1980.

/13/ R. Klemm, "Range and depth estimation by line arrays in shallow water", SIGNAL PROCESSING 3, (1981), pp. 333-344.

/14/ E. Bodewig, "Matrix Calculus", North Holland Publishing Company (1959), p. 218.

/15/ H. Cox, "Interrelated problems in Estimation and Detection, I, II", Proc. of the NATO Advanced Study Institute on Signal Processing, (1968), Enschede, The Netherlands.

/16/ R. Klemm, "Horizontal Array Gain in Shallow water", SIGNAL PROCESSING , 2 (1980), pp. 347-360.

/17/ F.B. Jensen and M.C. Ferla, "SNAP, the SACLANTCEN normal mode acoustic propagation model", SACLANTCEN SM-121, 1979.

/18/ C.S. Clay, "Effect of a slightly irregular boundary on the coherence of waveguide propagation", J. Acoust. Soc. Amer., Vol. 36, 1964, pp. 833-837.

/19/ W.A. Kuperman and F. Ingenito, "Attenuation of the coherent component of sound propagating in shallow water with rough boundaries", J. Acoust. Soc. Amer., Vol. 61, No. 5, pp. 1178-1187, May 1977.

/20/ R.J. Urick, "Principles of Underwater Sound", McGraw-Hill, 1975.

/21/ Klemm, R., "Low error bearing estimation in shallow water", to be published in IEEE-Trans. AES.

/22/ Cox, H., "Resolving power and sensitivity to mismatch of optimum array processors", Journal of the Acoust. Soc. of Am., Vol. 54, No 3 (1973), pp. 771-785.

GEOMETRIC ASPECTS IN ARRAY PROCESSING

by
José M. F. Moura
CAPS, Complexo, Instituto Superior Técnico
Av. Rovisco Pais, P-1096 LISBOA CODEX Portugal
Teleph: (351-1) 572399

ABSTRACT

The theory of array processing is usually directed to nonmoving , particular array shapes, like the linear, rectangular, disc, crossed arrays. These are geometric figures whose description is incorporated into the problem in an expedite way. In practice, due to causes beyond the control of the designer, the actual array departs from these simple geometries. Colinearity is difficult of maintaining in very long arrays. Uncertainty may arise concerning the exact location of the array sensors. If the array is a towed flexible aggregate, its steady state shape may significantly depart from a closed type description. The array may move, through irregular paths, eg. random accelerating. These motions are best modeled as stochastic dynamics.

The paper addresses the above questions, proposing and developing a new point of view for the description of the array shape. This is then coupled to the general situation of moving arrays and moving sources. The description uses ordinary differential equations, or more generally, stochastic differential equations. Instead of describing the geometry in an integrated way, like saying that it is linear, the array is assumed to be a given curve in space. Curves are studied in geometry, which provides parameterized descriptions. For example, trajectories of moving objects are normally parameterized by time. Curves may admit alternative parameter descriptions. The work concentrates on using the arc length ℓ as the description parameter for the curve. Specific applications might advantageously use others. The ℓ-description dualizes the t-description of point trajectories. Also, it provides an interpretation whose generalization to models with random uncertainty on the sensors location, remains intuitive. Because the description is in terms of stochastic differential equations, it is recursive. It is emphasized that this recursiveness is both with respect to the ℓ and t parameters.

To extend the array processing techniques to moving line arrays, a rigidity assumption is made. The array is considered nondeformable. The dynamics are then the motions of a rigid body - a problem of Classical Mechanics. The full complete model presented encompasses the following: i) irregular array shapes; ii) stochastic uncertainty on the sensors' locations; iii) randomly moving arrays; iv) stochastic source motions.

The paper presents examples that illustrate the techniques studied. Array processing and localization problems are cast in the framework of recursive estimation theory. For a specific problem the receiver is discussed, one of its blocks designed and its performance studied.

NATO ASI Series, Vol. F1
Issues in Acoustic Signal/Image Processing and Recognition
Edited by C. H. Chen

1. INTRODUCTION

The class of problems to be considered is the following. Noisy measurements of a wavefield are given

$$r(t,\ell),\ t \in T,\ \ell \in L, \tag{1}$$

where t is the time parameter that takes values on the observation interval T, and ℓ is the space variable that takes values on a set of coordinates L. The wavefield sample (1) is to be processed in order to obtain a certain objective: i) detection of the presence or absence of a target; and/or ii) determination of its position and dynamics. Figure 1 represents a schematic for the latter problem.

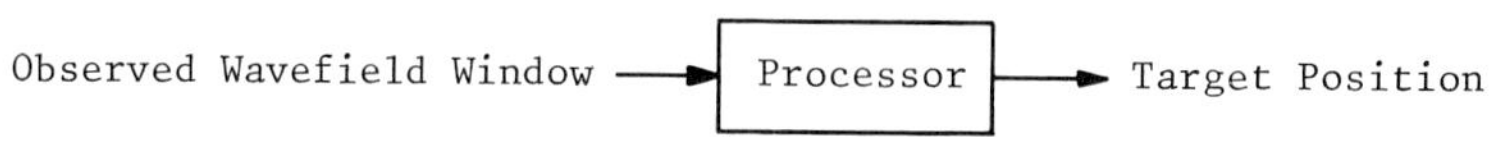

Figure 1. Problem Schematic

Relating to the processor, three issues are relevant: i) its design; ii) its implementation; and iii) its performance. With respect to the processor implementation, emphasis is here on its algorithmic aspects (most suitable for a digital technology), discussing its recursive or nonrecursive nature. The performance of the processor is considered in terms of the analysis of its predicted behavior, either via error bounds or simulated studies.

In a structured approach, the design stage is heavily dependent on the assumptions underlying the physics of the problem. To model the reality requires the mathematical definition of the following components:

i) the received wavefield $r(t,\ell)$ which is decomposed as

$$r(t,\ell) = s(t,\ell) + w(t,\ell),\ t \in T,\ \ell \in L \tag{2}$$

where $s(t,\ell)$ is the information bearing signal and $w(t,\ell)$ is the disturbing noise;

ii) the propagating channel between the observer and the target;

iii) the relative geometry.

Figure 2 summarizes these points.

The signal $s(t,\ell)$ is assumed to be of the directional type (propagating wave)

$$s(t,\ell) = s_o(t - \tau(t,\ell)) \tag{3}$$

where $\tau(t,\ell)$ is the travel time delay. In an active system, this includes both the processor/target and the target/processor travels, while for a passive system it is just a one way delay. The disturbing noise $w(t,\ell)$ is isotropic noise, given by the superposition of plane waves with uniform statistical level that propagate from all directions. The representation of general wavefield structures is an important problem in itself. The reader is referred to [1], where this topic is extensively dealt with in the underwater acoustics context. Equations (2) and (3) assume a channel of the delay type, where the disturbances are additive. Multipath and multiplicative noises are not herein explicitly considered. Much of what is stated is however also relevant in these more realistic conditions. The discussion proceeds in the above simpler context, so not to clutter the main point of the paper.

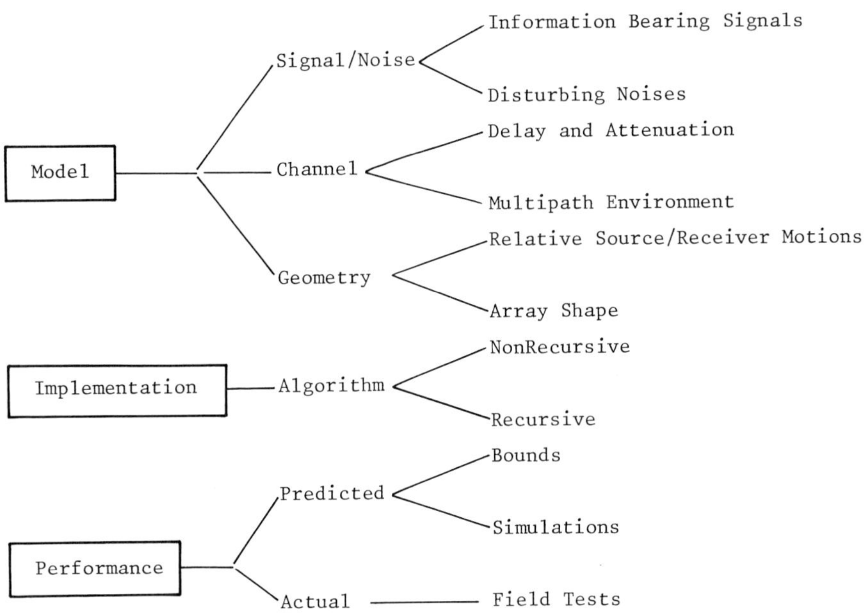

Figure 2. Brief Summary of Relevant Issues

From equation (3), it is seen that geometry comes into the problem through the delay function $\tau(t,\ell)$. For passive applications, with a homogeneous channel, the time delay is (see figure 3)

$$\tau(t,\ell_1, \ell_2)= R(t-\tau(t, \ell_1, \ell_2), \ell_1, \ell_2)/c \qquad \ell_1 \in L_1, \ \ell_2 \in L_2 \tag{4}$$

where $R(.,.,.)$ is the separation between the source point ℓ_2 and the receiver sensor ℓ_1, c is the medium propagation speed, and L_1 and L_2 are the sets defining the loca-

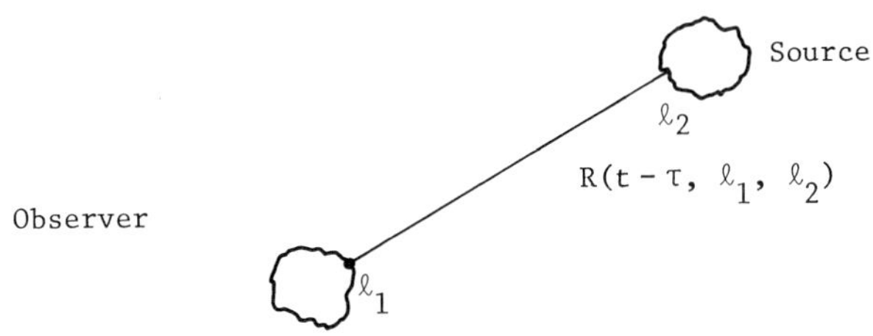

Figure 3. Source/Sensor Separation

tions of the observer sensors and the (spread) source. In the sequel, the following geometric simplifications are considered: i) the geometry is planar; ii) the source is point wise; iii) the observer's array is a line array (notice, not necessarily geometrically linear). Also, under conditions of practical significance, equation (4) is simplified to

$$\tau(t,\ell)= R(t,\ell)/c \quad t \in T, \quad \ell \in L \tag{5}$$

where $L = L_l$ stands for the set of locations defining the array. Equation (5) shows that the delay τ is a version of the source/receiver separation R. While the above arguments have been directed to passive systems, they can be modified to apply to active systems. Then, with no conceptual distinction, what is described next also concerns active systems.

2. GEOMETRIC CONSIDERATIONS

At the receiving end, as shown by equation (3), the propagating signal is modulated by the delay τ. On the other hand, figure (3) and equation (5) show that the delay is a function of the relative geometry, between the source and the receiver. To describe this geometry, it is further required the knowledge of:

i) the source shape and motions;
ii) the receiving array shape and motions.

The domains of interest for the delay function $\tau(t,\ell)$ have been given as T and L. The observations T are an interval on the real line. For example, for a centered time window of total duration T

$$T = [-T/2,\ T/2]. \tag{6}$$

In (6), notation is abused to use simultaneously letter T to designate the interval and its measure. The definition of the domain L may not be so simple a task. The principal point of the paper is precisely to provide a suitable definition for L. It is then shown how to use this definition to generalize the mathematical models commonly considered in array processing and localization problems, making them more conform to the physical world. These generalizations include the following contexts:

i) irregular array shapes, which depart from the more usual linear, circular or ring arrays.

ii) nondeterministic arrays, incorporating into the problem uncertainty about the array sensor's location.

The more general view point is obtained by considering the line array as a curve in space, and providing for it a suitable parameterized description. Recursive signal processing algorithms, eg. of the Kalman-Bucy type, can be applied. The point here is that recursivity is also over the space ℓ parameter.

3. PARAMETERIZED DESCRIPTION OF A LINE ARRAY

A line array is nothing but a curve in space. From Differential Geometry, it is well known that curves (or lines) are sets of points, with a predefined direction. The cartesian coordinates are given as

$$\begin{aligned} X &= \alpha(u) \\ Y &= \beta(u) \end{aligned} \qquad U_i \leq u \leq U_f \tag{7}$$

where α and β are continuous functions. If γ is the position vector of the general running point P, ie.,

$$\gamma = P - 0 \tag{8}$$

see figure 4, then

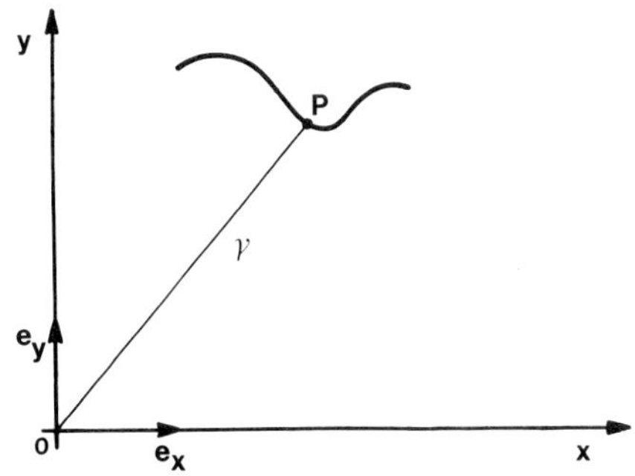

Figure 4. Curve on the plane

$$\gamma = \alpha(u)\, e_X + \beta(u)\, e_Y \tag{9}$$

where e_X and e_Y are the unit vectors along the X and Y-axes. This set of ideas can be given a more formal definition. A trajectory is an application

$$\begin{aligned} F:\ U=[U_i,\ U_f] &\longrightarrow \mathbb{R}^2 \\ u &\longmapsto (\alpha(u),\ \beta(u)) \end{aligned} \tag{10}$$

of the interval $U \subseteq \mathbb{R}$ on the real line into the plane R^2, or defined by

$$\gamma = F(u) \qquad u \in U \quad . \tag{11}$$

Sufficiently smooth conditions on α and β are assumed. The set of smooth trajectories on the plane is represented by T. The running point $P(u)$ describes a curve when u runs from U_i to U_f. Since different parameterizations are possible for the same curve, one can introduce in T an equivalence relation R between trajectories. A curve is then an equivalence class in the set of equivalence classes T/R.

Example 1: The two sets of equations

$$\begin{aligned} X &= \cos\theta \\ Y &= \sin\theta \end{aligned} \qquad \theta \in [0,2\pi]. \tag{12}$$

and

$$\begin{aligned} X &= \cos 2\theta \\ Y &= \sin 2\theta \end{aligned} \qquad \theta \in [0,\pi] \tag{13}$$

represent the circle of radius 1, of center at the origin, described clockwise, and starting at (1,0). The domain U of the parameter θ is part of the definition of the trajectory.

Of all possible parameterizations of a curve, one of special interest uses the arc length of the curve as a parameter. For an arc, its length is defined as

$$\ell(u) = \int_{u_a}^{u} ||d\gamma(u)/du||\, du\ . \qquad (14)$$

The arc length is independent of the parameterization being used. Under general conditions, a curve does admit a representation in terms of its arc length ℓ. For further comments on curves, see any introduction to the subject [2], [3]. Of further interest on the study of plane curves are the Frenet - Serret formula, the intrinsic equations of a curve, and the curvature of a line.

4. DIFFERENTIAL DESCRIPTION FOR A LINE ARRAY

The advantage of parameterizing the description of a line array, as done in the previous section, lies on the fact that the array constraint set L defining the array geometry can now be described in very compact terms. According to the specific trajectory chosen (in the sense of the previous section), the array constraint set L corresponds to the interval U on the real line, see equation (10). In example 1, equation (12), $L=[0,2\pi]$. Of course, the burden lies now on finding the particular set of equations describing the array shape (eg. equations (12) or (13)). But this is entirely dual to the point of view taken when describing the motions of an object. Having realized it, this dualism is further extended to obtain a differential description for the array.

Example 2: For the parameterization of equation (12), taking derivatives with respect to θ, it results

$$dX/d\theta = -Y$$
$$dY/d\theta = X \qquad \theta \in [0,2\pi],\ X(0)=1,\ Y(0)=0. \qquad (14)$$

Equation (14) describes the circular array in the so called state variables framework. In vector notation,

$$x = [X\ Y]^T$$
$$dx/d\theta = Ax,\quad x(0)=[1\ 0],\quad \theta \in [0,2\pi] \qquad (15)$$

with

$$A=\begin{bmatrix} 0 & -1 \\ 1 & 0 \end{bmatrix}\ .$$

Example 3: Alternatively, in polar coordinates (ρ,ϕ), the same directed circle can be represented by

$$d\rho/d\theta = 0$$
$$d\phi/d\theta = 1 \qquad \rho(0)=1,\ \phi(0)=0,\ \theta \in [0,2\pi] \qquad (16)$$

In terms of the arc length ℓ,

$$d\rho/d\ell = 0$$
$$d\theta/d\ell = 2\pi/L \qquad \ell \in [0,L]\ ,\ \rho(0)=1,\ \theta(0)=0 \qquad (17)$$

Example 4: The linear array, along the X-axis, from -L/2 to L/2 is described by

$$\begin{aligned} dX/d\ell &= 1 \qquad X(0)= -L/2 \\ dY/d\ell &= 0 \qquad Y(0)= 0 \end{aligned} \qquad \ell \in [0,L] \tag{18}$$

or, in vector form, by

$$dx(\ell)/d\ell = b, \qquad x(0)= [-L/2 \quad 0]^T, \quad \ell \in [0,L] \tag{19}$$

where $b= [1 \; 0]^T$.

All the above examples, have provided very elementary state variable descriptions. More irregular array shapes may require higher order derivatives in the state vector. This is analogous to the point of view of describing the dynamics of a target by differential equations derived from Newton's laws. More elaborate dynamics may require accelerations, or acceleration rates. Staying within the class of linear system descriptions for the array, the more general linear form is given by

$$\begin{aligned} dx(\ell)/d\ell &= A(\ell)\, x(\ell) + B(\ell)\, v(\ell) \\ x(0) &= x_o, \quad \ell \in L= [L_i, L_f] \end{aligned} \tag{20}$$

where $x(\ell)$ collects the coordinates (eg. cartesian X and Y, or polar R and θ), and possibly its derivatives with respect to the parameter ℓ, up to a given order. The matrices $A(\ell)$ and $B(\ell)$, of suitable dimensions, may be ℓ-dependent. In a system context, the vector $v(\ell)$ is referred to as the input. It provides an additional degree of freedom for the description (20). Remark that, depending on the coordinate system selected and the choice of system matrices A and B, and input v, equation (20) is capable of describing a very wide set of array shapes. This set includes, although not confined to, the geometrically linear array. The same shape may have a linear description in one coordinate set (choice of state variables), but require a nonlinear differential equation in another system. Other shapes may require all together a nonlinear differential description. For the sake of completeness, in state variable framework, this takes the form of

$$\frac{dx(\ell)}{d\ell} = f(x(\ell),\, v(\ell), \ell), \quad x(\ell_o) = x_o, \quad \ell \in L, \tag{21}$$

where f is the dynamics vector of the description.

5. ARRAYS WITH UNCERTAINTY ON THE LOCATION OF THE SENSORS: STOCHASTIC ARRAYS

In many applications, due to factors beyond the control of the processor designer, or due to the conditions of actual operation, it is not realistic to assume that the observer's array shape and location are a priori completely known. A more realistic attitude is to assume that the sensors of the array are randomly displaced about a nominal geometric shape. The differential description presented in the previous section supplies a structure which aptly handles this situation. Moreover, the ensuing estimation problem fits the framework of recursive filtering techniques, opening to the fields of array processing and localization a vast body of recursive algorithms. It is to be emphasized that recursiveness is meant on the spatial diversity of the antenna, ie. with respect to the arc length ℓ.

The proposed stochastic model assumes that in equation (20), or in equation (21), the input vector $v(\ell)$ is a sample function of a stochastic white Gaussian process. The white random process is an abstract entity that models wideband processes. Wideband refers here to the wavenumber contents of the process (Fourier dual dimension of the space-domain), not to its frequency spectrum. Also, the initial condition x_o is modeled as a random variable with given a priori probability density function see [4], [5]. A more rigorous treatment of these stochastic models requires that care must be taken. In particular, equation (21) has to be restricted to

$$\begin{aligned} dx(\ell) &= f(x(\ell),\ell)d\ell + g(x(\ell),\ell)\, d\eta(\ell) \\ x(\ell_o) &= x_o \qquad \ell \in L \end{aligned} \tag{22}$$

where f(.,.) and g(.,.) satisfy smooth and growth conditions guaranteeing the existence and uniqueness of solution for (22). Equation (22) is a formal writing, that should be interpreted in terms of a suitably defined stochastic integral (eg. in the sense of Ito). Notice also that it is written in terms of differentials. Formally, comparing equations (21) and (22)

$$v(\ell) = d\eta(\ell)/d\ell$$

where $\eta(\ell)$ is now a Brownian motion (or Wiener process). The linear class is readily obtained as

$$\begin{aligned} dx(\ell) &= A(\ell)\, x(\ell)\, d\ell + B(\ell)\, d\eta\,(\ell) \\ x(\ell_o) &= x_o \qquad \ell \in L\,. \end{aligned} \tag{23}$$

<u>Example 5</u>: A suitable model for a nominally linear array aligned along de X-axis with random displacements is

$$d\,x(\ell) = \begin{bmatrix} 0 & 0 \\ 1 & 0 \\ 0 & 0 \\ 0 & 1 \end{bmatrix} d\eta(\ell),\quad x(-L/2) = [-L/2\ 1\ 0\ 0]^T,\quad \ell \in L \tag{24}$$

where the state vector $x(\ell) = [X(\ell)\ \ dX(\ell)/d\ell\ \ Y(\ell)\ \ dY(\ell)/d\ell]^T$ includes the position coordinates and its derivatives with respect to the arc length parameter, $\eta = [\eta_1\ \eta_2]$.

Equation (25) models more appropriately the uncertainty sensor position problem than others that have been used in the litterature.

6. DYNAMICS OF A RIGID ARRAY

Section 5 has proposed a stochastic differential equation description for the array shape. When the array is static, equations (22) or (23) are used to yield an appropriate model. This section considers the generalization of the previous theory to the problem where the line array is moving. The shape of the array is a spatial constraint on the array. In section 5, this spatial constraint has been assumed to be time invariant. However, moving flexible arrays, eg. towed cabled arrays of hydrophones, deform their shape. A complete theory of moving flexible arrays requires the study of random fields. A narrower view point is taken here. It is assumed that the motions do not modify the array shape, which remains time invariant. In practical terms, the present theory applies only to problems where either the array is rigid or, if flexible, its shape has reached a steady state pattern. From the theory view point, the array is then considered nondeformable.

One of the more interesting problems in Classical Mechanics is the study of the dynamics of a rigid body. For a curvilinear array moving in a plane, rigidity implies that its entire configuration relative to a reference point P of the antenna is determined by the orientation of a single radius γ at P, see figure 5. Because the

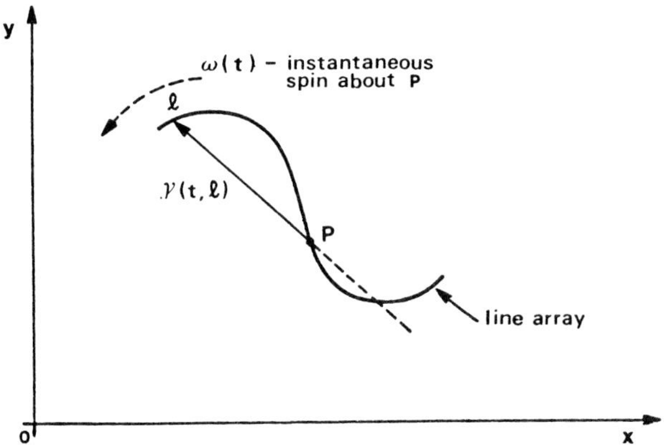

Figure 5. Configuration of Array Relative to P

line does not distort

$$d\|\gamma\ (t,\ell)\|\,/\,dt = 0, \tag{25}$$

only one degree of freedom remains. Alternatively, (25) implies

$$(\gamma,\dot{\gamma}) = 0 \tag{26}$$

where the dot stands for time differentiation, and(,) for the inner product. The relative position vector γ of particle ℓ of the array, and its instantaneous relative velocity $\dot{\gamma}$ are orthogonal. What is being said is a consequence of a deep result of Euler: The instantaneous motion of a rigid body relative to any of its particles is always a pure spin [6]. The free vector $\omega(t)$ whose magnitude is equal to the angular rate of rotation and which is directed parallel to the axis of a rotation in a right-handed sense is called the spin. Euler's theorem can be written analytically as the vector product

$$\dot{\gamma}(t,\ell) = \omega(t) \wedge \gamma(t,\ell). \tag{27}$$

For the planar motion of the line array under study: i) the modulus of the relative velocity of any point ℓ is directly proportional to $\|\omega(t)\|$ and to the distance $\|\gamma\|$ from the point to the reference P; ii) the rotational frequency $\|\omega(t)\|$ is independent of ℓ; iii) in general, $\|\omega(t)\|$ does vary with t; and iv) the spin direction remains invariant, being normal to the plane of the geometry.

To completely caracterize the dynamics of the line array, it is further required to specify the motions of the reference P.

From figure 6,

$$\Gamma(t,\ell)= \Gamma_P(t)+ \gamma(t,\ell) \ . \tag{28}$$

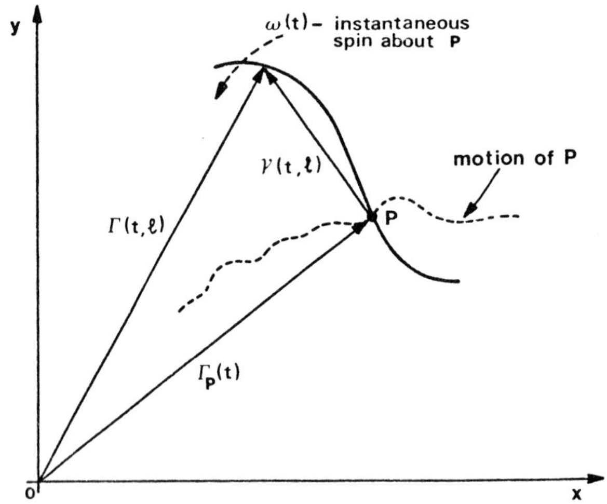

Figure 6. Complete Array Motions Caracterization

Time-differentiation of (28) leads to

$$v(t,\ell)= v_P(t) + \omega(t) \wedge \gamma (t,\ell) \tag{29}$$

where $v= d\Gamma/dt$ and $v_P= d\Gamma_P/dt$. In the sequel, to simplify the notation, the arguments of the functions will oftenly be omitted. Equation (29) shows that the motion of any sensor ℓ of the line array is conceptually decomposed as the sum of the dynamics of the reference and the instantaneous rotation of ℓ relative to P. In practical applications, to enhance performance, the receiver may maneuver. To correctly model these and other situations, higher order derivatives may be required. To obtain a differential rule, figure 7, indicates three frames: frame-β is the absolute external fixed

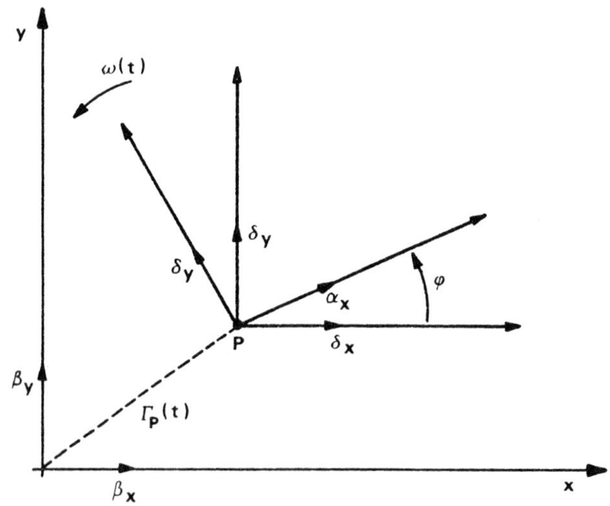

Figure 7. Three Reference Systems:
Absolute-β; Translational-δ; Rotational-α

reference; frame-δ follows the motion of P, remaining parallel to β; frame-α is solidary with the solid antenna, having a spin ω with respect to δ. By $(\dot{})_{\alpha,\beta,\delta}$ is meant the time differentiation of a given entity in frame α,β,δ respectively. Then [6], the relationship between derivatives in frames α and δ is:

$$(\dot{})_\delta(.) = \{(\dot{})_\alpha + \omega \wedge\}(.) \quad . \tag{30}$$

To obtain differentiation with respect to frame β, just add the corresponding time order derivative of vector $\Gamma_P(t)$. Differentiation of (29) leads to

$$a = a_P + \dot{\omega} \wedge \gamma + \omega \wedge (\omega \wedge \gamma), \tag{31}$$

where the acceleration vectors are $a = d^2\Gamma/dt^2$ and $a_P = d^2\Gamma_P/dt^2$.

With reference to α, the relative position vector γ is

$$\gamma = \rho(\ell)\left[\cos\theta(\ell)\ \alpha_X + \sin\theta(\ell)\ \alpha_Y\right] \tag{32}$$

where $\rho(\ell) = ||\gamma||$, and $\theta(\ell)$ provides the directional cosines associated with γ with respect to α. The matrix of the directional cosines of versors α_X and α_Y

$$\Phi(t) = \begin{bmatrix} \cos\phi\ (t) & \sin\phi\ (t) \\ -\sin\phi\ (t) & \cos\phi\ (t) \end{bmatrix} \tag{33}$$

satisfies

$$d\Phi(t)/dt = \Phi(t)\ I^*\ \omega(t) \quad , \tag{34}$$

where $I^* = \begin{bmatrix} 0 & 1 \\ -1 & 0 \end{bmatrix}$, and by an abuse of notation, $\omega(t)$ represents not the spin vector itself, but the ϕ angle rate of change.

The spatial-differential description of the array involves ℓ-differentiation of $\Gamma(t,\ell)$, or equivalently (see equation (28)) of $\gamma(t,\ell)$. Because of the factored aspect of the ℓ and t dependence in equation (32), the ℓ and t derivatives are decoupled. This simplification provided by the rigidity assumption dispenses the analysis of more sophisticated models requiring the apparatus of partial differential equations and random wavefields.

The following example summarizes the required steps to completely model the motions of a rigid line array.

Example 6:

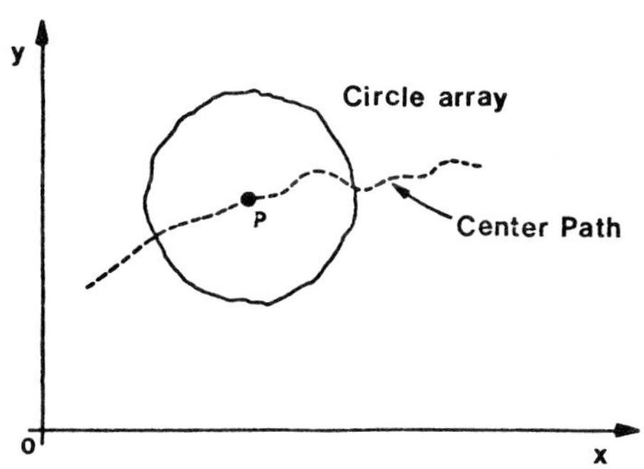

Figure 8. Distorted Moving Circle Array

Generalizing equation (17) of example 3, to a distorted circular array, let the spatial description be given by

$$\begin{aligned} dp(\ell) &= A\, p(\ell)\, d\ell + B\, d\, u_p(\ell) \\ p(0) &= [R\ 0\ 0\ 2\pi/L]^T \end{aligned} \qquad \ell \in [0,L] \tag{35}$$

where $p=[\rho\ d\rho/d\ell\ \theta\ d\theta/d\ell]^T$, $A=\begin{bmatrix}0100\\0000\\0001\\0000\end{bmatrix}$, $B=\begin{bmatrix}0&0\\1&0\\0&0\\0&1\end{bmatrix}$ and $u_p(\ell)=[u_\rho(\ell)u_\theta(\ell)]^T$.

The center P of the array is assumed to follow a nominal linear path with random accelerations. In cartesian coordinates

$$\begin{aligned} d\, x_P(t) &= A\, x_P(t)\, dt + B\, du_x(t) \\ x(0) &= [0\ \dot{X}_0\ 0\ \dot{Y}_0]^T \end{aligned} \qquad t \in [0,t] \tag{36}$$

where $x_P(t)=[X_P\ \dot{X}_P\ Y_P\ \dot{Y}_P]^T$, A and B are as in equation(35) and $u_x(t)=[u_X\ u_Y]$.
The angle rate $\omega(t)$ is a stochastic process modeled as

$$\begin{aligned} d\Omega\ (t) &= C\ \Omega\ dt + b\ d\ u_\Omega \\ \Omega(0) &= [\omega_0\ 0]^T \end{aligned} \tag{37}$$

where $\Omega(t)=[\omega(t)\quad \dot{\omega}(t)]$, C is in companion form and $b=[0\ 1]^T$.

Integration of (35) constructs for each ℓ the values of ρ and θ to be substituted in equation (32). Equation (37) provides the rotation $\omega(t)$ that placed in (34) yields the cosine directions for the versors of frame α. Hence γ results. Equation (36) directly supplies the position vector Γ_P. In practice, however the noise inputs to the different equations (35)-(37), as well as possibly some or all of the initial conditions are not known. The merit of the above equations is then to lay down a model that can be used to derive estimation algorithms that are recursive both on ℓ and t.

7. SOURCE MOTIONS

Since the source has been assumed to be pointwise, to study its motions is a task similar to the modeling of the reference point P of the array, see the previous section. If $x_S(t)$ is the position state vector, including all required time derivatives of the position coordinates, this model is of the general type

$$\begin{aligned} d\, x_s\ (t) &= f(x_s,t)\ dt + g(x_s,t)\ d\, u_s\ (t) \\ x_s(t_0) &= x_0\ . \end{aligned} \tag{38}$$

Specific vector functions f and g result for particular trajectories.

8. ESTIMATION ALGORITHMS

As seen in section 1, equations (3) and (4), the geometry modulates the signals through the delay $\tau(t,\ell)$. From figure 9, one can see

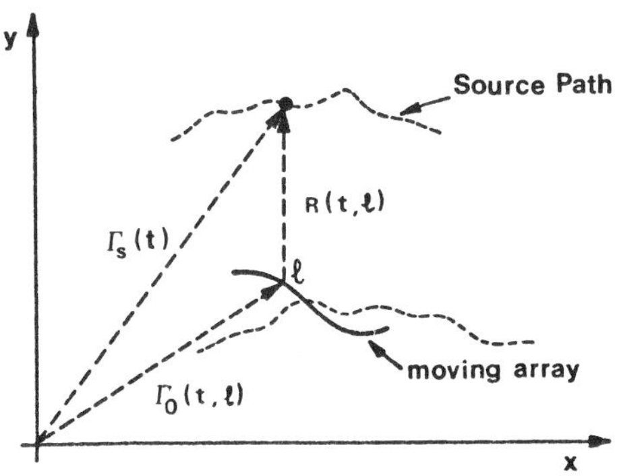

Figure 9. Source/Receiver General Geometry

that

$$\tau(t,\ell) = \|\Gamma_s(t) - \Gamma_o(t,\ell)\|/c \tag{39}$$

where $\Gamma_s(t)$ and $\Gamma_0(t,\ell)$ are the source and observer position vectors. Sections 5 to 7 have developed recursive models for these vectors. The emitted signal $s_o(t)$ may also be modelled as a stochastic process satisfying a differential equation of the type of (38). With this generality, the resulting delay estimation problem has not a practical finite dimensional solution. As remarked in [4] and [5], conceptually, the receiver is seen as a cascade of two blocks, see figure 10. The first block is a

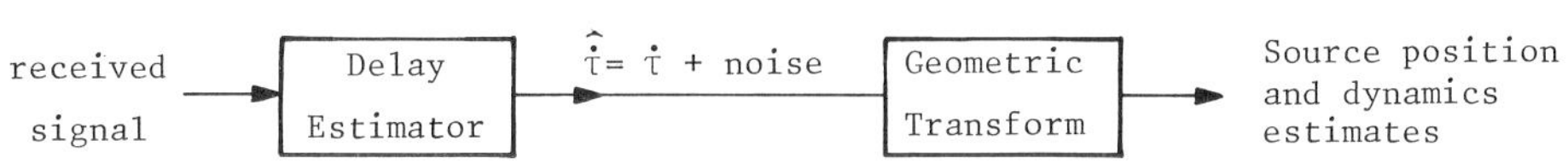

Figure 10. Conceptual Decoupling of the Receiver

(spatial and/or time) Doppler estimator. With a discrete array this is equivalent to differential delay estimation. Its structure is highly dependent on the assumptions made on the signal $s_o(t)$ and the receiving noise. The second block achieves the geometry demodulation from the noisy estimates of the Doppler (or relative delays). The models of the previous sections are the basis for the design of this block. One should notice that there is an intrinsic difference between the spatial aspects and time aspects of the second block. The spatial processing is equivalent to a smoothing problem, while the time processing is a filtering problem. In other words, the data record corresponding to the total array length L is used to estimate the geometry with respect to any sensor ℓ of the array, while only the time record up to the present epoch t is used to estimate the present geometry. So, the second block is a spatial smoother/temporal filter.

The time recursiveness of the present theory has been applied to the positioning problem in [7], [8]. See the latter paper and references there in for several analytical bounds for the mean square error performance.

Example 7

This example presents results for the performance of the spatial algorithm, when processing data from an hyperbolic distorted type array. In polar coordinates, let the spatial state vector

$$x(\ell)= [R(\ell)\ \ dR/d\ell]^T \tag{40}$$

where R is the source/receiver separation. The relative/source receiver geometry is assumed to be stationary. The geometry constraints are assumed to be

$$dx(\ell)=\begin{bmatrix}0 & 1\\ a & 0\end{bmatrix} x(\ell)\, d\ell + \begin{bmatrix}0\\ 1\end{bmatrix} du(\ell)$$

$$x(0)= [R_o\ \ \dot{R}_o] \quad , \quad \ell \in [0,L] \tag{41}$$

The input $u(\ell)$ is a Brownian motion of variance q. As noted above, the second block processes the (spatial) Doppler to yield the geometry demodulation (in this case, the state vector estimates which provide the source range). The input to the second block of figure 10 is then

$$dr(\ell)= H\, x(\ell)\, d\ell + dw(\ell) \qquad \ell \in [0,L] \quad , \tag{42}$$

where $H= [0\ \ 1]$. If the array was a discrete array, the measurements (42) would correspond to the available estimates of the relative delays between the sensor elements. The "spatial" Brownian motion $w(\ell)$ has variance σ. The time aspects of the problem are not considered here. As mentioned before, for a stationary configuration, they affect the structure of block one of figure 10. They also affect the performance of block 2, but this through the equivalent signal to noise ratio (parameter σ). Also, equation (42) assumes that the Doppler estimators provide good estimates, so that the equivalent output noise $w(\ell)$ (input to block 1) is adequately modeled by a Brownian motion. If the Doppler structure is not operating in a linear region, then bias will affect the output of the geometry demodulator. This bias may induce a significant degradation in performance.

To study the meansquare error performance associated with the estimates at the right end of the array $\ell=L$, a Kalman-Bucy filter, working on the spatial dimension, was designed. The error variance is obtained by integrating the associated Riccatti equation. This integration is in the space dimension. The Kalman-Bucy filter here processes the output of each sensor one at each iteration, in a recursive fashion. The gains of the combiner are given in terms of the error variance computed by the Riccatti equation. Figure 11 shows the Riccatti predicted mean square error performance and the average results for 50 Monte Carlo runs, for the range state variable. Figure 12 displays equivalent curves for the range (spatial) rate. The interesting behavior to be noted is that, due to the uncertainty sensor location, the improvement in performance is significantly slowed down, after an initial transient. Curves of this type may be used to dimension the array size L. Figure 11, indicates that for a prespecified output range error standard deviation, the corresponding $L= \ell_{acq}$.

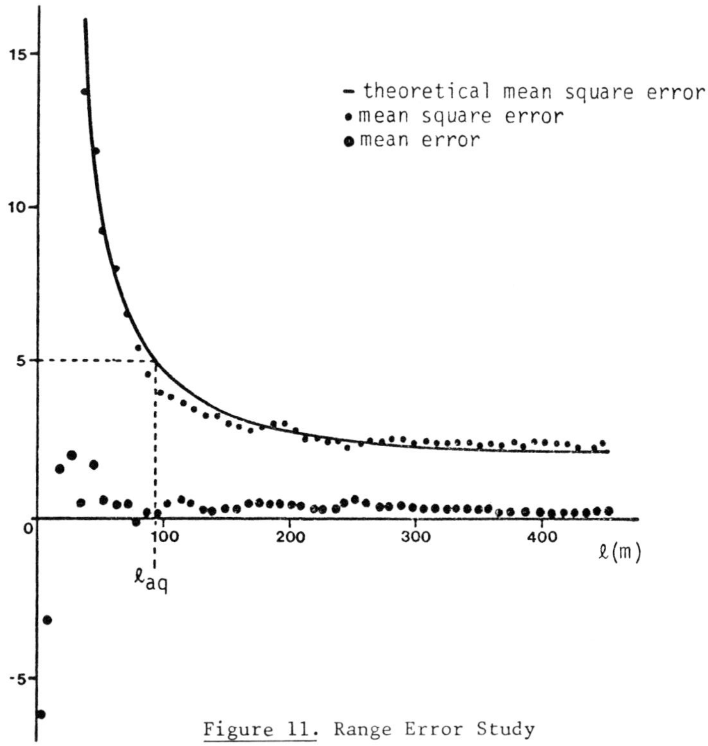

Figure 11. Range Error Study

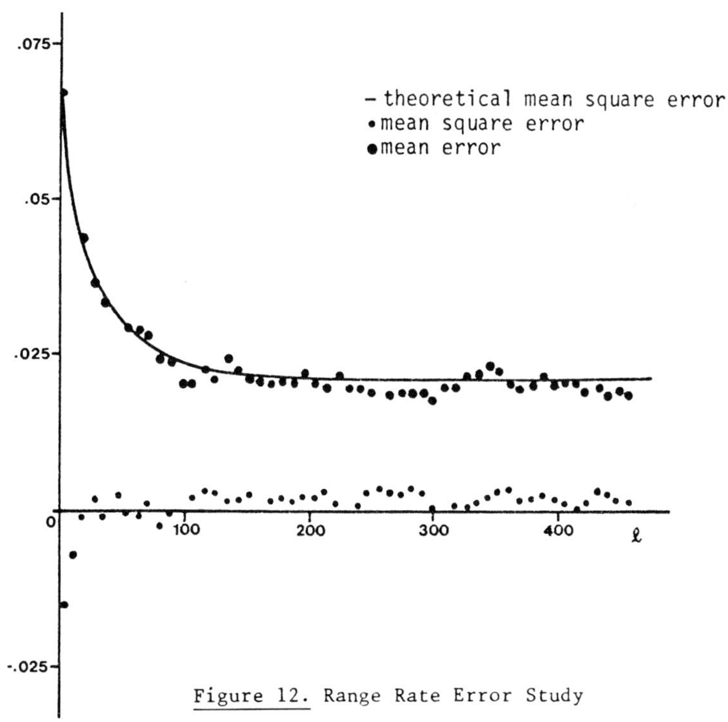

Figure 12. Range Rate Error Study

9. CONCLUSIONS

The paper has presented a novel perspective to describe the geometric constraints in array processing and localization problems. The approach extends to the spatial aspects, namely to the array description, the stochastic differential models that have been widely used in modeling temporal stochastic processes. The class of models studied encompasses combination of i) irregular line array shapes; ii) stochastically distorted line arrays; iii) stochastic motions of rigid arrays; iv) stochastic source motions.

Generalization with no added extra effort to extended line sources is possible. Several examples have illustrated the applicability of the techniques. In a forthcoming paper, the structure of the receiver will be explored in greater detail, with emphasis on the recursive spatial smoothing algorithm.

BIBLIOGRAPHY

[1] - Arthur B. Baggeroer, "Space/Time Random Processes and Optimum Array Processing", NURDC Technical report, San Diego, CA, USA, 1973.

[2] - F.R. Dias Agudo, "Lições de Análise Infinitesimal", Escolar Editora, Lisboa, 1972.

[3] - Karrel Rektorys, ed., "Survey of Applicable Mathematics", english translation by the MIT Press, Cambridge, Mass,1969.

[4] - José M.F. Moura, "Modeling, Algorithmic and Performance Issues in Deepwater Ranging Sound Systems", Underwater Acoustics and Signal Processing, ed. Leif Bjørno, D. Reidel, 1981.

[5] - José M.F. Moura, "Recursive Techniques for Passive Source Location", IEEE ICASP-82, May-1982, pp.403-406, Paris, France.

[6] - Rodney Hill, "Principles of Dynamics", Pergamon Student Editions, Pergamon Press, 1964.

[7] - José M.F. Moura, H.L. Van Trees, and A. B. Baggeroer, "Space/time Tracking by a Passive Observer", Proceedings of the 4th Symposium on Nonlinear Estimation Theory and Its Applications, San Diego, CA, Sept., 1973.

[8] - José M.F..Moura, "The Hibrid Algorithm: A Solution to Acquisition and Tracking", Journal of Acoustical Soc. of Am., vol. 69, nº 6, pp. 1663-1672, June 1981.

"Adaptive Range Tracking of Underwater Maneuvering Targets Using Passive Measurements"

Richard L. Moose
Department of Electrical Engineering
Virginia Polytechnic Institute and State University
Blacksburg, Virginia 24061

ABSTRACT

This paper examines the problem of adaptively tracking in range and velocity a maneuvering target using passive time delay measurements. The target is free to maneuver in velocity and to make depth changes at times unknown to the observer.

Tracking is accomplished by using the basic polar model of target and observer motion previously developed and reported upon [1],[2]. Now however, a nonlinear prefilter has been added to the tracking system. This leads to two major benefits: the first, is that the need for extended Kalman filters is completely eliminated which gives the passive tracking system a much greater degree of "robustness" than it previously had. The second benefit is a decoupling of depth estimation from the polar range estimator, which considerably reduces the computational level of the adaptive tracking system.

I. INTRODUCTION TO TARGET TRACKING

During the past several years much effort has been spent in the development of sophisticated digital filtering algorithms for tracking maneuvering targets. A common method has been to model the target dynamics in a rectangular coordinate system which results in a linear set of state equations, but forces the measurements to be nonlinear functions of the state variables. With this model an extended Kalman filtering algorithm is frequently used both to provide current state variable estimates and, by a one-step prediction process, to linearize the next measurement vector. This method works moderately well until the target makes an abrupt change in its trajectory in response to pilot or missile-guidance program commands. In this situation the velocity and position estimates can, and often do, diverge from the true unknown values. The inherent problems of this approach can lead to large bias errors and sometimes complete filter divergence.

Earlier work on the maneuvering target tracking problem includes Jazwinski's limited memory filtering [3], in which the filter gains are prevented from decaying to zero. Another technique, described by Thorp [4], involves switching between two Kalman filters in response to a detected maneuver. A third approach, due to Singer [5], models the target trajectory as a response of the target model to a time-correlated random acceleration. With this method additional state variables are used to generate the correlated forcing functions which, in turn, increase the

Sponsored by ONR under Contract No. 0014-77-C-0164.

NATO ASI Series, Vol. F1
Issues in Acoustic Signal/Image Processing and Recognition
Edited by C. H. Chen

dimension of the Kalman filtering algorithm. In this manner the technique provides the filter with statistical information concerning target maneuvers based on an assumed range of possible accelerations. Singer's method was subsequently extended by many others.

Parallel to the effort was the method of modeling major changes in target trajectories by a semi-Markov process. An application of this approach to tracking maneuvering targets in two-dimensions by Moose [6] was successfully extended by Gholson and Moose [1] to three-dimensional tracking.

The general approach which uses the "adaptive semi-Markov maneuver model" of [6] and [7] implies a discretization of possible vehicle accelerations or velocities. The estimation algorithm then views the maneuvering vehicle as if it is responding to commands which are modeled by a semi-Markov process, i.e., a random process with a finite number of "states" (commands) which are selected according to the transition probabilities of a Markov process. A semi-Markov process differs from a Markov process in that the duration of time in one state prior to switching to another state is itself a random variable. Incorporating the semi-Markov concept into a Bayesian estimator was done in [6] and [8]. This estimation algorithm provides a substantial improvement in filter stability, which means that large bias errors are prevented from being built up due to unmodeled target accelerations. An important aspect of this adaptive estimation algorithm is its elimination of a "growing memory" which is prevalent in many adaptive filters. It has been shown that by combining the concepts of Singer and Moose, et. al. that the mean values required to prevent filter divergence is greatly reduced. This combination is illustrated in Figure 1.

The target trajectory is generated by the random selection of an input time-correlated Gaussian process whose mean value u_i is applied to the target plant dynamics for a random duration of time. This input disturbance process lasts until a new input u_j is randomly chosen from among a finite set of n possible inputs. With this model as a background and using an appropriate choice of state variable equations to represent target dynamics, either submarine or aircraft, it is possible to develop an "optimal" (in the minimum mean-square error sense) tracking filter that adaptively learns, then quickly adjusts itself for each major alteration of target trajectory.

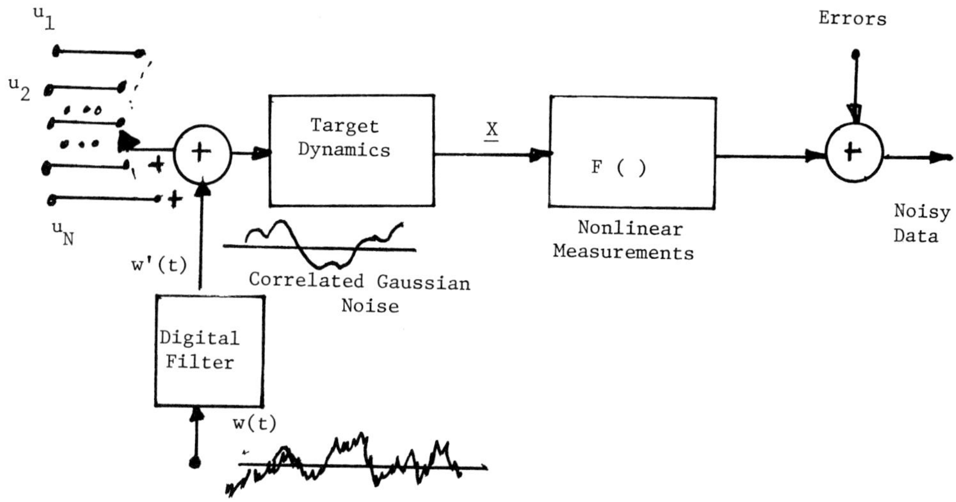

Figure (1). Maneuvering Target Motion Model

II. LINEAR POLAR STATE VARIABLE MODEL

Referring to Fig. 2, the polar coordinate system (ρ, z_{so}) is attached to and moves with the observer. If (x_o, y_o, x_s, y_s) are the horizontal position coordinates of the observer and source, respectively, in a fixed (w.r.t. ocean floor) rectangular coordinate system, then

$$\rho = [(x_s - x_o)^2 + (y_s - y_o)^2]^{\frac{1}{2}} \tag{1}$$

Equation (1) is linearly expanded as follows:

$$\begin{aligned} \rho_{k+1} = \rho_k &+ \left.\frac{\partial\rho}{\partial x_s}\right|_k (x_{s_{k+1}} - x_{s_k}) + \left.\frac{\partial\rho}{\partial y_s}\right|_k (y_{s_{k+1}} - y_{s_k}) \\ &+ \left.\frac{\partial\rho}{\partial x_o}\right|_k (x_{o_{k+1}} - x_{o_k}) + \left.\frac{\partial\rho}{\partial y_o}\right|_k (y_{o_{k+1}} - y_{o_k}). \end{aligned} \tag{2}$$

By modeling the source dynamics in the X and Z directions as a linear drag model (drag coefficient α) with a forcing function consisting of a Singer correlated Gaussian random process (time constant = $1/a$), the following state variable model can be shown to emerge from (2) for a sample interval T.

$$\begin{bmatrix} \rho \\ \dot{\rho} \\ w'_{s_\rho} \end{bmatrix}_{k+1} = \begin{bmatrix} 1 & A & B \\ 0 & E & F \\ 0 & 0 & e^{-aT} \end{bmatrix} \begin{bmatrix} \rho \\ \dot{\rho} \\ w'_{s_\rho} \end{bmatrix}_k + \begin{bmatrix} C & (A - T) \\ A & (E - 1) \\ 0 & 0 \end{bmatrix} \begin{bmatrix} u_{s_\rho} \\ V_o \cos B_{so} \end{bmatrix}_k + \begin{bmatrix} D \\ G \\ J \end{bmatrix} w_{s_{\rho_k}} \tag{3}$$

where $A = (1 - e^{-\alpha T})/\alpha$, $B=[1 + (ae^{-\alpha T} - \alpha e^{-aT})/\alpha - a)]/(\alpha a)$, $C = (\alpha T - 1 + e^{-\alpha T})/\alpha^2$, $D = [T + (\alpha A - \alpha J)/(\alpha - a)]/(\alpha a)$, $E = e^{-\alpha T}$, $F = (e^{-\alpha T} - e^{-\alpha T})/(\alpha - a)$, $G = (J - A)/(\alpha - a)$, and $J = (1 - e^{-aT})/a$.

V_o is the observer's horizontal velocity and B_{so} is the relative bearing of the source with respect to the observer. A similar state model exists for the vertical (Z) direction. Note that the state equation for $\dot{\rho}$ in (3) is arrived at by applying the linear expansion (2) to the time derivative of (1).

Depth Channel Model

Using a discretized version of the basic linearized drag model of

$$\begin{bmatrix} d_T + \alpha \dot{d}_T = w'_z \\ \dot{w}'_z + a\, w'_z = w_z \end{bmatrix}$$

and letting d_T be the target depth the following state model is presented.

$$\begin{bmatrix} d_T \\ \dot{d}_T \\ w'_z \end{bmatrix}_{k+1} = \begin{bmatrix} 1 & A & B \\ 0 & E & F \\ 0 & 0 & e^{-aT} \end{bmatrix}_k \begin{bmatrix} d_T \\ \dot{d}_T \\ w'_z \end{bmatrix} + \begin{bmatrix} D \\ G \\ J \end{bmatrix} w_{z_k} \tag{4}$$

The state model is also subject to the constraint of a constant velocity observer.

III. TRACKING GEOMETRY

The tracking geometry of Figure 2 has been investigated by Hassab [9] and shown to yield two different time delays τ_1 and τ_2. Here τ_1 is the difference in time between the direct path and the surface reflected path from source to observer, and τ_2 the direct and bottom bounce time difference. For *long* ranges i.e.: $r^2 >> d_o^2$, then $r = \rho$ the polar range, and

$$\tau_1 = \frac{2\, d_o d_T}{\rho C} \quad . \tag{5a}$$

If, in addition, we have the "shallow water ranging situation," i.e. that (d_k^2/r^2) is << 1. For example, if $r \geq 2.25\ d_k$, our error of approximation is less than 1.4%, thus τ_2 becomes

$$\tau_2 = \frac{2\, d_k (d_w - d_T)}{\rho C} \tag{5b}$$

where d_w = total ocean depth

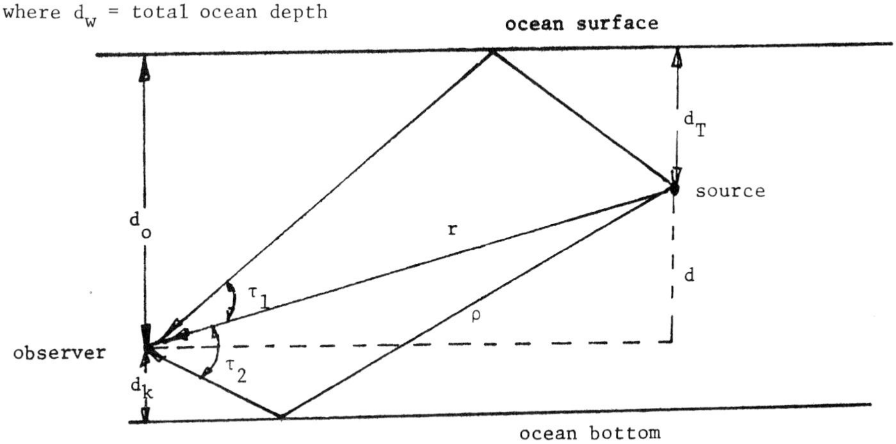

Figure (2). Two Dimensional (Range and Depth) Tracking Geometry

Solving equations 5a and 5b for target depth d_T then substituting into (5a) to get target polar range. ρ we find that:

$$\rho = \frac{b_o}{\tau_1 + a_o \tau_2} \tag{6a}$$

$$d_T = \frac{\tau_1 d_w}{\tau_1 + a_o \tau_2} \tag{6b}$$

where $a_o = d_o/d_k$, $b_o = 2\ d_o d_w/C$, and C equals the sound velocity in sea water. In reality, we do not have τ_1 and τ_2 given to us but only the noisy set of measurements

$$z\tau_1 = \tau_1 + v_1$$

$$z\tau_2 = \tau_2 + v_2$$

where v_1 and v_2 are Gaussian random processes with zero mean and variances σ_1^2 and σ_2^2 respectively. This results in the noisy set of measurements

$$z_{d_T} = \frac{z\tau_1 d_w}{z\tau_1 + a_o z\tau_2}$$

$$z_\rho = \frac{b_o}{z\tau_1 + a_o z\tau_2} \tag{7}$$

IV. NONLINEAR PREFILTER DESIGN AND ANALYSIS

In equation (7) we see that a noisy measurement (z_ρ) of polar range (ρ) is found by combining the weighted time delay measurements and dividing them into the geometry coefficient (b_o) previously defined. This process is shown in figure (3). The output (z_ρ) consists of true range (ρ) plus non zero mean and non Gaussian additive noise (v_ρ).

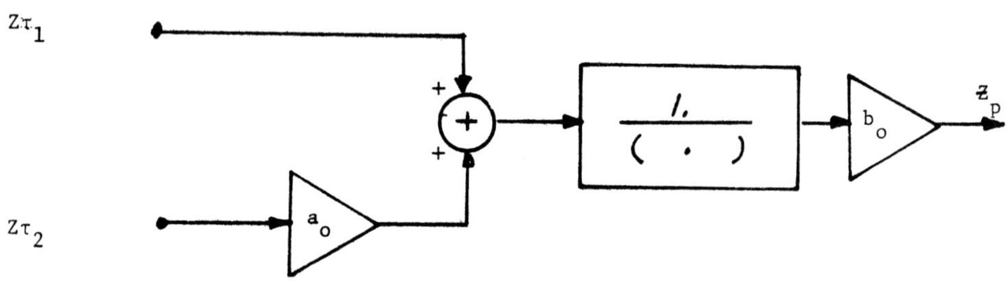

Figure (3). Nonlinear Prefilter for the Target Range Measurement

After a statistical analysis the following probability density function for v_ρ was determined

$$p_2(v_\rho) = \frac{\rho\ \tau_T}{\sigma_T\sqrt{2\pi}\ (\rho + v_\rho)^2} \exp\left[\frac{-\tau_T^2\ v_\rho^2}{2\sigma_T^2(\rho + v_\rho)^2}\right] \text{ where } \sigma_T^2 = (\sigma_1^2 + a_o^2\ \sigma_2^2). \tag{8}$$

The nature of this density function is shown in the following figures for the target and observer at fixed depths, but closing in range.

It appears that the density function $p_2(v_\rho)$ given by equation (8) is of a form that does not appear in complete sets of integral tables, nor does it seem likely that there is a closed form for the expected value ($v_\rho = -ax/(x+k)$), or for the expected value of $v_\rho^2 = a^2x^2/(x+k)^2$ where x is $N(0,\sigma_\rho^2)$. Thus one must use *numerical integration or simulation techniques* in order to get an approximation for the mean and the variance of measurement error (v_ρ). Examining the density function, we see from Figure (4) that it is a smooth continuous curve. As target range closes the mean and variance becomes smaller and the density becomes more and more symmetrical approaching a near Gaussian like density.

Table (1) shows that v_ρ has a non-zero mean and a variance which are nonlinear functions of range and SNR. These statistics must be altered for they are not compatible with the Kalman filters that will process the range measurements. For example, v_ρ must be zero mean. This is achieved by a table look-up of the measurement noise mean based on the tracker's previous estimate of ρ and the SNR. This mean is then subtracted from the current measurement z_ρ producing a zero mean noise process. A similar table look-up method is used to select the proper variance required by the Kalman filters. Simulations have shown that the values in Table (1) are sufficient to insure measurement compatibility with the Kalman filters.

	$\sigma_{1,2}$ = 1ms		$\sigma_{1,2}$ = 5ms	
Range	Mean	Var	Mean	Var
10,000	3.8	.209E + 05	62.	.549E + 06
20,000	23.	.336E + 06	487.	.102E + 08
30,000	72.	.172E + 07	1650.	.605E + 08
40,000	163.	.552E + 07	3518.	.192E + 09
50,000	310.	.137E + 08	5588.	.427E + 09
60,000	527.	.290E + 08	6841.	.780E + 09
70,000	831.	.555E + 08	7068.	.127E + 10
80,000	1238.	.972E + 08	6204.	.190E + 10

Linearized Measurement Error Means and Variances vs Range

Table I

A limiter is required at the output of the nonlinear prefilter. This is due to the fact that a density function of the form given by equation (8) has a large variance and an occasional bad data point is observed. The limits are quite loose and were set at $\hat{\rho}_k \pm 10^5$.

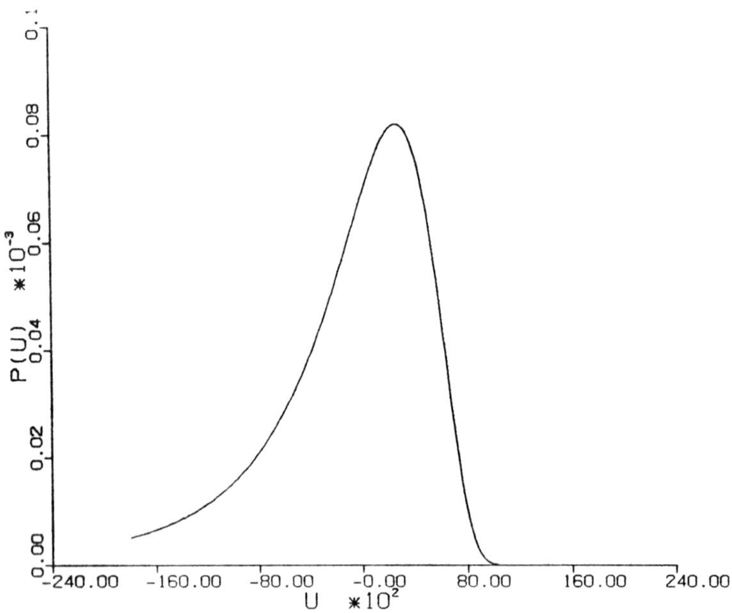

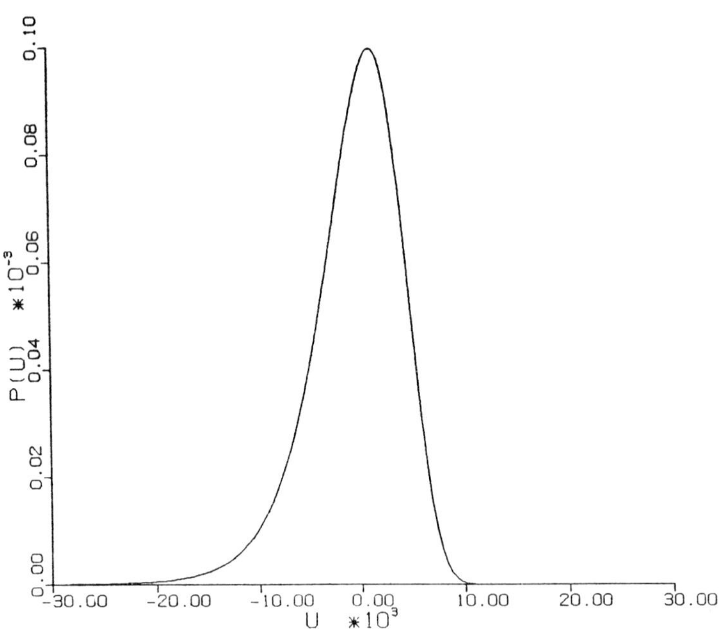

Figure (4). Measurement Error Density Function for Closing Range

V. STATE ESTIMATION AND ADAPTIVE TRACKING SYSTEM STRUCTURE

5.1 Introduction

In this section, we will discuss the basic estimator system structure. We will make use of the linearized measurements containing the nonstationary, non-Gaussian, statistics that were previously discussed in section (4) of the paper.

Referring to Figure (5), the nonlinear time delay measurements $z_{\tau 1}$ and $z_{\tau 2}$ are fed into the nonlinear prefilter. This unit develops a linearized measurement of target range (ρ). The errors in measuring these target parameters are both non-Gaussian and nonstationary depending upon the geometry of the tracking situation. As target range closes, or opens the mean value, and variances of these errors change, and must be accounted for by the stored means and variances of Table (1).

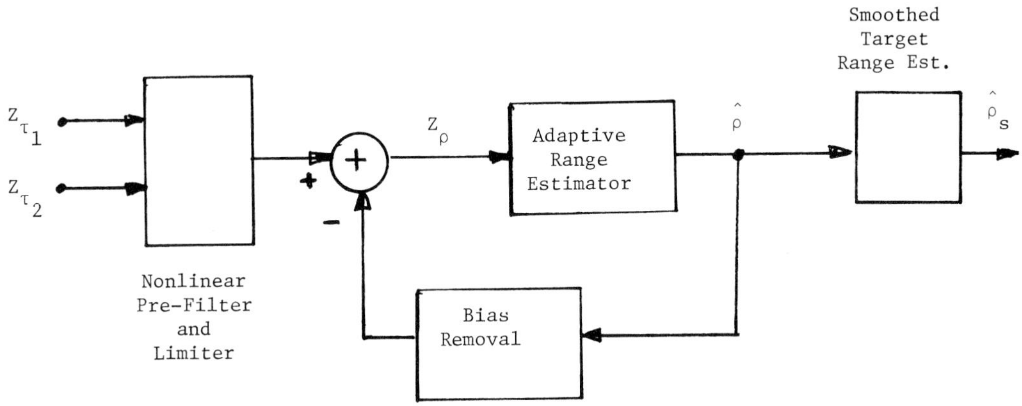

Figure (5). Basic Range Estimation Structure

In Figure (6) a discrete time system model is presented showing the development of noisy time delay measurements $z_{\tau_1}(k)$ and $z_{\tau_2}(k)$. The upper portion shows the generation of $\underline{x}_{k+1} = [\Phi\, \underline{x}_k + \Gamma\, u_k + \underline{\psi}\, w_k]$ where Φ, Γ, and $\underline{\psi}$ are discussed in eq. (3). The deterministic input u_k is unknown to the tracking filter and serves to generate large scale target maneuvers in velocity. A measure of randomness in target trajectory ($x_1 = \rho_k$ and $x_2 = \dot{\rho}_k$) is generated by applying a Gaussian random input $\underline{w}_k$ to the simulated target, exponentially correlated, forcing function component as state vector $x_3 = w_\rho(k)$.

Once ρ_{k+1} is generated, it is acted upon in a nonlinear manner to generate τ_1 and τ_2, which when added with the Gaussian random measurement errors v_1, v_2, produce the noisy time delay measurements z_{τ_1} and z_{τ_2}.

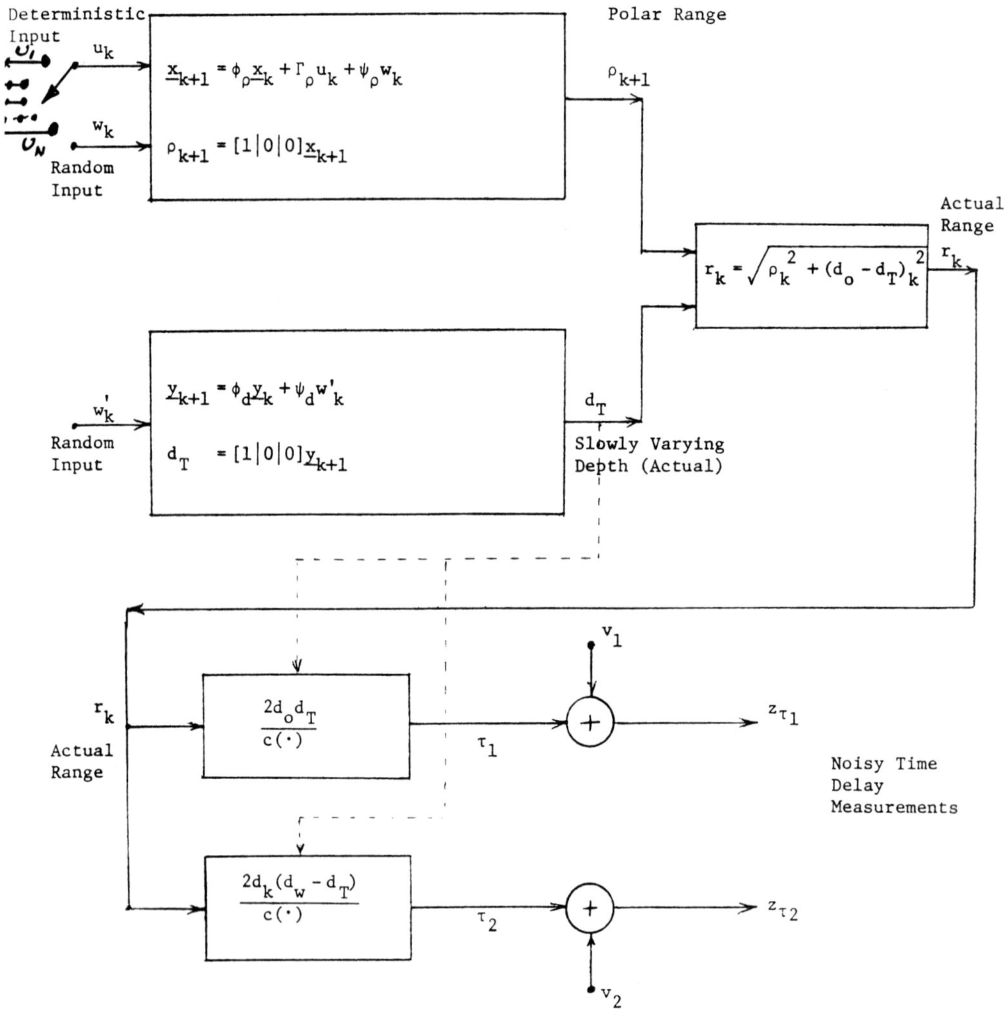

Figure (6). Data Generation

5.2 The Polar Range Adaptive State Estimator

The heart of the adaptive filter summarized in this paper is in the forming of the total estimate of the target states, from a weighted sum of state estimates conditioned on the N possible discrete input levels $u_\rho^{(i)}$. Consider the state model of equation (3). This state model views the target input acting in the polar direction as being derived from a time correlated Gaussian density having a mean value u_ρ. Next consider a series of N such Gaussian curves with displaced mean values $u_\rho^{(i)}$, i = 1, 2, ..., N and partially overlapping "tails" outlined in Figure one. If a bank of N Kalman filters is formed, each filter based on the state equations of equation (3) with the deterministic input $u_\rho^{(i)}$ being a different one of these N mean values, then a series of N estimates is obtained, each conditioned on a different Gaussian curve. Next a weighted sum of these estimates is found to be the total unconditioned estimate of the target states given by the optimal mean-squared estimate

$$\underline{\hat{x}}(k+1) = \sum_{i=1}^{N} \underline{\hat{x}}^{(i)}(k+1)\; W_i(k+1). \tag{9}$$

Now as the target executes a series of evasive maneuvers in the polar channel, the changing input necessary to produce these maneuvers is viewed as randomly switching among the N Gaussian curves. By applying semi-Markov statistics to this switching process a series of N probabilities W_i, i = 1, 2, ..., N is generated where

$$W_i \equiv \Pr\{\text{target input is being derived from the Gaussian curve whose mean value is } u_\rho^{(i)}\} \qquad \text{OR,}$$

$$W_i(k+1) = \Pr\{u_\rho(k) = u^{(i)} \mid Z(k+1)\} \tag{10}$$

and

$\hat{x}^{(i)}(k+1) = E\{x(k+1) \mid u(k) = u^{(i)}, Z(k+1)\}$ represents the ith Kalman filter conditioned on $u_\rho = u^{(i)}$. For example, filter (i) is given by

$$\hat{x}^{(i)}(k+1) = \Phi\hat{x}^{(i)}(k) + \Gamma u^{(i)} + K(k+1)\,[z(k+1) - H\Phi\hat{x}^{(i)}(k) - H\Gamma u^{(i)}] \tag{11}$$

where

$$K(k+1) = M(k+1)\, H^T [HM(k+1)\, H^T + R]^{-1},$$

$$M(k+1) = \Phi P(k)\, \Phi P(k)\, \Phi^T + \psi\, Q\psi^T$$

$$P(k+1) = [I - K(k+1)H]\, M(k+1).$$

The matrices Φ, Γ and ψ are used to denote the respective coefficient matrices in equation (3).

The weighting probabilities can be expressed in a vector recursive form as

$$\underline{W}_{k+1} = C_k \underline{P}_k \underline{\Theta}^T \underline{W}_k \tag{12}$$

where $\underline{P}_k$ is a time varying diagonal matrix whose elements have been previously computed in each of the (N) Kalman filters (the measurement residual), i.e.:

$$P_{ii} = N\,[m_i(k+1),\ V(k+1)].$$

The term C_k is a normalizing constant computed at each iteration to ensure that the sum of probabilities equals unity. The matrix $\underline{\theta}$ is a precomputed matrix whose elements contain statistical knowledge of the randomly switching plant inputs. In practice, the diagonal elements are nearly unity and the off diagonal terms are set equal.

Although it might appear from equation (9) that an entire Kalman filter algorithm is being executed N times (for each of the possible inputs) at each time iteration, such is not the case, since the process and measurement covariances Q and R are the same for each filter.

What differentiates the different target "states" is the discrete levels u_i; however, the target dynamics remain unchanged! The entire covariance, and gain analysis of the Kalman filter algorithm becomes identical for each state in a given channel and, consequently, need be executed just once rather than N times. The adaptive filter structure is shown in Figure 7.

VI. RANGE ESTIMATION RESULTS

In order to have a *benchmark* with which to compare the adaptive estimator's performance the following scenario was devised. A target at the extreme range of 100,000 was generated on a closing trajectory of 25 ft/sec at a depth of 600. The target's plant model was matched exactly with that of the filter. This was done to simulate and test the performance of the optimal (non-realizable) tracking filter. In addition, it also provided a test of the nonlinear prefilter and the adaptive state estimator whose weights were all set at zero for the unmatched filters and unity for the single matched filter.

Target tracking results are shown in figure (8) for additive noise of 5 m sec. Examining the figure we see the performance that one would expect out of the unrealizable filter. Note that the variance of measurement error decreases with decreasing range as one might expect.

The data simulation of Figure (6) made use of the following parameters: the maneuver time constant ($1/\alpha$) was chosen as 25 seconds, the Singer correlated acceleration time constant ($1/a$) in equation (3) was selected to be 40 sec, and the data rate was chosen to be one sample every 10 seconds. Although studies were made indicating a marginal increase in tracking performance, for sample intervals of

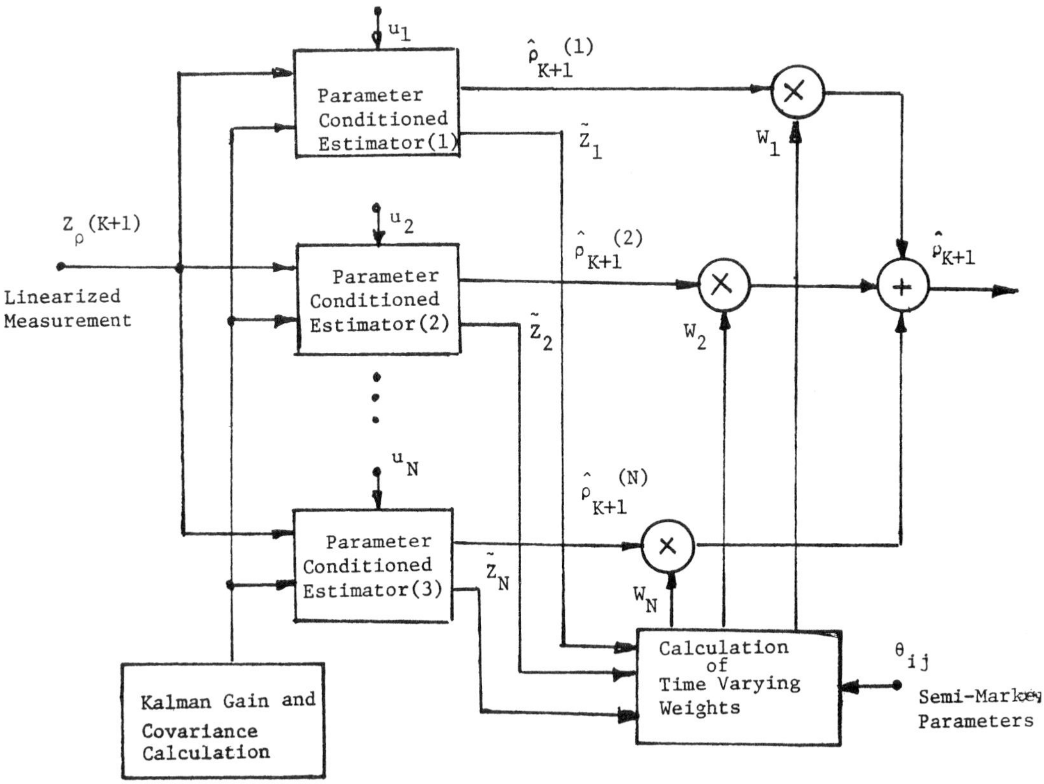

Figure (7). Adaptive Polar Range Estimator

2 and 5 seconds, the sample interval was retained at 10 seconds in order to reduce the computation burden for real world applications.

In the adaptive estimator of Figure (7), 6 levels of input u_i were chosen to span the expected target velocity range of +30 (ft/sec) for an opening target and -45 (ft/sec) for a closing target. If the velocity ranges are greater than this number, N could be increased as needed. The u_i's were chosen to model +30, +15, 0, -15, -30, -45 (ft/sec).

The next figure (9) shows the realizable adaptive filter tracking in the presence of low noise (1 m sec). The target makes a random step change in velocity at time k = 150, which corresponds to 1500 seconds into the sceanrio. The range estimate $\hat{\rho}$ tracks very well converging very quickly after the target maneuver.

In Figure (10) the weighting coefficients are shown to switch appropriately from 0 to near unity as required.

In Figure (11), we see the results of tracking a medium range (0-45k) target as it undergoes a maneuver at k = 140 and k = 340. The target is on a closing trajectory then reverses its velocity at 140 and 340. The estimate $\hat{\rho}$ tracks very well at close ranges and progressively gives noiser estimates as the targets range in-

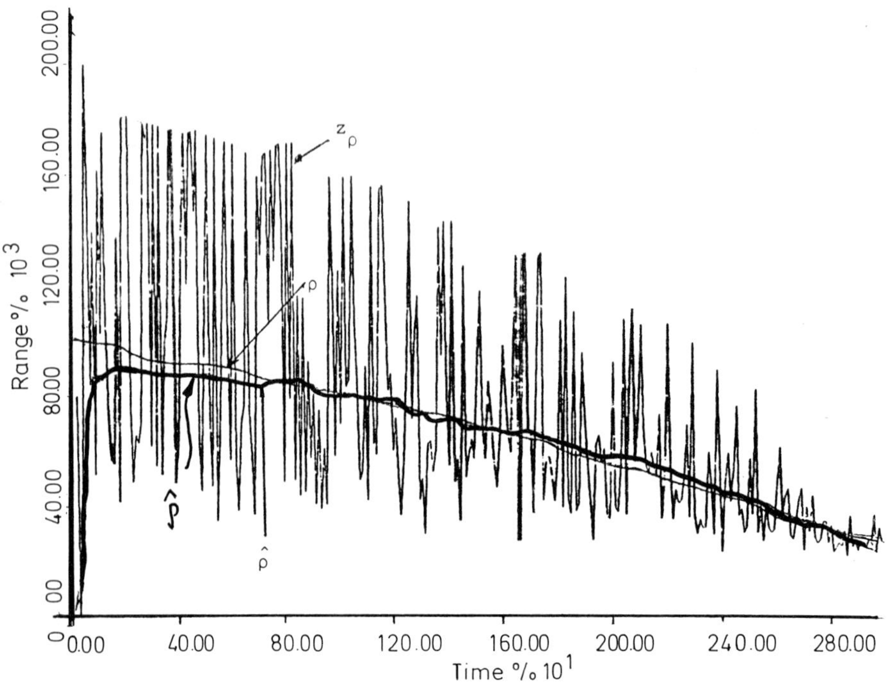

Figure (8). Range Estimate of Matched (unrealizable) Kalman Filter (Low SNR Case).

creases. This is due to the variance of the measurements, which are not shown, increasing with range. The additive time delay noise is $\sigma_1 = \sigma_2 = 3$ m sec. a considerable increase. Several conclusions can be drawn from Figure 11. The first is that we need to limit ourselves to close proximity targets, or smooth the estimate and weighting coefficients $\hat{w}_i(k)$, or operate in an environment with smaller additive noises which is totally unrealistic.

6.1 Modification to the Adaptive State Estimator

In order to operate in a noisy environment where the additive measurement errors attached to τ_1 and τ_2 are greater than or equal to 3 m sec, it becomes necessary to smooth both the weighting coefficients and/or the output estimate $\hat{\rho}_{k+1}$. It was found that a simple first order digital filter of the form

$$\tilde{\rho}_{k+1} = a\tilde{\rho}_k + b\hat{\rho}_{k+1} \tag{13}$$

was successful for good smoothing and small lag in tracking maneuvers. The term $\tilde{\rho}_{k+1}$ represents the smoothed output range estimate at t_{k+1} and $\hat{\rho}_{k+1}$ the "rough" filter input. Coefficients a and b were set at 0.8 and 0.2 respectively.

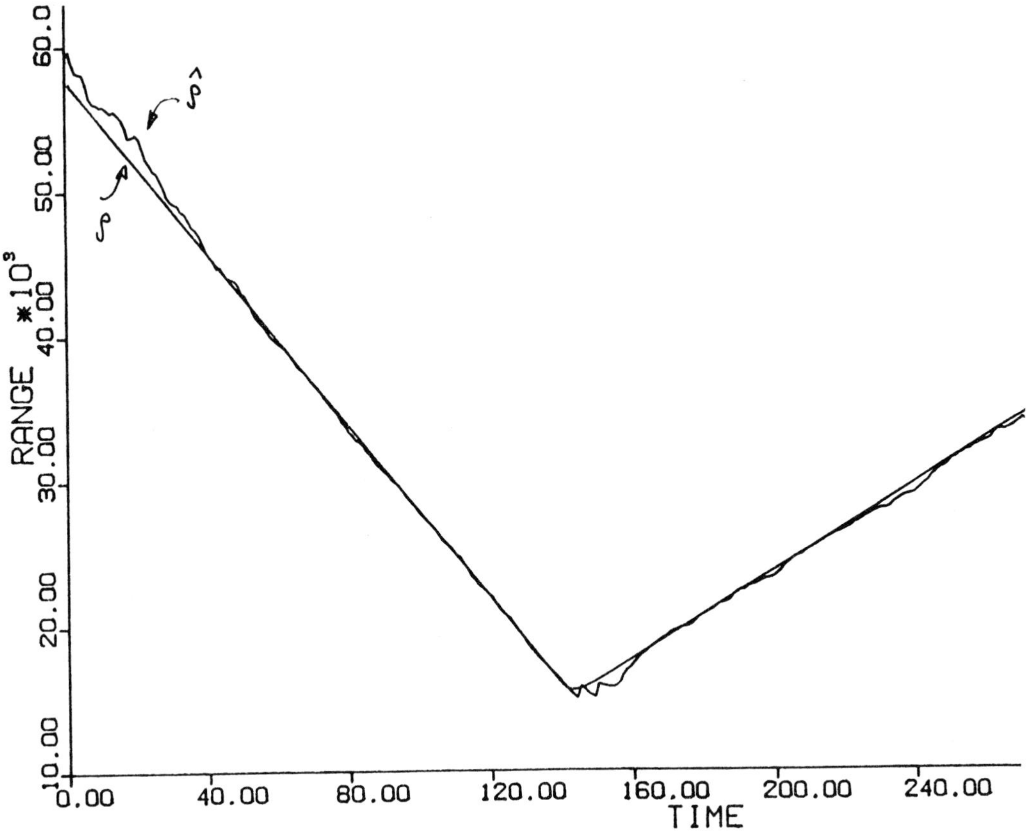

Figure (9). Target Tracking for the Low Noise Case Realizable Filter

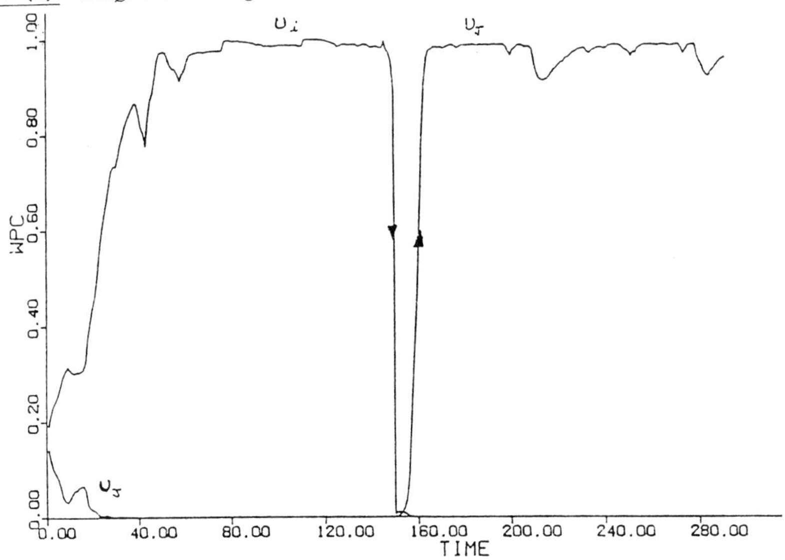

Figure (10). Plot of Weighting Coefficients vs Time for Target Maneuver

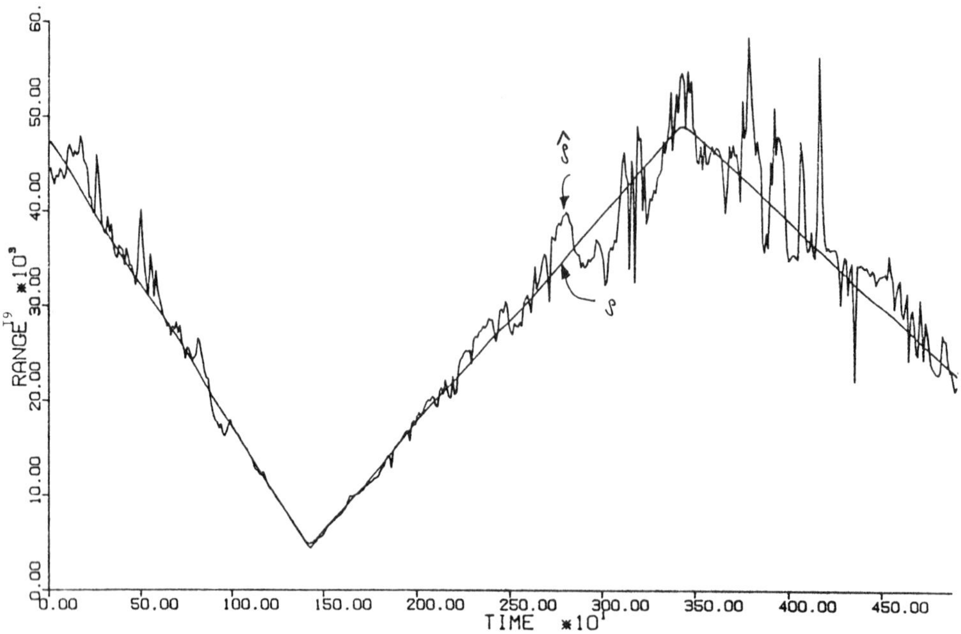

Figure (11). Target Tracking of Multiple Maneuvering Target Under Increased Measurement Error

In addition, to output smoothing, the weighting coefficients $w_i(k)$ $i = 1, 2, \ldots, 6$ were averaged as follows:

$$\tilde{w}_i(k) = \frac{1}{N} \sum_{j=0}^{N-1} w_i\,(k - j) \tag{14}$$

Here $\tilde{w}_i(k)$ represents the time averaged output of the moving window averager equation (14). Numerous simulations were performed as N was changed from 10 to 30. Some results follow.

Figures (12 and 13) illustrate the performance of the filter's range and velocity estimates and the adaptive nature of the weighting functions. The figures also show how the number of averaged data points $\tilde{w}_i(k)$ effects all of the above mentioned parameters. The scenario used to generate figures (12, 13) starts the target at a range of 50,000 ft. and a velocity of -30 ft/sec (-means toward observer) and changes the targets velocity to +30 ft/sec after 2,500 seconds.

VII. CONCLUSION

An adaptive state estimator has been developed and extensively tested to track a target making random large scale maneuvers in velocity. The target/observer scenario is constrained to the vertical plane in the ocean environment. This was intentionally done so as not to compete with well established bearing tracking programs.

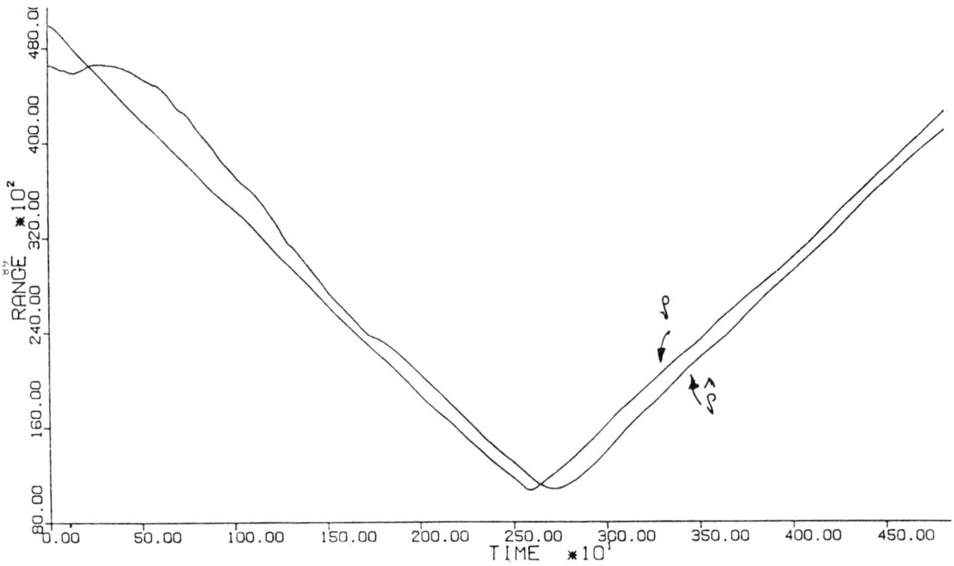

Figure (12). Estimate of Range for Approaching/Opening Target (20 points) Averaged

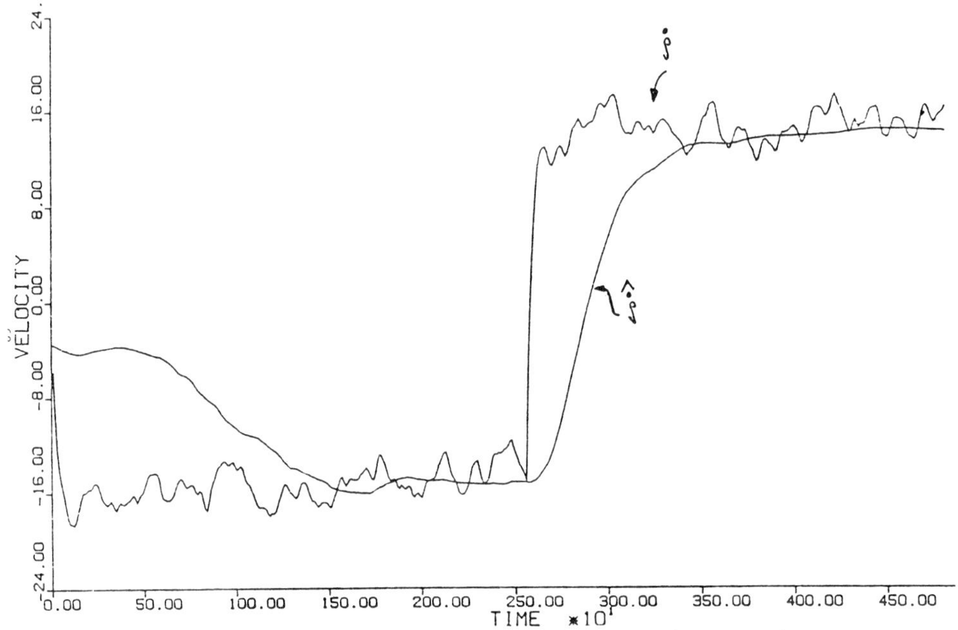

Figure (13). Estimate of Velocity for Maneuvering Target (20 points) Averaged

The adaptive estimator made use of a nonlinear prefilter to *uncouple* the state variables that model target motion in both depth and range. An additional benefit was the elimination of all extended Kalman filters in the tracking system. This results in a more *robust* tracker and significantly fewer computations. The cost of doing this, is that the linearized measurements contain nonstationary and non-Gaussian measurement errors.

System inputs to the tracking system consists of noisy time difference measurements of bottom/direct, and surface/direct multipath time delays. The adaptive tracker pre-filters the noisy multipath measurements in a nonlinear operation and then transmits the new linearized range measurements into its filtering channel. The range channel is fairly complex. Since the target is free to make major random velocity changes it required six Kalman filters and an adaptive weighting technique to span the expected range of all target velocities. Computationally this was quite easily done since each filter was only third order, and had the same Kalman gain and covariance matrix which only required one basic computation common to all six filters. The filters differed only in the deterministic input u_i $i = 1, 2, \ldots, 6$ built into each. As the target changed range in 5k increments new means and variances were programed into the filter bank thus compensating for the effect of nonstationary linearized measurement errors.

Overall tracking results seem quite good, especially in the low signal to noise ratio cases. An on-going effort is underway to study techniques of averaging data to Gaussianize the measurement errors.

BIBLIOGRAPHY

[1] N. H. Gholson and R. L. Moose, "Maneuvering target tracking using adaptive state estimation," IEEE Trans. Aerosp. Electron. Syst., May 1977.

[2] R. L. Moose, H. F. VanLandingham, D. H. McCabe, "Modeling and Estimation for Tracking Maneuvering Targets," IEEE Trans. on AES, vol. AES-15-No. 3., pp. 448-456, May 1979.

[3] A. H. Jazwinski, "Limited memory optimal filtering," IEEE Trans. Automat. Contr., vol. AC-13, Oct. 1968.

[4] J. S. Thorp, "Optimal tracking of maneuvering targets," IEEE Trans. Aerosp. Electron. Syst., vol. AES-9, July 1973.

[5] R. A. Singer, "Estimating optimal tracking filter performance for manned maneuvering targets," IEEE Trans. Aerosp. Electron. Syst., July 1970.

[6] R. L. Moose, "Adaptive estimator for passive range and depth determination of a maneuvering target (U)," U. S. Naval J. Underwater Acoustics, July 1973.

[7] R. A. Howard, "System analysis of semi-Markov processes," IEEE Trans. Mil Electron., vol. MIL-8, pp. 114-124, April 1964.

[8] D. H. McCabe and R. L. Moose, "Passive Source Tracking using Sonar Time Delay Data," IEEE Trans. on Acoustic Signal Processing, June 1981.

[9] J. C. Hassab, "Passive Tracking of a Moving Source by a Single Observer in Shallow Water," Journal of Sound and Vibration (1976) 44 (1).

Bearing Estimation using Random Arrays

Melvin J. Hinich
Virginia Tech

Abstract

A sonobuoy field is a random two dimensional array if the signals from the sensors are coherently processed. The sensor position coordinates must be estimated to a fraction of a wavelength in order to coherently process the signals, which is a separate processing task since the field has a changing random pattern due to drifting of the buoys. This paper reviews the statistical properties of the maximum likelihood bearing estimator computed from coherent summation of output signals from randomly deployed sensors. It is shown that the approximate mean-square bearing error for a small random array favorably compares with that for a square array of equally spaced sensors with about one-half the aperture. The random array is not subject to the aliasing (grating lobes) problem of an equally spaced array, and a great deal of array gain can be achieved from a large array of randomly placed sensors. Serious consideration should be given to tracking the locations of drifting sonobuoys so as to coherently process them as an array.

NATO ASI Series, Vol. F1
Issues in Acoustic Signal/Image Processing and Recognition
Edited by C. H. Chen

Bearing Estimation using Random Arrays

Melvin J. Hinich

Introduction

A sonobuoy field is a random two dimensional array if the signals from the sensors are coherently processed (Thorn, Booth, and Lockwood [1]). The sensor position coordinates must be estimated to a fraction of a wavelength in order to coherently process the signals, which is a separate processing task since the field has a changing random pattern due to drifting of the buoys. It is technically feasible to track the pattern, but the processing gain from such an effort must be perceived as sufficiently great to justify such an effort. This paper reviews the statistical properties of the maximum likelihood bearing estimator computed from coherent summation of output signals from randomly deployed sensors. Stochastic properties of the array response function are discussed by Thorn, et. al. and Steinberg [2].

The signal received at a sensor from a distant acoustic source in a horizontally homogeneous ocean is a plane wave plus ambient noise (Clay and Medwin, [3]). Using complex variable notation, a single frequency plane wave at point (x_1,x_2) and time t is

$$p(t,x) = A \exp[i2\pi f_o(t-(x_1 \cos\theta_o + x_2 \sin\theta_o)/c)]$$

where θ_o is the horizontal direction of propagation with respect to the x-axis, c is the phase velocity, f_o is the wave's frequency (the wavelength is $\lambda_o = c/f_o$) and A is the complex amplitude. This amplitude will be assumed to be a constant for the expositon of coherent processing, but actual signals will have some amplitude and phase modulation which reduces the signal coherence across the array. A reduction of coherence results in an increase in the mean square error of bearing computed from any of the coherent processing methods discussed next.

1. Coherent Processing

The most widely used coherent processing technique is delay-and-sum beamforming. Suppose that the signal received by the kth sensor in a horizontal planar array is $s(t,\underline{x}_k) = p(t,\underline{x}_k) + \varepsilon(t,\underline{x}_k)$ where $\varepsilon(t,\underline{x}_k)$ is a Gaussian noise field at time t and point (x_{1k},x_{2k}). The usual approach for beamforming is to compute the beam output signal $B(t,\theta_\ell) = \Sigma_{k=1}^{M} s(t+\tau_{k\ell})$ for a set of delays $\tau_{k\ell} = (x_{1k}\cos\theta_\ell + x_{2k}\sin\theta_\ell)/c$ for sensors $k=1,\ldots,M$ and beam (look) angles $\ell=1,\ldots,N$. The energy in beam θ_ℓ over a sampling window of length T is $|B(\theta_\ell)|^2$ where $B(\theta_\ell)$ is the filtered beam signal in a band about f_o of bandwidth 1/T. If the plane wave signal is broadband, then the filter passband is adjusted to the bandwidth of the signal.

Levin [4] shows that the maximum likelihood estimator of the propagation angle θ_o is the $\hat{\theta}_o$ in the interval $0 \leqslant \theta < 2\pi$ that maximizes $|B(\theta)|^2$. Define

$$\hat{\sigma}_1^2 = M^{-1} \sum_{k=1}^{M} (x_{1k} - \overline{x}_1)^2, \qquad \hat{\sigma}_2^2 = M^{-1} \sum_{k=1}^{M} (x_{2k} - \overline{x}_2)^2 \tag{1.1}$$

where $\overline{x}_1 = M^{-1}\Sigma_{k=1}^{M} x_{1k}$ and $\overline{x}_2 = M^{-1}\Sigma_{k=1}^{M} x_{2k}$. Set $\Sigma_{k=1}^{M}(x_{1k}-\overline{x})(x_{2k}-\overline{y}) = 0$

by a proper rotation of the coordinate system. Then if $\rho M\hat{\sigma}_1^2$ and $\rho M\hat{\sigma}_2^2$ are large, where ρ denotes the power signal-to-noise ratio in a narrow band about f_o, then the mean-square error of $\hat{\theta}_o$ is given by the approximation (Hinich [5])

$$E(\hat{\theta}_o - \theta_o)^2 \simeq (\lambda_o/2\pi)^2(2\rho M)^{-1}(\hat{\sigma}_1^{-2}\sin^2\theta_o + \hat{\sigma}_2^{-2}\cos^2\theta_o)\ \mathrm{rad}^2. \qquad (1.2)$$

This approximation also holds for a discrete set of beam angles and an estimator $\hat{\theta}_N$ that maximizes $|B(\theta_\ell)|^2$, provided that the error $|\theta_{\ell+1}-\theta_\ell|$ is small for all ℓ. In many applications, however, the bias due to the discreteness of the θ_ℓ is larger than the square-root of the right hand side of (1.2). It is important in these applications to interpolate between $\hat{\theta}_N$ and its adjacent angles.

A simple way to interpolate is to use the frequency-wavenumber equivalence with beamforming (Hinich [6]). Suppose that $s(t,\underline{x}_k)$ is passed through the same narrow band filter as was used to produce $B(\theta)$. For example, the fast Fourier transform (FFT) can be used to filter discretely sampled signals to yield the filtered output

$$s(\underline{x}_k) = N^{-1}\Sigma_{n=0}^{N-1} s(n\tau,\underline{x}_k)\exp(-i2\pi mn/N), \qquad (1.3)$$

where $T = N\tau$ is the sampling period, τ is the sampling interval, and m is the integer part of f_oT. Then

$$B(\theta_\ell) = \sum_{k=1}^{M} s(\underline{x}_k)\exp[i(\kappa_{1\ell}x_{1k} + \kappa_{2\ell}x_{2k})] \qquad (1.4)$$

where $\kappa_{1\ell} = (2\pi/\lambda_o)\cos\theta_\ell$ and $\kappa_{2\ell} = (2\pi/\lambda_o)\sin\theta_\ell$. The FFT can be used to compute (1.4) for as fine a grid as desired by imbedding $\{\underline{x}_k\}$ in a square lattice of equally spaced points, and adding zeros to expand the two dimensional array.

From now on it will be assumed that the bearing interpolation is sufficiently refined and ρM is sufficiently large so that expression (1.2) holds for $\hat{\theta}_o$. This expression will now be used to analyze the ex-ante root mean-square error of $\hat{\theta}_o$, rmse $\hat{\theta}_o$, when the sensor locations are randomly determined.

2. Gaussian Random Arrays

Suppose that the x_{1k} and x_{2k} are M independent realizations from two independent Gaussian random variables X_1 and X_2. With no loss of generality center the origin of the coordinate system so that $X_1 \sim N(0,\sigma_1^2)$ and $X_2 \sim N(0,\sigma_2^2)$, where σ_n^2 is the variance of X_n for n=1,2. To further simplify the analysis, let $\sigma_1 = \sigma_2 = \sigma$.

It then follows that $\sigma^{-2}\Sigma_{k=1}^{M}(x_{1k} - \bar{x}_1)^2$ and $\sigma^{-2}\Sigma_{k=1}^{M}(x_{2k} - \bar{x}_2)^2$ are independent chi-square χ^2_{M-1} variates with M-1 degrees of freedom. Denote $\beta_T = \Pr(\chi^2_{M-1} > T^2)$, e.g. $0.99 = \Pr(\chi^2_{15} > 5.2)$ using a standard chi-square distribution table. By independence, the probability that

$$\Sigma_{k=1}^{M}(x_{1k} - \bar{x}_1)^2 > (\sigma T)^2 \quad , \quad \Sigma_{k=1}^{M}(x_{2k} - \bar{x}_2)^2 > (\sigma T)^2$$

is β_T^2, and thus

$$\Pr\{M^{-1}(\hat{\sigma}_1^{-2} \sin^2\theta + \hat{\sigma}_2^{2} \cos^2\theta) < (\sigma T)^{-2}\} > \beta_T^2 \ . \tag{2.1}$$

Applying (2.1) to (1.2) and ignoring the approximation error,

$$\text{mse}\ \hat{\theta}_o < (\lambda_o/8\pi\rho^{1/2}\sigma T)^2\ \text{rad}^2 \tag{2.2}$$

with probability greater than β_T^2. For example if $\beta_T = 0.99$, $M = 16$, $\sigma = 6\lambda_o$, and $\rho = 1/10$ (-10 db), then $T^2 = 5.2$ and from (2.2), rmse $\hat{\theta}_o < 0.53°$ with probability greater than 0.98.

It is useful to compare this result with the mean-square error of $\hat{\theta}_o$ for a square lattice array whose $M=N^2$ sensors are at the points $\{md,nd:m,n=1,\ldots,N\}$. Setting $L = (N-1)d$ for the length of the square's sides,

$$\sum_{k=1}^{M} (x_{1k} - \bar{x}_1)^2 = \sum_{k=1}^{M} (x_{2k} - \bar{x}_2)^2 = ML^2/12. \tag{2.3}$$

Thus from (1.2) the mse $\hat{\theta}_o$ for the <u>random array is less than</u> that for the square lattice array (with probability greater than β_T^2) if

$$L < (12/M)^{1/2}T\sigma. \tag{2.4}$$

For example if $M=16$ and $\beta_T = 0.99$, then the bound is $L < 1.98\ \sigma$. If $M = 36$ and $\beta_T = 0.99$, $T^2 = 18.5$ and $L < 2.48\ \sigma$.

If the square is centered on the origin rather than at $((N+1)/d,(N+1)/d)$, the probability that a sensor is in the square is $[\Pr(|Y| < L/2\sigma)]^2$ where $Y \sim N(0,1)$. Thus if $L = 4\sigma$, this probability is 0.91. Since $(0.99)^2 = 0.98$, it follows from the examples above that the small Gaussian random array has approximately the same bearing accuracy as a square lattice array of about one-half the aperture.

This comparison with the lattice array has been made very conservative by the selection of $\beta_T = 0.99$. The rmse $\hat{\theta}_o$ for a small random array will often be as small as a lattice array of equal aperture.

3. Conclusion

It is shown that the approximate mean-square bearing error for a random array favorably compares with that for a square array of equally spaced sensors with about one-half the aperture. The random array is not subject to the aliasing (grating lobes) problem of an equally spaced array, and a great deal of array gain can be achieved from a large array of randomly placed sensors. Serious consideration should be given to tracking the locations of drifting sonobuoys so as to coherently process then as an array.

References

1. J. V. Thorn, N. Booth, and J. C. Lockwood, "Random and Partially Random Acoustic Arrays," *J. Acoust. Soc. Am.* 67, 1277-1285 (1980).

2. B. D. Steinberg, *Principles of Aperture and Array System Design* (Wiley, New York, 1976), Chap. 8.

3. C. S. Clay and H. Medwin, *Acoustical Oceanography* (Wiley, New York, 1977), Chap. 9.

4. M. J. Levin, "Least-Squares Array Processing for Signals of Unknown Form," *Radio Electron. Eng.* 29, 213-222 (1965).

5. M. J. Hinich, "Estimating Signal and Noise using a Random Array, *J. Acoust. Soc. Am.* 71, 97-99 (1982).

6. M. J. Hinich, "Frequency-Wavenumber Array Processing," *J. Acoust. Soc. Am.* 69, 732-737 (1981).

HOW VHF ACOUSTICS CAN HELP PROCESS UNDERWATER ACOUSTIC SIGNALS

by H. GAUTIER and P. TOURNOIS

THOMSON-CSF, DASM - BP 53 - 06801 CAGNES/MER Cedex

- France -

ABSTRACT

Surface Acoustic Wave (SAW) techniques allow to build analog Fourier transformers and convolvers able to process wideband (up to 100 to 500 MHz) limited duration (10-100 μs) signals. The processing speed is equivalent to 10^9 to 10^{10} floating point operations per second.

They exhibit 10-12 bit (60-80 dB) dynamic range and a floating point type accuracy of 6 to 8 bits. The major drawback is the lack of memory and this is why it is very often useful to place digital interfacing circuits around the SAW processor to be able to employ the digital memories.

Addition of these circuits limit the useful bandwidth to a few tens of MHz but adds a lot of flexibility : One can process non simultaneous signals and signals with time, bandwidth characteristics very different from those of the analog acoustic devices. Underwater acoustic signals are a good example.

Several applications belonging to the sonar field are presented. They require high computation speeds like those quoted earlier when real time operation and small volume and power are needed. Fast Fourier analysis can be used to perform beamforming. This requires fast multiplexing of the various input signals. Multi-channel spectrum analysis is another application where input memories perform time scale compression. In both instances, two-dimensional schemes exist and are discussed. Finally correlation techniques are briefly presented. They always require digital compatibility and offer interesting solutions to the problem of signature analysis for instance.

These are not new concepts but SAW devices and compatible digital circuits have only been recently available. Few implementations have been tried in the underwater acoustics field but it seems already clear that interest exists when information rate is large enough : a lower bound can be set around 200 kword/s to-day.

NATO ASI Series, Vol. F1
Issues in Acoustic Signal/Image Processing and Recognition
Edited by C. H. Chen

INTRODUCTION

Hardware implementations of signal processors may rely on various technologies of which the major ones are the digital, sampled analog (CCD), optical, microwave, and VHF acoustics. Under the last label are gathered the techniques which make use of propagation phenomena of high frequency surface acoustic waves on the surface of piezoelectric cristals. The abbreviation SAW is generally employed.

The natural field of application of these acoustic techniques is radar and telecommunication, but it can be widened to include sonar by making digitally compatible hybrid sub-assemblies.

To show this point, this paper reviews the relevant material concerning SAW devices and leading to the concept of hybrid modules and presents in a rather schematized way examplesof applications to underwater signal processing.

The format chosen is that of lecture notes and the reader interested by more formal presentations and a more extended list of references may refer to publications [1] and [2] .

1.- SAW TECHNOLOGY APPLIED TO SIGNAL PROCESSING - COMPONENTS AND ANALOG PROCESSORS

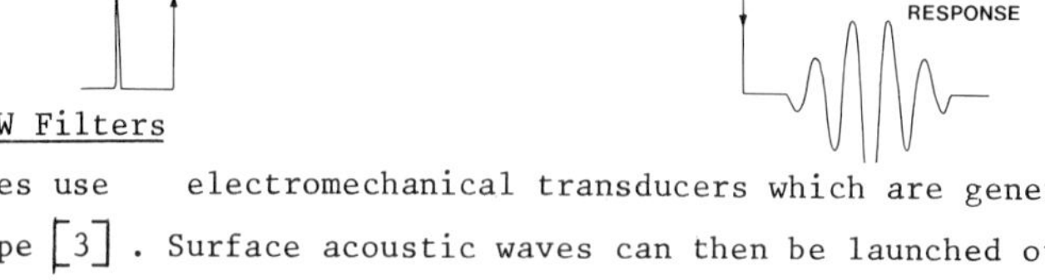

Figure 1. SAW Filters

All SAW devices use electromechanical transducers which are generally of the interdigital type [3] . Surface acoustic waves can then be launched or detected on piezoelectric substrates like Quartz or Lithium Niobate. Relatively slow propagation velocity (∿3500 m/s) allows time delay of tens of μs. Standard photolithographic process limits the line resolution to 0.6 μm corresponding to a maximum center frequency of 1,5 GHz. Higher frequencies are obtained by using faster waves or harmonic operation. Various transducer configurations exist ; varying the finger overlap as schematized in the figure allows to weight the impulse response and is the basis of the design of SAW filters.

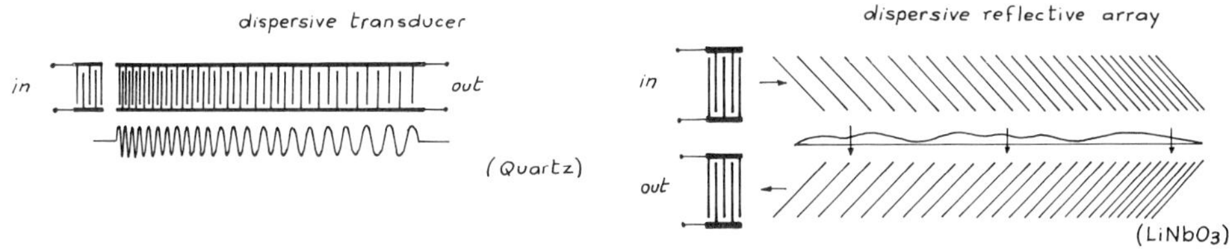

Figure 2. Dispersive delay lines

The first major family of SAW devices applied to signal processing is that of dispersive delay lines. These are filters with a linear FM chirp impulse response with duration T and bandwidth B. Two techniques exist : one employs dispersive transducers [3] generally laid down on quartz substrates and applies to modest T x B product devices ; the other employs dispersive reflective arrays [4] generally engraved on $LiNbO_3$ substrates, and applies to higher performance devices. A posteriori phase correction of the transfer function is possible by means of a correction film deposited between the two arrays.

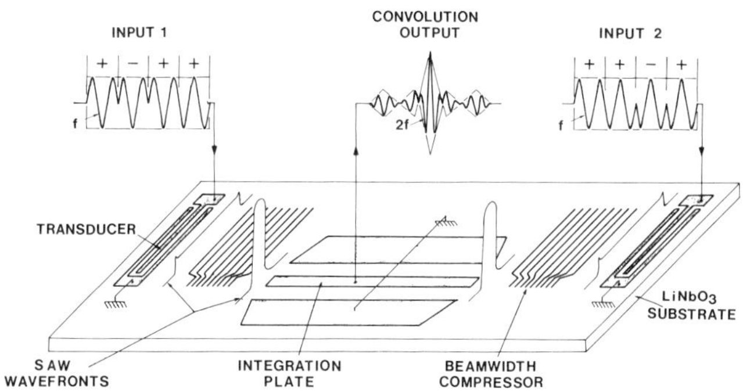

Figure 3. Piezoelectric Convolver

The other important device for signal processing application is capable of performing the convolution of any two signals with limited duration and bandwidth. The two signals are carried by two contra directive waves with identical frequencies f, and are multiplied together using the intrinsic material non linearity. Charges generated at frequency 2 f are spatially in phase and can be simply collected by a metallic plate. Input beamwidth compressing techniques [5] are required to enhance the power density and hence the efficiency.

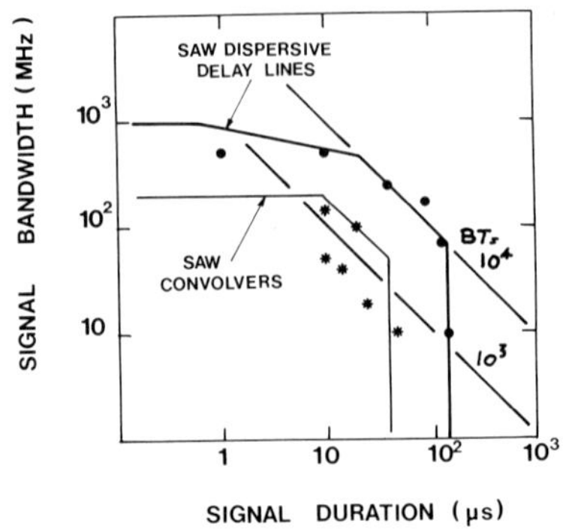

Figure 4. Time and bandwidth of SAW devices

Acoustic component are characterised by the duration and bandwidth of the allowed input signals. The linear devices (mainly dispersive delay lines) are limited to duration around 200 μs and bandwidth around 1 GHz ; the max. T x B product reaches 10^4. The performances of non linear components (convolvers) are more modest in a ratio of some 5 to 1.

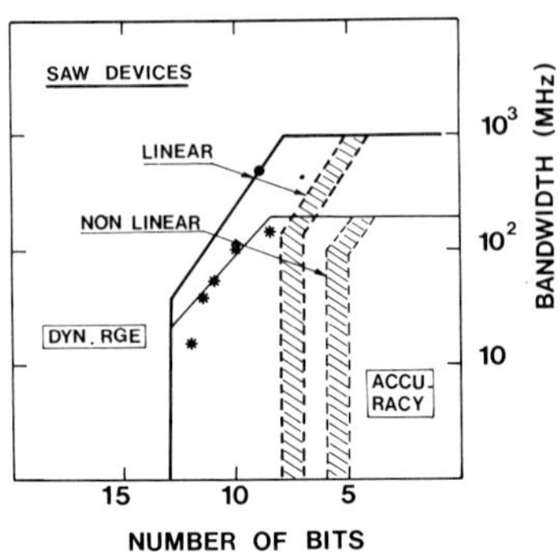

Figure 5. Dynamic range and accuracy

These two notions are independent in analog devices and one may associate two different equivalent number of bits for comparison with digital systems. Dynamic range varies from 60 to 80 dB corresponding to 10-13 bits. As for accuracy best devices show 0.1 dB amplitude errors and 1° r.m.s. phase errors allowing for 35-45 dB side-lobe suppression. One can estimate this to be equivalent to 6-8 bit floating point type accuracy. Convolvers are somewhat less performing.

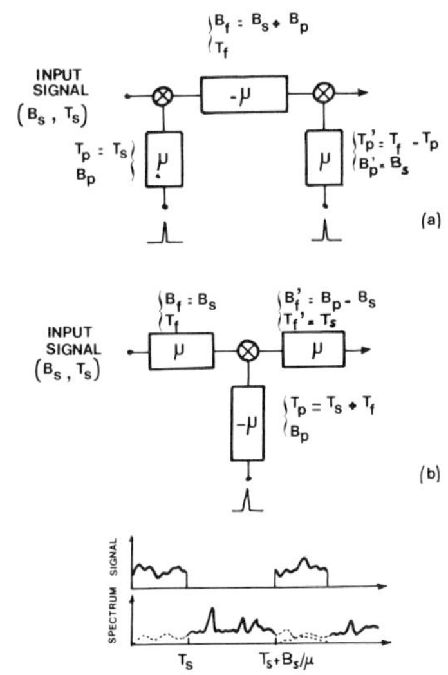

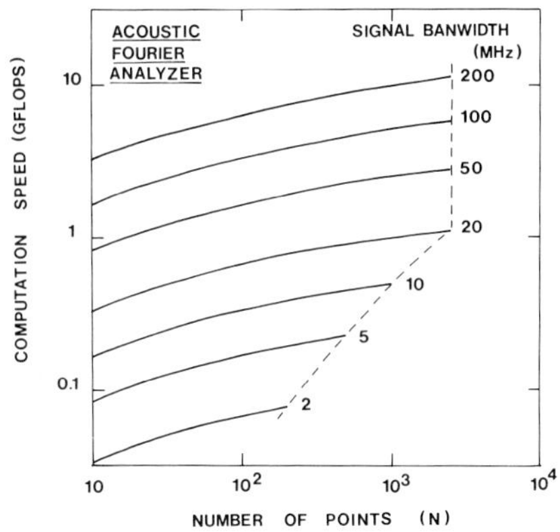

Figure 6. Acoustic Fourier Transform modules employs one of the two dual chirp transform schemes recalled above ; where each "μ" box represents a dispersive filter with specified T x B characteristics. The transform of a signal slot of duration T_s appears in the following time slot of duration B_s/μ ; one often chooses $\mu = B_s/T_s$ so that input and output rates are identical, and the circuits must be doubled to provide 100 % probability of intercept. To be used as a spectrum analyzer weighting in added to the premultiply filter (a-scheme) or preferably to the output filter (b-scheme).

Figure 7. Equivalent computation speed of an AFT

When $B_s/\mu = T_s$, $N = B_s T_s$ independent points (frequency lines) are computed in a time T_s. By comparison to the FFT algorithm, a complex"butterfly" circuit would have been operated $\frac{N}{2}\log_2 N$ times. As such a butterfly performs 10 elementary operations ; the equivalent number of floating point operations performed per second would have been $(5N\log_2 N)/T_s$. This results is plotted above. Several GFLOPS speeds are readily obtainable.

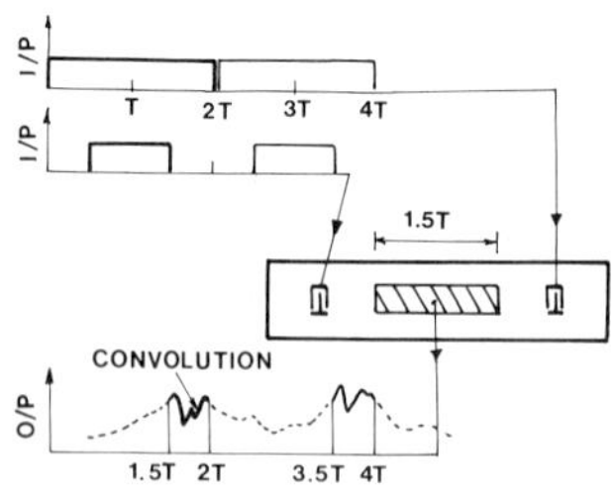

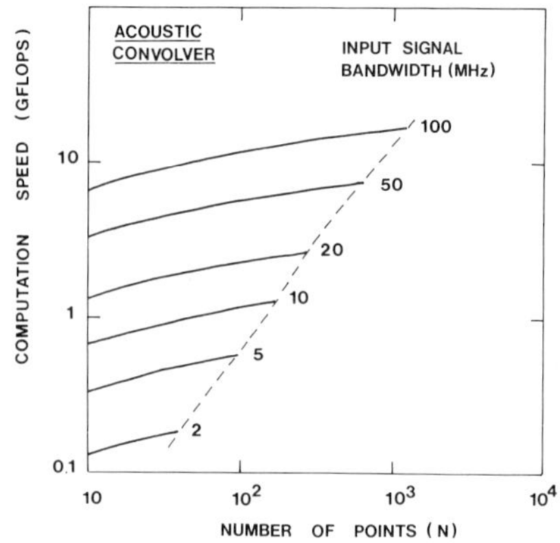

Figure 8. Convolution Modules are built around acoustic convolvers . In a typical operation signals of bandwidth B and duration T and 2T resp. are convolved. N = B x T true convolution points are computed in a time of the order of 2 T.

Figure 9. The equivalent speed of a convolver can again be expressed in FLOPS by estimating the number of operations an FFT implementation would require. Even for modest input signal bandwidths (10-20 MHz) the equivalent computation speed is 1 to 3 GFLOPS.

2.- DIGITAL PROCESSORS USING SAW DEVICES

The intrinsic characteristics and performances of the SAW devices presented above clearly show that these devices are well fitted for wideband short duration VHF-like signal processing. This is typically the case in radar and (some) telecommunication systems. It is therefore of no surprise to find most applications of the SAW technology in these fields where it brings small volume, speed and reliability (low power dissipation).

But SAW processors as such lack flexibility because no real store element exist. Memories are indeed most useful
(i) to process non simultaneous signals
(ii) to acquire a large number of signals and perform multiplexing
(iii) to be able to vary the time scale by changing the memory read-out speed.

Since the only available memories are digital, coupling the analog acoustic components to digital circuits must be performed and the question of compatibility of technology is raised. It concernes mainly the digital-to-analog interfacing circuits (A/D and D/A converters) and the memories. The question has been addressed in some detail elsewhere [2] . And the performances bounds that can be set to day are :
- Operating bandwidth below 30 to 50 MHz ; limited by the operation of the A/D converters.
- T x B product (independent points) below 1000 or 2000 ; limited by the acoustic devices.
- Depth of coding below 8-10 complex bits ; limited by the accuracy of the converters and the performances of the acoustic modules.

Hybrid SAW-digital modules have been developped and typical examples are described next.

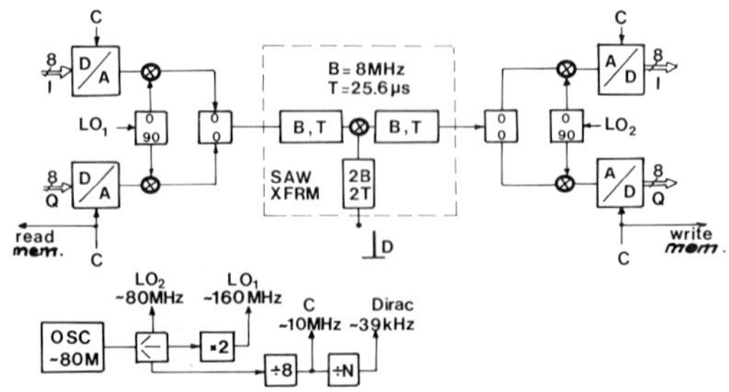

Figure 10. Digital I/O Fourier Transform Module

One of the modules actually built is schematized here. Standard quadrature modulators and demodulators as well as commercial D-to-A and A-to-D converters are employed to interface with the digital inputs. I and Q signals are coded over 8 bits and clock frequency is 10 MHz. The SAW Fourier Transformer can analyze 8 MHz bandwidth 25,6 μs duration signals. Although not represented in the figure, the transformer has a 100 % probability of intercept, and a logarithmic output amplifier is used to compress the output dynamic range by 24 dB and keep a full coding over 8 bits.

Nota : Existing standard technology would allow to multiply sampling rate and number of samples by 2 to 4.

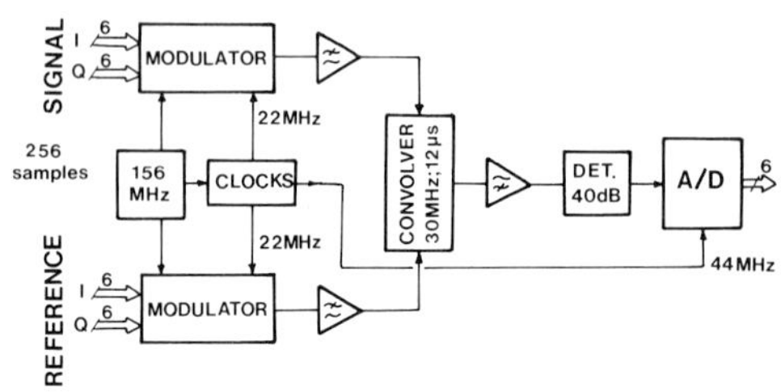

Figure 11. Digital I/O convolution/correlation module.

One such actual processor is according to the above schematic. It is compatible with an input strobe rate of 22 MHz and coding over 2 x(5 bits + sign).
The maximum signal length corresponds to 256 samples. As such it is able to perform convolution or correlation provided one input signal is fed time reversed.

Nota : The modules described in this page were built under DRET contract

		FOURIER TRANSFORMER	CONVOLVER
INPUT	# bits	8	6 → 8
	Coding	complex	complex
	Clock	10 MHz	20 MHz
	# samples	256	256
SAW	Bandwidth	8 MHz	30 MHz
	Time	256 μs	12 → 20 μs
OUTPUT	# bits	8	6
	Coding	complex	module
	Clock	10 MHz	40 → 60 MHz
TOTAL	Volume	2 ℓ	2 ℓ
	Power	48 W	40 W

Figure 12. Characteristics of the hybrid processors described in figures 10 and 11. The volume and power quoted are those of demonstration prototypes ; and their low values are to be noticed.

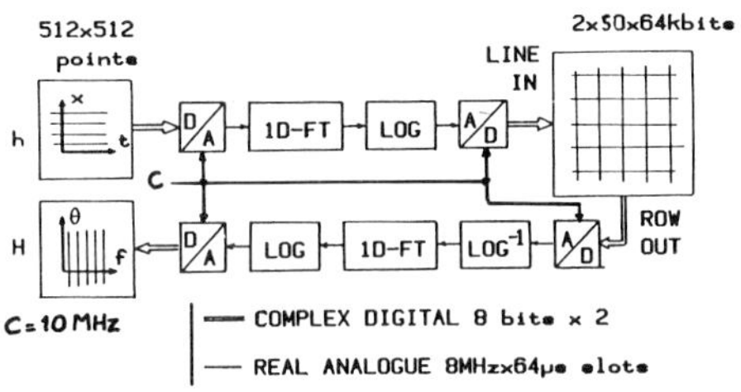

Figure 13. Two-dimensional Fourier Transform can be implemented by means of two one-dimensional Fourier transformers and an intermediary memory where the information is written-in along lines and read-out along rows. Using 10 MHz clocks, 4 MHz base band TV-like images can be processed in real time. The estimated volume for this processor is 4 liters and the power dissipation would be around 175 Watts of which 100 W are required by the memory. One major problem with such an implementation is the number of bits to carry along : indeed each 1-D Fourier transformer supplies a large processing gain which would require addition of 9 more bits for full coding. Starting with 8 one would end up with 26 ! This not being consistent with the dynamic range of the SAW processor, data reduction or compression must be performed along the chain ; and it will highly depend on the application.

III.- APPLICATIONS TO UNDERWATER SIGNAL PROCESSING

Underwater acoustic signals are typically low data rate signals (in the kHz or 10 kHz per channel) but there may exist a very large number of channels (100 to 1000 for example). The global data rate is there relatively high (1 to 10 MHz or more) and fast processors (around 100 MFLOPS) are required when real time operation is needed. Solutions based on SAW and digital technologies are applicable in different cases and they lead to small volume, limited power dissipation hardware. Examples are described next which make use of Fourier Transform or Spectrum Analysis operators.

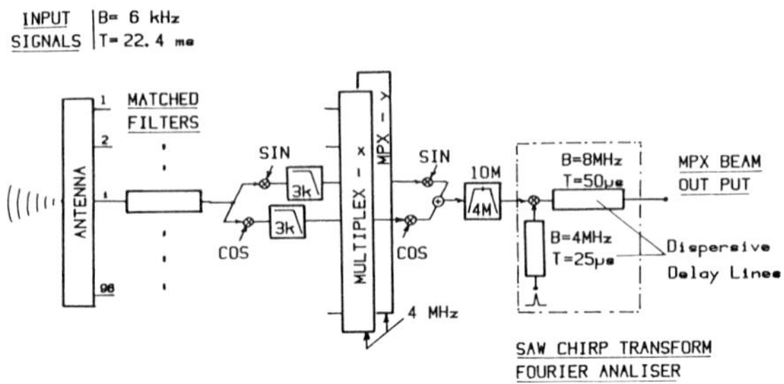

Figure 14. Application of Fourier Transform to Beamforming in a Sea bottom Imaging Sonar

One of the many applications of spectrum analysis is beamforming. In the case of sea-bottom imaging sonars, antennas are generally linear arrays of periodically spaced transducers. The relative bandwidth of these active sonars is rather narrow so that beamforming may be achieved by Fourier transformation of the signal made of the time multiplexed sequences of complex (amplitude and phase) samples of all transducer signals. Correct sampling of each transducer requires that the sampling frequency be much larger than the operation bandwidth.

Shown in the figure is the example of a sonar system with 96 receiving transducers and one illuminator. FM pulses with 22.4 ms duration and 6 kHz bandwidth are transmitted. After reception, matched filtering is performed by CCD means . Complex demodulation follows : the real and imaginary signals of each channel are next sampled and time multiplexed at a 4 MHz rate ; complex modulation of a carrier frequency generates 4 MHz bandwidth 25 μs duration signals which can easily be analyzed by a SAW module.

The output is the beamformed signal corresponding to a given range cell. The next sampling of the transducers 50 - 100 μs later yields the information concerning the next range cell. It is worth noting that SAW processing of sonar signals does not here require any time compression or storage. Interfacing is simply achieved through fast sampling of the many parallel inputs.

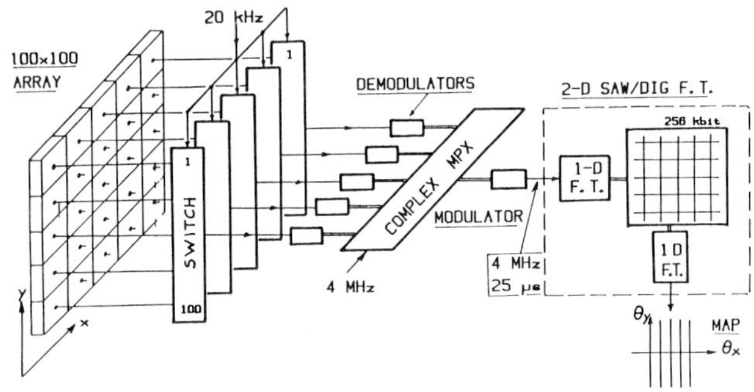

Figure 15. 2D Beamforming in imaging sonars

The antenna is composed of a square matrix of narrow-band transducers with one illuminator. Two-dimensional beamforming is to be performed in a time corresponding to the system depth of focus to limit the background reverberation noise. A high resolution sonar with a 100 x 100 element antenna might, for instance, require a depth of focus of 5 to 8 m and integration time should then be limited to 5 ms. The maximum scan rate would then be 200 images/s.

Beamforming may again be performed by Fourier analysis of the fast sampled inputs (see fig. 14). A set of 100 parallel low-speed multiplex switches connects each line of transducers to complex demodulators. Fast 4 MHz multiplexing and HF modulation results in 4 MHz bandwidth, 25 μs duration line signals which are inputs to a SAW-digital 2D-Fourier transformer. This transformer requires a 256 kbit memory size with a 200 ns access time and 5 MHz clock rate converters.

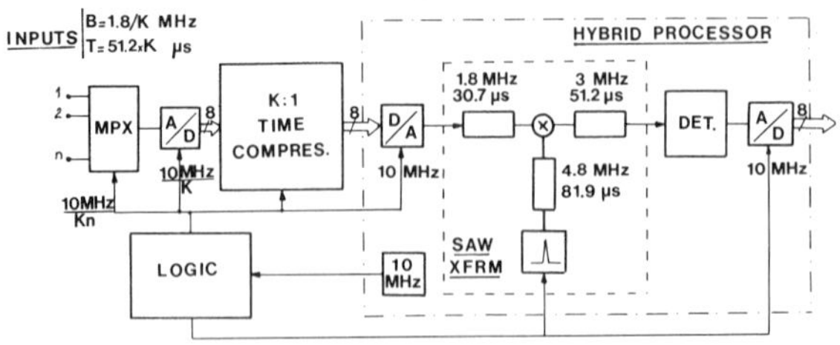

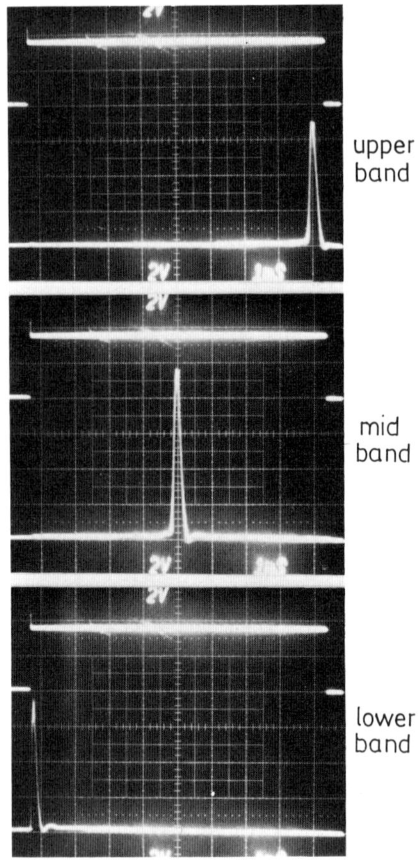

<u>Figure 16.</u> <u>Multichannel Spectrum Analyzer</u>

Real time spectrum analysis of several hundred sonar channels (or beams) can be performed using hybrid SAW processors. One system was developed to analyze preformed beam signals and resolve 75 frequency lines per channel.

The system schematized above shows that input beam signals are sampled, time multiplexed and stored to be then read-out K-times faster. This time compression is required to adapt the signal characteristics to those of the acoustic components and allow for a single processor to process a large number of beams. The acoustic analyzer employs the "T" scheme. The first filter also rejects spurious frequencies due to sampling and the last filter also includes weighting for sidelobe reduction. The digital I/O processor itself fills 1 liter and dissipates 32 Watts. Real operation demonstrated full 8 bit compatibility w.r.t dynamic range, accuracy, distorsion. Typical output signals are shown to the right.

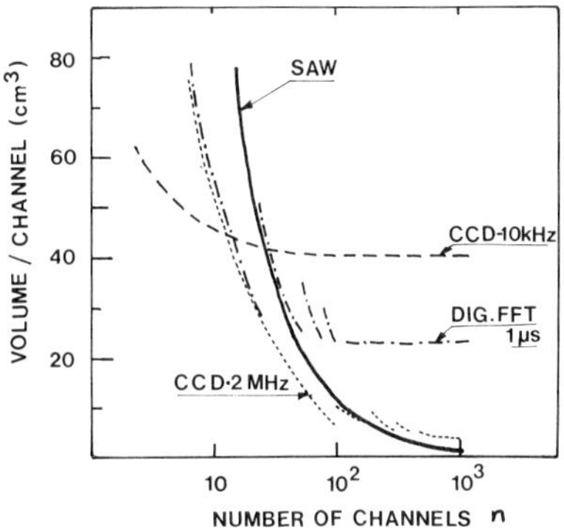

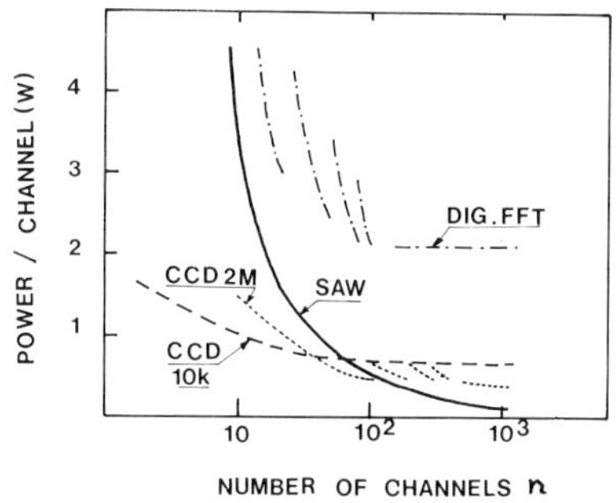

Figure 17. Multichannel spectrum analyzer - Discussion

The maximum number of channels the processor of fig. 16 can process in real time is equal to K. In addition the input compression memory size must equal 2nBT bytes when input signals have a bandwidth B, duration T and are sampled at Nyquist rate. If we let BT $\approx$ 100 and B $\approx$ 2 kHz, K must equal 10^3 and the processor can analyze up to 1000 channels in real time.

With these assumptions volume and power dissipation per channel were estimated and plotted as function of number of channels n. Similar plots are drawn for competitive technologies namely CCD filter with 10 kHz and 1 MHz clocks, and digitals processors built around 1 µs cycle time butterfly circuits.

Whereas the all digital solution should be discarded, the low frequency non multiplexed CCD circuits are best for $n < 10$; multiplexing higher clock rate CCDs are preferable when $10 < n < 100$ and the hybrid SAW processor becomes interesting when $n > 100$

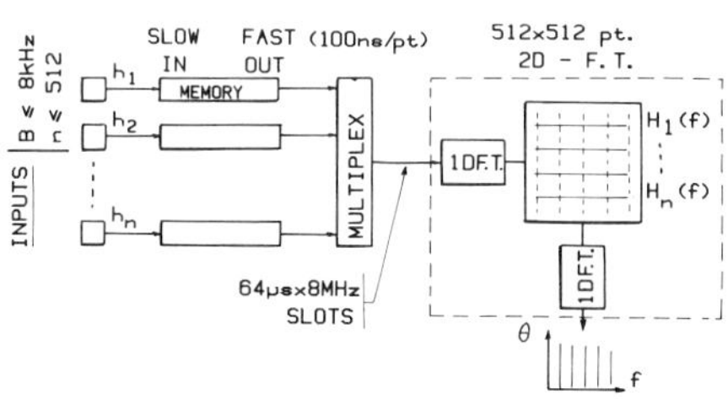

Figure 18. Frequency against bearing map in passive listening sonar

In this case hydrophones h_i, i = 1 to n, are periodically spaced along an axis $\vec{x}$. The frequency f against bearing θ map, $H(\theta,f)$ can be computed by Fourier transformation of the signal $h(x,t)$. The input signal bandwidth being wide, frequency analysis is performed first. Beams are next formed in each frequency cell. Signals are time compressed and serial multiplexed before being fed into a 2D acoustic Fourier Transformer. In this way up to 500 channels can be analyzed and 500 frequency lines resolved. Owing to the high speed of the analyzer, the maximum input bandwidth is then 8 MHz.

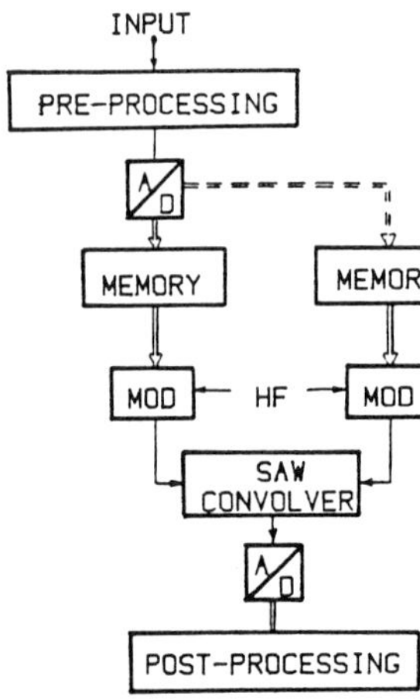

Figure 19. General One-dimensional Correlation Receiver

Incoming signals are stored in digital memories and successive signal slots are later compared to previously stored or library signals. For that purpuse, the memory contents modulate a carrier and the resulting signals are combined in a convolver. The output is stored and/or further processed.

Most application to-day belong to the radar and telecommunication fields ; the convolver is used either as a calculator to compute a convolution function, or as a programmable filter, when one input is considered as a reference which can be changed or optimized at will. This can be the basis for adaptive filtering.

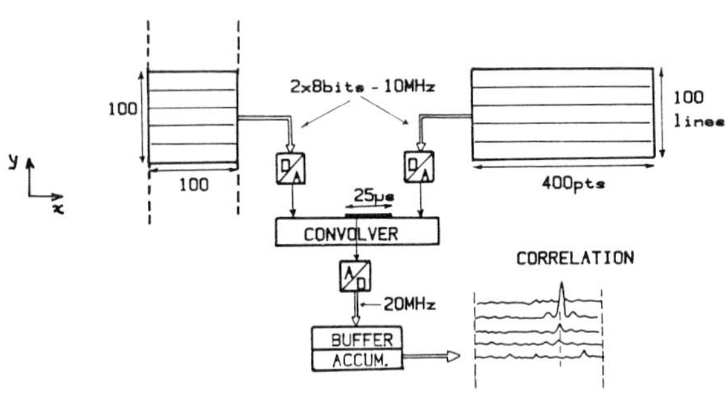

Figure 20. Principle of a 2-D map Correlation System

Since the 2-D correlation is not a separable function, it is not possible to perform it by means of two cascaded 1-D processors. Solutions exist like that depicted here which requires one convolver and one digital accumulator. The convolver performs the integration over the x-axis and integration along the y-axis is obtained by accumulating the adequate number (100 here) of line correlations.

CONCLUSIONS

SAW technology can be used to build ultrafast processors ; and basic possible operations are Fourier Transform and Convolution. In addition it is obvious that to be easy to apply in system, digital compatibility is necessary : this leads to hybrid structures.

Although limited by both technologies, the performances are already quite attractive : bandwidth around 20 MHz, T x B product around 1000, equivalent speed of 1 GFLOPS together with volume of a few liters and power dissipation in tens of watts.

The price one has to pay for such high Speed/(Power x Volume) ratios is a limited equivalent number of bits : around 8 complex bits today. This is enough in a number of cases like those described here. Other applications relate to radar, telecommunication and image processing.

The most useful operation appears to be Fourier Transform for beamforming and spectrum analysis in one or two dimensions. Also mentionned was convolution which may lead to programmable (and adaptive) filtering.

Improvements in technology act on two grounds ; (i) speed of the interfacing circuits, hence bandwidth ; this will allow to make fuller use of the SAW device capabilities ; (ii) accuracy of the analog processors.

These relatively new concepts are experimented slowly but important stepping stones are being laid down. Actual modules exist ; they are evaluated by system engineers and they are compared to other technologies. Ranges of applicability begin to appear like a minimum computation output rate of some 2×10^5 independent points per second.

REFERENCES

[1] H. GAUTIER and P. TOURNOIS - "Very Fast Signal Processors as a Result of the Coupling of SAW and Digital Technologies" - IEEE Trans. Micr. Th. and Tech., MTT - 29, 404-409 (1981)

[2] H. GAUTIER - "Digital Processors using SAW Devices" - Proc. 1981 IEEE Ultrasonics Symp. 206-219 (1981)

[3] M.B.N. BUTLER - "SAW Devices for Signal Processing" - IEE Conf. Pub. n° 180 21-33 (1979)

[4] R.C. WILLIAMSON - "Properties and Applicationsof Reflective Array Devices" - Proc. IEEE, 64, 702-710 (1976)

[5] Ph. DEFRANOULD and C. MAERFELD - "A SAW Planar Piezoelectric Convolver" - Proc. IEEE 64, 748-751 (1976)

[6] T.W. BRISTOL - "Review of Spectrum Analysis with SAW chirp Transforms and Filter banks" - IEE Conf. Pub. n° 180 - 226-231 (1979)

CHARACTERIZATION OF THE UNDERWATER CHANNEL
APPLICATION TO COMMUNICATION

G. JOURDAIN
Professeur à l'INPG
CEPHAG - BP46 - 38402 Saint-Martin-d'Hères (France)
Laboratoire Associé au CNRS (LA 346)

ABSTRACT

This paper is a synthesis of the CEPHAG's last years studies on the submarine acoustic channel characterization and its application to optimal communication. The methodology used for the characterization is recalled and some experimental results are given. More recently we have used some of these models to construct and value an optimal binary communication scheme. The more advanced results concern the fading multipath case and the frequency dispersive channel. Other results, concerning the estimation of data transmitted in a multipath channel, are also mentioned.

1 - INTRODUCTION

This paper synthetizes the last years CEPHAG's studies on the submarine acoustic channel characterization and its application to optimal communications. Our approach is a systemic one : the channel is characterized by its action on the transmitted signals. We have developed a whole methodology of active characterization and several test signals in order to learn the channel action.

We limit this paper to the bipunctual case : the signal is transmitted from a point A to a point B. This elementary situation appears in all cases (detection, source imaging, communication,...). The most adapted model to take many situations into account is a linear filter : in this filter's parameters, some physical knowledge of the channel can be used, it is possible to distinguish the well-known part of the channel from the other fluctuating part, and it is especially the most convenient model to imagine some optimal communication system over the so-modelled channel.

This paper has two parts : in the first one, we study the active characterization of the channel. We show both theoretical and experimental results. We have applied these methods in many different physical situations : vertical or horizontal propagation, long or short range, mobil or fix "points" AB, and so on... In the second part we use the knowledge of channel characterization to construct an optimal communication system and calculate its performance. This second part has been especially developed for two years and we show some recent results, expecially theoretical ones, up to now. More details about all the results can be found in the references given at the end of the paper.

NATO ASI Series, Vol. F1
Issues in Acoustic Signal/Image Processing and Recognition
Edited by C.H. Chen

2 - CHARACTERIZATION OF THE SUBMARINE CHANNEL

2.1 - Introduction

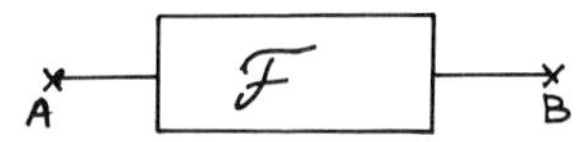

The most general model to represent the channel AB is a randomly time varying filter $\mathcal{F}$. The hypothesis of linearity is often true experimentally -as long as the emitted signals are not too strong-. But the "fluctuating" aspect of the transmission AB can be represented either by randomness, or by time variation, or by both. This fluctuating aspect appears as soon as the transmission AB distorts in any way the emitted signal and this distorsion cannot be represented only by additive noise.

The most important point in this active characterization is to know the time scale of interest. For example, during one communication digit, the channel will be possibly constant, whereas it is time varying all along a communication period.

The important fluctuation phenomena are essentially in a.s.m. (acoustic submarine propagation) : fading, multipath and reverberation. The following experimental result (Fig. 1) shows an example of fading transmission. It gives the variation of the instantaneous power $A^2(t)$ versus t, received at B, when emitting in A different simultaneous monochromatic waves (ν_j given in the schema) for horizontal and vertical transmission.

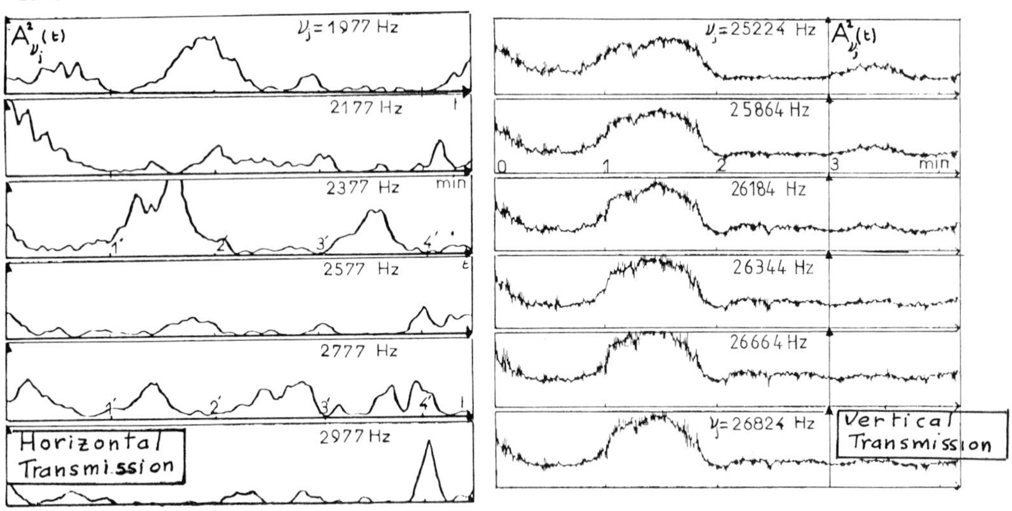

Fig. 1 . Instantaneous power versus t (from (4))

We see that the two cases are very different :

i) The fading is much slower in the vertical case.

ii) The variations are decorrelated from one frequency to another in the horizontal case, whereas they are very correlated in the vertical case.

The figure 2 gives an example of multipath transmission : it is obtained when emitting large WT signals and estimating the square impulse response of the channel (§2.3.2).

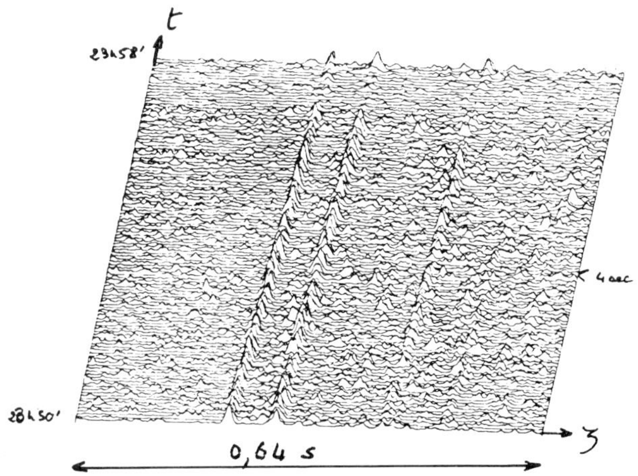

Fig.2 . Time evolution of the square of the impulse response of the channel (from [3])

These two examples exhibit the important parameters of the a.s.m. transmission. The time evolution of the fading or of the multipath will lead to the choice of one model, or another. For example fast fading will be represented by a time varying $\mathcal{F}$. Multipath can be taken into account either by a time dispersive $\mathcal{F}$, or a random $\mathcal{F}$, or both.

2.2 - The theoretical models for a randomly time varying filter

For the most general case, the formulas are given below :

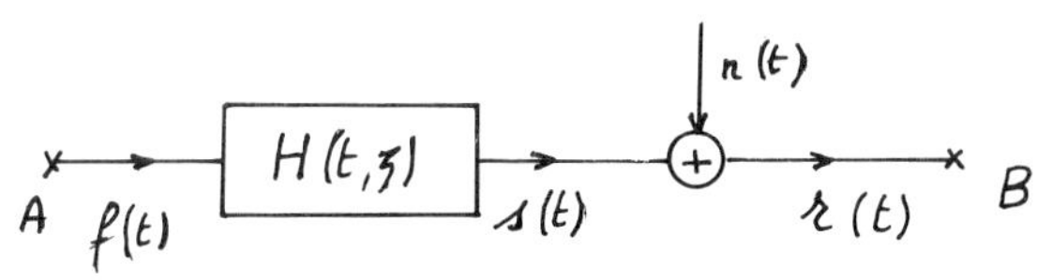

Fig. 3

The deterministic signal f(t) is transmitted through the randomly time varying filter $\mathcal{F}$ characterized by its random bitemporal impulse response $H(t,\zeta)$. The received signal is a random one s(t) :

$$s(t) = \int_R H(t,\zeta)\, f(t-\zeta)\, d\zeta \qquad (1)$$

n(t) is the additive noise. As the filter is often assumed gaussian, its statistical properties are only :

i) The first moment $E\{H(t,\zeta)\}$ which corresponds to the determinitic part of the channel (specular component).

ii) The second moment of the centered pure random part $H_r = H - E\{H\}$, which is the covariance of H_r. It is often assumed (see [3]) that the channel is WSSUS (Wide Sense Stationary Uncorrelated Scatterers). So the covariance of H_r is :

$$E\{H_r(t,\tau)\, H_r(t',\tau')\} = \Gamma_{H_r}(t-t',\tau)\ \delta(\tau-\tau') \tag{2}$$

and the channel is characterized by its scattering function $S(\nu,\tau)$ which is the Fourier Transform versus $t-t'$, of $\Gamma_{H_r}(t-t',\tau)$; $S(\nu,\tau)$ expresses the signal energy dispersion over the time-frequency plane.

This definition of scattering function is often related to small bandwidth signals. In this case the model is generally related to the complex amplitudes of the emitted and received signals, relative to the center frequency ν_0 of the signals. This may be the case in many situations. But it is not always the best model. So we had proposed another random time varying model, the reverberating model [1] defined by the following input-output equation :

$$s(t) = \iint_{R^2} G(\lambda,\tau)\ f(\lambda t-\tau)\ d\lambda\ d\tau \tag{3}$$

in which $G(\lambda,\tau)$ characterizes the reverberating filter. A new scattering function $D(\lambda,\tau)$ is also defined by means of the covariance of G_r :

$$E\{G_r(\lambda,\tau)\ G_r(\lambda',\tau')\} = D(\lambda,\tau)\ \delta(\lambda-\lambda')\ \delta(\tau-\tau') \tag{4}$$

Essentially the difference between these two kinds of models are that in the first case the frequential spreading of the channel is independent of the center frequency. In the second model it is proportional to the center frequency (it corresponds to a true Doppler effect).

2.3 - General Processing for choosing a model

We have developped [1], [3], a methodology in two steps to estimate the random channel H_r (after estimating the mean value).

2.3.1. - Test with polychromatic waves. This enables :

i) to see the importance of the frequential spreading (i.e. of the time modulation phenomenon) and its time scale,

ii) to study the comparative amplitude and phasis evolution [5],[7],[2],

iii) to test the WSSUS hypothesis (by studying the correlation between the received signals : see fig. 1). This can be used for testing the WSSUS hypothesis for the classical time varying filter, and also for the reverberating model, and we can choose the model. We have given different examples in [2],[4].

2.3.2 - Processing large product WT signals [3],[6].

By emitting large product WT signals and by calculating the interambiguity between the received signal and a copy of the emitted signal, it is possible to obtain an estimation of the scattering function. According the first or the second chosen model, we process translation or compression interambiguity, and we estimate $S(\nu,\tau)$ or $D(\lambda,\tau)$.

Two examples of estimated scattering functions : $S(\nu,\tau)$ in different experimental conditions are given below :

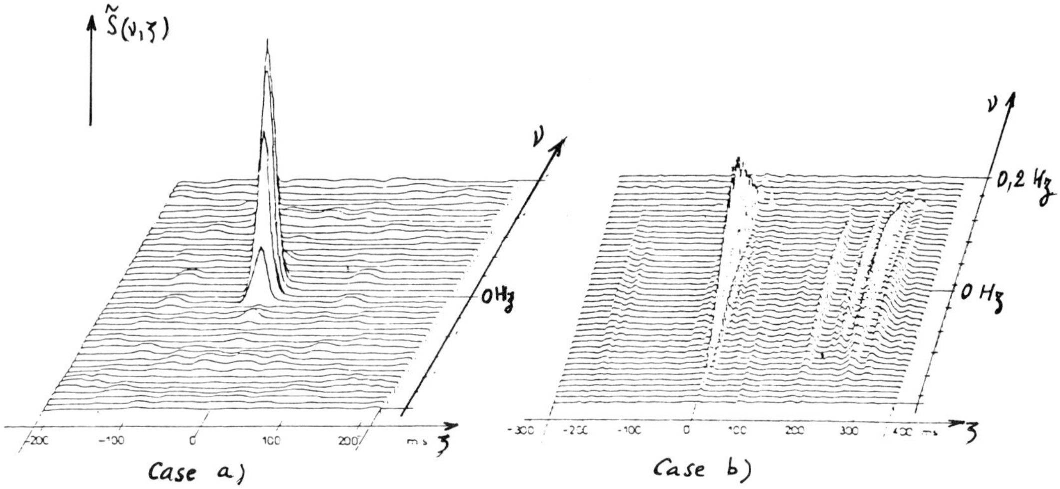

Fig. 4 . Examples of estimated scattering functions

The first one (case a) is a one path transmission. The scattering function exists only at the origin of the time-frequency plane. The temporal fading of the mean transmitted power is measured by the amplitude of the peak. This corresponds to the theoretical formula :

$$S(\nu,\tau) = M\,\delta(\nu)\,\delta(\tau) \qquad (5)$$

This case will be called the "one path Rayleigh model" (as the received signal is assumed to be gaussian).

The second example (case b) corresponds to a true time and frequency random dispersive transmission. The time dispersion (versus τ) exhibits N multiple paths (5 or 6 important paths). The mean power transmitted over each path is not the same. The frequential dispersion (versus ν) is not very important. It has often this monotonic evolution and we will model it with a Lorentzian form.

So the theoretical formula corresponding to the case b is now :

$$S(\nu,\zeta) = \sum_{i=1}^{N} q_i\, S(\nu)\, \delta(\zeta-\zeta_i), \qquad S(\nu) = \frac{2k}{k^2 + (2\pi\nu)^2} \tag{6}$$

k measures the frequential spreading of the channel, and ζ_i are the transmission paths.

We will now apply this characterisation of a.s.m. channel to construct an optimal system and particularly we will use the models (5) and (6).

3 - OPTIMAL COMMUNICATION

3.1 - Introduction

The channel is now assumed modelled, for example it has been modelled as a WSSUS channel, with a given (estimated) scattering function. The aim is to construct the optimal receiver using the channel knowledge. Naturally many communications schemes can exist. Two cases are studied here corresponding to two different kinds of problems.

The first case is to study the optimal receiver corresponding to one elementary digit transmission. The channel model is assumed complex enough (WSSUS with different possible scattering functions). This case corresponds to a situation where much security is necessary and where few information is transmitted : monoship sonar, remote control...

The second case is on the contrary, the case of a whole data transmission, and here, the important point is the data flow. In this case the channel will be considered as a deterministic one during each bit duration, but with multipath. This case corresponds for example to remote measurement, image transmission, ... , where the data rate is more important than the error probability.

3.2 - First case : communication through a randomly dispersive channel

We study here the binary symetric case. The deterministic orthogonal signals $f_i(t)$, i = 0 or 1, are transmitted through the random dispersive channel characterized by its scattering function $S(\nu,\zeta)$.

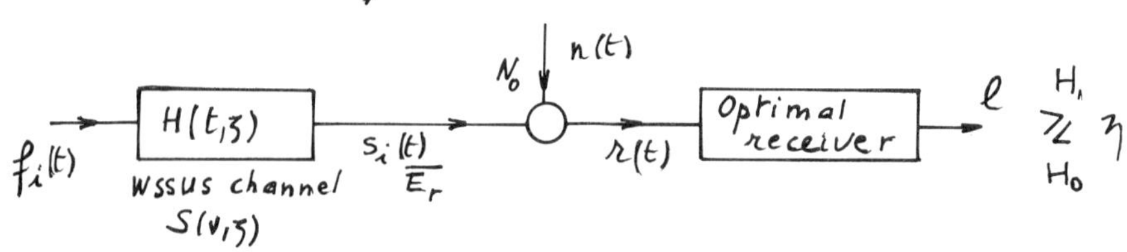

Fig. 5 . The communication model

Some noise n(t) is added to the received signal $s_i(t)$, (i = 0 or 1). N_0 is the power spectral density of n(t). The mean received signal power is $\overline{E_r}$ and the received SNR is $\overline{E_r}/N_0$. $s_i(t)$ and n(t) are assumed gaussian.

All the quantities correspond to the signal complex amplitudes (bandpass case). The problem is to construct the optimal receiver which calculates the sufficient statistic ℓ and decides in the best way between the 2 hypothesis. The criterion is the error probability P(e) and we want to calculate this error.

This problem can be simple or not, depending on the scattering function of the channel.

3.2.1 - The Rayleigh case

The simplest model for $S(\nu,\zeta)$ is the one path Rayleigh model, given by (5) and it leads to the simplest solution which is well known. The optimal receiver is a matched filter followed by an envelope detector and the performance is given by :

$$P(e) = \frac{1}{2 + \overline{E_r}/N_0} \tag{7}$$

It depends only on the SNR because the channel is not truly "dispersive".

It is well known that to decrease this error probability, explicit diversity techniques can be used. We will speak of "K diversity transmission paths".

An important known result is : given a transmitting energy per binary digit, and assuming the same SNR on each diversity path, there exists an optimal K_0 such that :

$$K_0 \sim \frac{1}{3} \frac{\overline{E_r}}{N_0}$$

We have applied this diversity technique in ASM long range propagation [8],[10]. Frequential diversity was used. The results were very satisfactory and in good agreement with the theoretical diversity formulas. So, on one hand, the Rayleigh model was well adapted to this transmission ; and on the other hand, the choice of the "diversity paths" was good.

3.2.2 - The general WSSUS case

When the scattering function is not so simple, both the receiver structure and the calculation of the error probability become more complex, and P(e) depends now on the form of f(t). We give some new recent results with essentially two kinds of scattering functions.

a) With a fading multipath channel

Let us choose a special form of (6) :

$$S(\nu,\tau) = \sum_i q_i \; \delta(\tau - L_i)\,\delta(\nu) \qquad (8)$$

The channel is a pure multipath dispersive channel. In this case we have succeeded to obtain the structure of the optimal receiver and also to calculate exactly the performance. The optimal receiver [9],[13] has been shown to possess as many adapted filters as channel paths (the Fig. 6 shows the case of 5 paths) and the optimal receiver is a quadratic combination Q of the outputs of these matched filters.

Particularly, we have shown that this quadratic form is greatly simplified if the emitted signals are well chosen : i.e. if their correlation function becomes zero for each delay path.

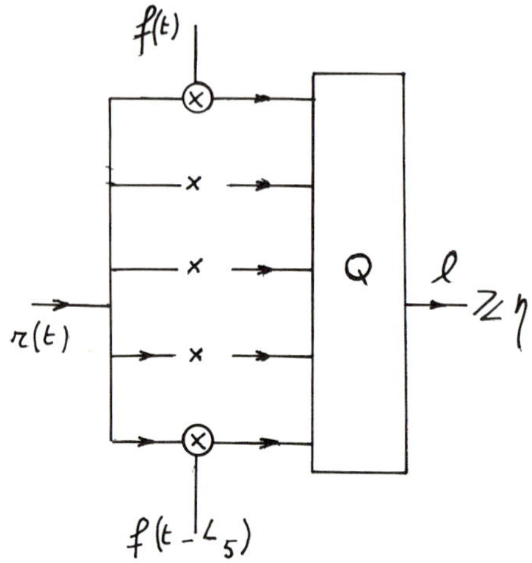

Fig. 6
The o.r. with a 5-path channel

The exact calculation of P(e) has been exhibited [9]. The important results are :

i) The best performance is obtained by distributing the transmitted mean power over the channel paths in order that the SNR on each path is about equal to 3 : it is an implicit diversity system (which can be related to the explicit diversity techniques) [15].

ii) For the signal design problem, the important parameter is the bandwidth of $f_i(t)$ and the best choice is recalled just above concerning the signal correlation function.

b) With a frequency dispersive channel

The scattering function is given by (6) in which we have both multipath and frequency dispersion measured by k.

This case is the more complex one because :

i) To construct the optimal receiver, an "internal state model" is needed for the channel, which was not useful up to now. In the case of a Lorentzian frequency dispersion, as above, a first order state model is sufficient. The optimal receiver is also obtained by means of an internal state model of the same oder as the channel one. It is now a variable gain Kalman filter.

Fig. 7 below gives the structure of the optimal receiver for 5 Lorentzian frequency dispersive paths. [9]

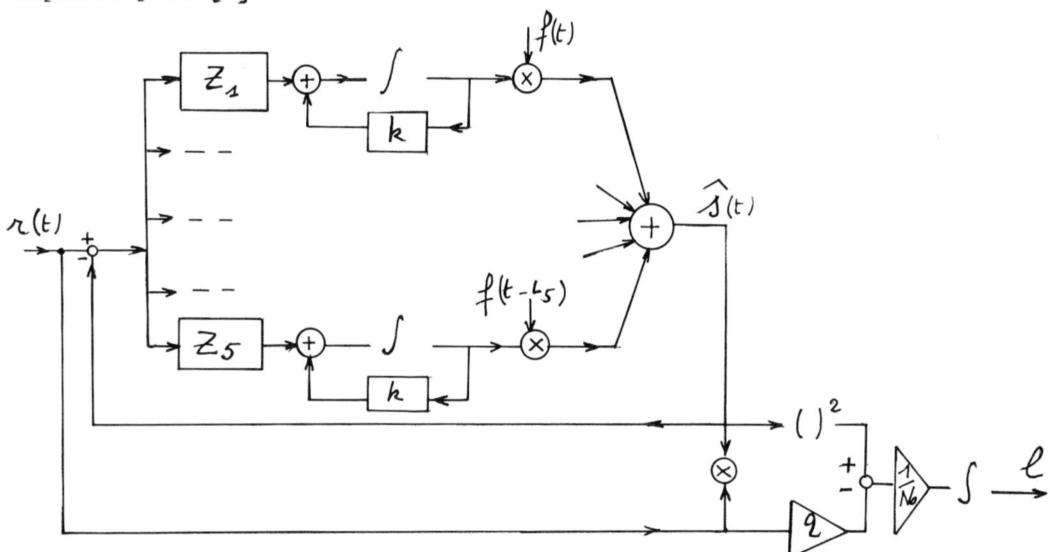

Fig. 7

The o.r. with a 5-path channel and frequency dispersive channel

ii) It is no longer possible to calculate exactly the error probability. The authors propose either approximate formulas for P(e), or bounds. Recently we have exhibited a new approximate formula [12] which is more adapted to this dispersive transmission case. This new formula exhibits well the diversity concept, and enables to look at the signal design problem in a more precise way. The important results are that in the frequency dispersive case :

- The important parameter is the signal duration, or rather an effective signal duration :

$$T_u = \left[\int | f(t) |^4 dt \right]^{-1} \tag{9}$$

We have tested different signal envelopes f(t) (see [11]) with the same T_u and the results for P(e) were nearly the same.

- The best performance is obtained when managing T_u in order that :

$$\frac{\overline{E_r}}{N_0} \cdot \frac{1}{N} \sim 3.4 \, , \qquad N = 1 + k \, T_u \tag{10}$$

where 1/k is the "channel correlation time" (inverse of the frequency dispersion). It is still an implicit diversity system where N plays as the diversity number of the multipath case.

3.3 - Second case : optimal sequential transmission with a multipath channel

Now the problem is to estimate data transmitted through a multipath ASM channel. The channel model is a deterministic one : so this means that there is a learning period in order to estimate the multipath structure, then a long communication period during which the channel is assumed to be nearly constant ; after that there may be another learning period and another communication period.

The problem is stated in Fig. 8 below :

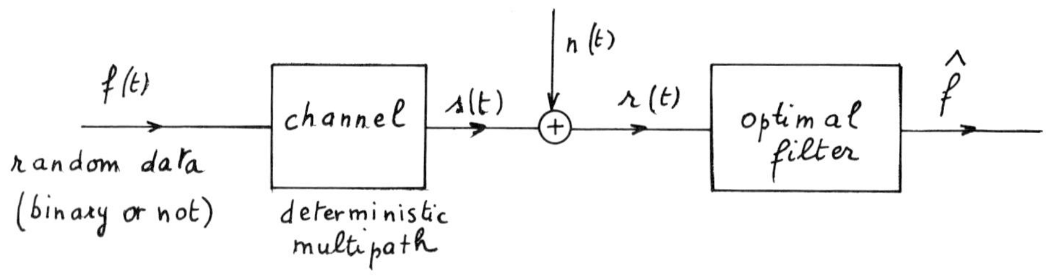

Fig. 8 . Model for the sequential transmission

The data f(t) are assumed to be random, either binary or not. The channel is a known multipath filter. The transmitted signal s(t) is added to gaussian noise n(t) and the aim is to use the knowledge of the channel to construct the optimal filter which estimates the transmitted data with a minimum quadratic error ε^2 :

$$\varepsilon^2 = E\left\{|f - \hat{f}|^2\right\} \tag{11}$$

So the problem is now an optimal estimation problem. It is a Wiener problem. It is well known that, with a strong SNR, the optimal filter becomes an "inverse" filter. So this is also naturally related to deconvolution problems and evidently, to equalization system in telecommunications. Different currently works about this problem are developed in our laboratory and will be soon published. We compare particularly transversal and recursive structures of the optimal filter for a given multipath channel.

Fig. 9 shows an example of experimental inverse filtering. The column a shows the channel response $H^2(\xi)$ with an evident multipath structure. The column b shows the action of the optimal inverse filter which "gathers" the paths.

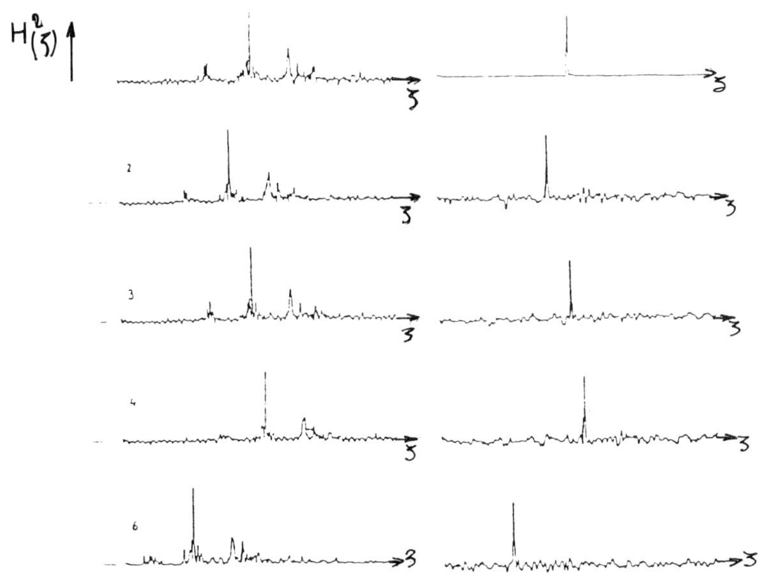

Fig. 9 . Example of inverse filtering

Actually we have shown that the optimal filter has a mixed transversal and recursive structure [16] and that in some cases it is very easy to elaborate.
The last example given below (Fig. 10) is a simulated experiment.

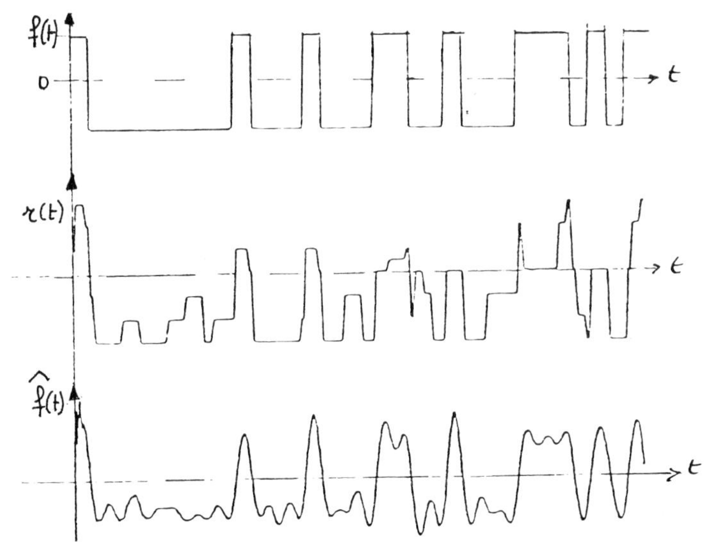

Fig. 10 . Inverse filtering after a 3-paths channel

The data f(t) are binary and transmitted through a three-paths channel. r(t) is the received signal (with strong SNR). After the optimal inverse filter, the data $\hat{f}(t)$ are nearly well recovered.

4 - CONCLUSION

The methodology of an active characterization of a randomly time varying channel has been recalled here. Actually these techniques have become classical for us, and we apply them in many different experimental conditions. The application of these models to optimal communications have begun. The important point is the good choice of the model and this is very related to the time scale of the communication. We have pointed out that in some configurations an optimal system is possible to construct and valued. We actually study in a parallel way some suboptimal simpler structures to be compared to the optimal ones.

5 - BIBLIOGRAPHY

[1] G. JOURDAIN : "Filtres linéaires aléatoires et non stationnaires. Modèles. Simulations et Applications", Thèse de Doctorat-d'Etat, INPG, USMG, Septembre 1976.

[2] J.Y. JOURDAIN : "Etude du Milieu Marin. Propagation acoustique à moyenne distance" Rapport final de la convention DRET n° 78/1093.

[3] G. JOURDAIN - J.Y. JOURDAIN : "Méthodes et techniques d'estimation de la fonction de diffusion", Colloque GRETSI, Nice 1977.

[4] G. JOURDAIN - J.Y. JOURDAIN : "Characterization of Submarine Acoustic Transmission Channels. Underwater and Signal Processing", NATO ASI, Copenhagen 1980.

[5] J.Y. JOURDAIN - P. LAMBERT - G. REVOL : "Altération de la phase dans une transmission ASM", GRETSI 1981.

[6] B. FAURE : "Estimation de la fonction de diffusion d'un milieu réel", Rapport CEPHAG n° 24/79.

[7] J.Y. JOURDAIN - P. LAMBERT : "Transmission acoustique de signaux monochromatiques" Extrait de la revue du CETHEDEC, 17e année, 2e Trimestre 1980.

[8] G. JOURDAIN : "Transmission FSK à voies diversité non identiques. Comparaison avec un codage correcteur d'erreur", Signal processing, Vol. 2, n° 2, Avril 1980, pp. 151-172.

[9] G. TZIRITAS : "Communication en milieu aléatoire et dispersif", Thèse de Docteur Ingénieur, INPG, Février 1981.

[10] J.P. HENRIOUX : "Etude d'une transmission acoustique sous-marine à grande distance utilisant un codage FSK", Thèse, INPG, 1979.

[11] H. CHERIFI : "Etude des performances du récepteur optimal pour la communication binaire symétrique dans un canal aléatoire et dispersif en fréquence", Rapport DEA, Juin 1981.

[12] G. TZIRITAS : "Approximation de la probabilité d'erreur des récepteurs à traitement quadratique", Colloque GRETSI, Nice, 1981.

[13] G. JOURDAIN - G. TZIRITAS : "Optimal receivers over randomly time and frequency dispersive channels : multipath case", Signal Processing, Eurasip, 1980.

[14] J.Y. JOURDAIN - G. LOUBET : "Filtrage inverse appliqué au regroupement des trajets multiples", Rapport CEPHAG n° 26/80.

[15] G. JOURDAIN - G. TZIRITAS : "Communication in a fluctuating channel. Models and use of explicit and implicit diversity", ICASSP, Paris, Mai 1982.

[16] J. MARTIN - G. JOURDAIN : "Etude d'égalisateur après transmission dans un canal à trajets multiples. Application au cas à deux trajets", Rapport DEA, 1981.

\- 0 -

MULTIDIMENSIONAL FILTERING OF ACOUSTIC RECORDS IN SEISMIC SIGNAL PROCESSING

Giovanni Garibotto [x]

CSELT - Torino - Italy

1. INTRODUCTION

In the paper the problem of acoustic wave propagation within a homogeneous medium is considered. Examples of application are referred in the field of acoustics and in seismic signal processing. The wave migration in the frequency domain corresponds to linear all pass phase filtering and phase specifications are derived for the ideal two-dimensional case. The resulting hyperbolic impulse response is symmetrical with respect to the time axis and two-dimensional half-plane recursive filters are proposed to solve the wave equation backwards in the time variable. Numerical examples are referred to analyse the practical constraints of the proposed processing system which is compared against parabolic approximations of the two-dimensional wave equation. In most application a general three-dimensional model of spherical waves is much more suitable. The proposed solution in the 3-D case is still based on recursive filtering so that it can be profitably used also for large amounts of data. The linear processing method described here is limited to homogeneous media so that it does not represent the best solution for the wave migration in geophysical prospecting, even if it has been proved an efficient method for preliminar estimation. On the other hand it seems particularly useful for some other acoustics problems such as impulse analysis of loudspeakers.

2. 2-D WAVE EQUATION

The acoustic propagation of an elastic wave in a homogeneous medium, characterized by an acoustic velocity c, is described by the differential equation :

$$c^2 \nabla^2 u = \frac{\partial^2 u}{\partial t^2} \tag{1}$$

In general this equation corresponds to spherical waves in the 3-D space (x,y,z).

[x] He is now with 3M Italia Ricerche S.p.A. - 17016 Ferrania (Savona) Italy

NATO ASI Series, Vol. F1
Issues in Acoustic Signal/Image Processing and Recognition
Edited by C. H. Chen

The problem is simplified here in two dimensions (x,z) and it is supposed that no variations occur in the orthogonal direction y. This reduced model is commonly used in most seismic applications, in which case the wave equation (1) can be written as:

$$c^2 \left(\frac{\partial^2 u}{\partial x^2} + \frac{\partial^2 u}{\partial z^2} \right) = \frac{\partial^2 u}{\partial t^2} \tag{2}$$

In the ideal situation depicted in Fig. 1 a single source at depth z_0 emits an impulse at time $t=t_0$, which propagates through the homogeneous medium up to the receivers located on the surface z=0. A data record is available as u(x,t,0) and it is requested to estimate the depth position z_0 of the source. By inverting the 2-D wave equation it is possible to evaluate the terms (t_0, z_0), corresponding to the maximum of the response $u(x,t_0,z_0)$. A solution can be found in the 2-D frequency domain. Let $U(\omega x, \omega t, z)$ be the 2-D Fourier transform, in the variables (x,t) of u(x,t,z), according to the transform relation:

$$U(\omega_x, \omega_t, z) = \iint_{-\infty}^{\infty} u(x,t,z) \exp\{-j(\omega_x x + \omega_t t)\} \, dx \, dt \tag{3}$$

By the theorem of differentiation, equation (2) becomes:

$$\frac{\partial^2 U}{\partial z^2} = -\left(\frac{\omega_t^2}{c^2} - \omega_x^2 \right) U \tag{4}$$

which is an ordinary second order differential equation in the variable z. The general solution $U(\omega x, \omega t, z)$ is evaluated as:

$$U(\omega_x, \omega_t, z) = A e^{jzk} + B e^{-jzk} \tag{5}$$

$$k = \sqrt{\omega_t^2/c^2 - \omega_x^2}$$

The first term in (5) corresponds to a wave which propagates in the negative direction of the depth variable (-z), and the second term is referred to a wave propagating in the positive direction (+z). In the following only upgoing waves are considered (since they are recorded on the surface, if there are no further reflection events) and the second term in (5) is neglected, B=0. Coefficient A corresponds to the Fourier transform $U(\omega x, \omega t, 0)$ of the record on the surface u(x,t,0) and it represents the boundary condition of the partial differential equation (4). In this hypothesis relation (5) becomes:

$$U(\omega_x, \omega_t, z_0) = U(\omega_x, \omega_t, 0) \exp\left\{ j z_0 \sqrt{\omega_t^2/c^2 - \omega_x^2} \right\} \tag{6}$$

Hence, starting from the surface record u(x,t,0) the function u(x,t,z) can be evaluated at any depth z_0 by computing equation (6) and inverse Fourier transform of $U(\omega_x, \omega_t, z_0)$. The frequency product (6) can be performed by linear filtering according to the frequency specifications:

$$H_{z_0}(\omega_x, \omega_t) = \begin{cases} \exp\left\{ j z_0 \sqrt{\omega_t^2/c^2 - \omega_x^2} \right\} & \text{on } S_1 \equiv \left\{ |\omega_t| > c|\omega_x| \right\} \\ \exp\left\{ - z_0 \sqrt{\omega_x^2 - \omega_t^2/c^2} \right\} & \text{otherwise} \end{cases} \tag{7}$$

The frequency function (7) can be further expressed as the product of two terms:

$$H_{z_0}(\omega_x, \omega_t) = P_{z_0}(\omega_x, \omega_t)\, F_{z_0}(\omega_x, \omega_t) \tag{8}$$

where $F_{z_0}(.)$ is responsible for the amplitude attenuation with zero-phase constraints:

$$F_{z_0}(\omega_x, \omega_t) = \begin{cases} 1 & \text{on } S_1 \\ \exp\left\{ - z_0 \sqrt{\omega_x^2 - \omega_t^2/c^2} \right\} & \text{otherwise} \end{cases} \tag{9}$$

and $P_{z_0}(.)$ is related to the phase specifications:

$$P_{z_0}(\omega_x, \omega_t) = \exp\left\{ j z_0 \sqrt{\omega_t^2/c^2 - \omega_x^2} \right\} \quad \text{on } S_1 \tag{10}$$

There are no phase constraints outside the frequency domain S_1 which is filtered out by the fan filter $F_{z_0}(.)$.

Unfortunately the phase function 10) cannot be realized by any causal recursive filter since its impulse response is defined for negative values of the variable t (see Appendix).

A solution to the problem can be obtained as follows. At first a causal recursive all-pass filter is designed according to the frequency specifications:

$$Q_{z_0}(\omega_x, \omega_t) = \exp\left\{-j z_0 \sqrt{\omega_t^2/c^2 - \omega_x^2}\right\} \qquad \text{on } S_1 \tag{11}$$

Hence the input data u(x,t,o) are processed along the negative direction of the time variable, starting from the last record t=T up to the first record of data t=0. The inverse Fourier Transform of (11) has dominant values along the hyperbole of equation (see Appendix)

$$c^2 t^2 - x^2 = z_0^2 \tag{12}$$

As a consequence the impulse response of the filter to be designed is defined in the triangular sector included between the asymptotes $x = \pm ct$ (see Fig. 2).
The design of filter $H_{z_0}(.)$ is considered in section 4 with reference to the amplitude specifications 9) and the phase constraints in 11).
Next section contains a brief discussion on some application problems of geophisics and acoustics in which the two-dimensional wave equations is profitably used.

3. EXAMPLES OF APPLICATION

The two-dimensional wave equation previously discussed plays an important role in many problems of physical science. In the following a synthetic description of some applications is referred.

I - Seismic signal processing.

A seismic record is a collection of signals from different sources [1]

a) Teleseismic signals which are characterized by an apparent velocity $v \geq 4$ Km/sec and, in the frequency-wave number domain (ω_t , ω_x), are included within the sector T of Fig. 3.
The bandwidth is limited to a few cicles/second and they represent the useful signal component to be detected.

b) Microseismic noise with an apparent velocity between 2.5 and 4 Km/sec approximately.

c) Low frequency incoherent noise without specific orientation.

Hence the signal to noise ratio can be significantly improved by filtering out the signal component having apparent velocity less than 4 Km/sec, by using a band-pass fan-filter as described before.

II. 2-D Wave Migration

One of the most important problems in geophysical application is the estimation of the reflecting layers in the earth. Then it is required to recover a space-depth profile (x,z) starting from the time section recorded on the surface u(x,t,o). The usual approximation consists in neglecting multiple reflections and in lumping the whole space between sources and reflectors into a homogeneous medium characterized by an average acoustic velocity c. [2] By using two-way travel time from source to receivers and defining midpoint coordinates in "stacked" sections[3] the reflected sound wave is the linear superposition of spherical waves generated at time zero at the different layers. In this case it is necessary to estimate the downward continued time sections u(x,o,z) by sampling the depth variable $z=n\Delta z$.
The processing scheme is depicted in Fig. 4.

III. Acoustic analysis of loudspeakers

The solution outlined in the previous section can be used for the impulse response analysis of a loudspeaker excited by white noise. The block diagram of Fig. 5 shows the experimental set-up for data collection. A linear array of microphones is placed at a fixed distance from the source. The recorded spherical wave is migrated towards the origin in order to estimate the position of the impulse as well as its spatial spread.

4. HALF-PLANE FILTER DESIGN

In order to solve the problem stated in section 2 it is necessary to design a band-pass fan-filter and an all-pass phase filter according to the frequency contraints (9) and (11) respectively. In both cases the specifications are symmetrical with respect to the time axis t and the designed filters should have an impulse response in the half-plane $t \geq 0$. Many efficient techniques have been recently proposed to design half-plane recursive filters with amplitude constraints. At the end of the section a general method for the design of half-plane all-pass phase filter is also presented. Anyway, in this particular application, we can take advantage of some additional design constraints which are peculiar of the problem at hand. In fact the hyperbolic impulse response to be approximated is defined in a triangular sector as depicted in Fig. 2. Hence a symmetrical mask of coefficients (Fig. 6b) is particularly suitable both for potential parallel processing and for the minimization of the computational complexity. Moreover a further simplification of the problem can be obtained by properly chosing the sampling interval Δt (after low-pass filtering the analog data to the required bandwidth). In fact for $\Delta t = \Delta x/c$ the hyperbole asymptotes in the space time domain (x, t) are orthogonal and the filter design can be carried out in the rotated first quadrant (x_1, x_2) of Fig.7. In this new domain the impulse response $h(m_1, m_2)$ is constrained to be

$$h(m_1, m_2) = 0 \qquad \text{for} \quad m_1 + m_2 \quad \text{odd} \tag{13}$$

and the locus of definition of the impulse response (12) becomes:

$$2\, x_1 x_2 = z_0^2 \tag{14}$$

A rotation through an angle β in the space domain determines an equal rotation in the frequency domain. The new variables (ω_1, ω_2) corresponding to the space variables (x_1, x_2) are related to the frequency wavenumber pair (ω_t, ω_x) by :

$$\omega_x = \frac{\sqrt{2}}{2} (\omega_2 - \omega_1)$$

$$\omega_t = \frac{\sqrt{2}}{2} (\omega_2 + \omega_1)$$

and the phase function in the rotated variables becomes:

$$\phi(\omega_1, \omega_2) = - z_0 \sqrt{2\, \omega_1 \omega_2}$$

and it is defined in the first quadrant of Fig. 8. The examples referred in the paper are based on this processing method by using the mask of coefficients shown in Fig. 6 c) [4] . A critical analysis of results is carried out in Section 5. In the following some techniques for the design of half-plane recursive filter design are briefly summarized.

Amplitude approximation

The problem of 90° fan-filter design with frequency specifications as in (9) has been recently addressed in many papers and different solutions have been proposed. Best results have been achieved using the following methods :

a) non-linear optimization [5] [6] with non-symmetric half-plane mask of coefficients as in Fig. 6 a).

b) two-dimensional linear prediction and non-linear optimization, with symmetric constraints on the coefficients [7] .

c) non linear optimization in the first quadrant and filter rotation [8] , using a symmetric mask of coefficients as in Fig. 6c). This solution allows the minimum number of coefficients in the space-domain. Fig. 9 illustrates the amplitude response of the designed fan filter.

d) two-dimensional wave filter design by rotation,[9] which minimizes the number of parameters to be optimized.

Phase approximation

There are just a few techniques available for the design of all-pass phase filters [10] [11] and none of them has been proved quite efficient for the solution of the problem stated here. An iterative linearized method in the first quadrant rotated domain is presented in details in [4] . In the following a generalization to symmetric half-plane filters is briefly referred. Let a phase function $\emptyset\ (\omega_1, \omega_2)$, be given. The input specifications at the frequency pairs $(\omega_{1K}, \omega_{2K})$ are:

$$P(\omega_{1K}, \omega_{2K}) = \exp\{j\,\emptyset(\omega_{1K}, \omega_{2K})\} = R_K + j\,I_K \tag{17}$$

It is required to design a 2-D all-pass phase filter with symmetric half-plane coefficients (Fig. 5), having a transfer function:

$$A(z_1, z_2) = z_1^{n}\,\frac{D(z_1^{-1}, z_2)}{D(z_1, z_2)} \tag{18}$$

and

$$D(z_1, z_2) = 1 + \sum_{n_1=1}^{n} \left\{ d(n_1, 0) + \sum_{n_2=1}^{n} d(n_1, n_2)(z_2^{n_2} + z_2^{-n_2}) \right\} z_1^{n_1} \tag{19}$$

the frequency response of the filter at $(\omega_{1K}, \omega_{2K})$ is:

$$A(\omega_{1K}, \omega_{2K}) = \exp\left\{-j\left[n\omega_{1K} + 2 \angle D(\omega_{1K}, \omega_{2K})\right]\right\} = \exp\{j\,\varphi_{AK}\} = X_K + j\,Y_K \tag{20}$$

n being the order of the filter.

The coefficients $d(n_1, n_2)$ in (14) can be determined by minimization of the error function :

$$J(d) = \sum_k |e_k|^2 = \sum_k \left[(R_k - X_k)^2 + (I_k - Y_k)^2 \right] \tag{21}$$

which can be expressed in terms of the phase differences as:

$$J(d) = 4 \sum_k \sin^2 \left\{ \frac{\phi(\omega_{1k}, \omega_{2k}) - \varphi_{Ak}}{2} \right\} \tag{22}$$

This is a non-linear problem which can be converted into a linear form, by minimizing a weighted error function in an iterative way, so that at step i the error function becomes:

$$J_i(d) = \sum_k | e_k D_{k,i} / D_{k,i-1} |^2 \tag{23}$$

where $D_{k,i-1}$ represents the frequency response of the denominator (19) whose coefficients have been determined at the previous iteration. The partial derivatives of the error (23) are equated to zero and a linear system of equation is solved at each step. Convergence is achieved when there is no significant change of the filter coefficients in the subsequent iterations; this corresponds to the minimization of the original unweighted error (22). Stability of the half-plane filter is tested at each step according to a recently proposed technique [12] . When the designed filter is found unstable a stabilization procedure in the complex cepstrum domain [13] is used, to guarantee a meaningful final result.

5. ANALYSIS OF RESULTS.

The purpose of this section is to evaluate the performance of the proposed processing method for the solution of the 2-D wave equation. The results discussed here have been obtained by phase filter design in the first quadrant and 45° rotation of the coefficient mask with the sampling constraint $\Delta t = \Delta x/c$. With reference to the phase function (11) an additional time delay has to be included in the frequency specifications to be approximated by a symmetric half-plane recursive filter. Using a fourth order section as in Fig. 6 c) a delay D of 2 samples in the t direction has been selected and the phase function (11) has been changed into :

$$Q_{\Delta z}(\omega_x, \omega_t) = \exp\left\{-j\left(D\omega_t + \Delta z\sqrt{\omega_t^2/c^2 - \omega_x^2}\,\right)\right\} \tag{24}$$

Of course by reducing the elemental step Δz, also the approximation error $J(d)$ in (22) is strongly reduced. In Table 1 some results of the error function are referred for different values of Δz. On the other hand, since the depth migration to $z + \Delta z$ requires the filtering of the entire data array u(x,t,z), the step Δz should be large enough to minimize the overall computation. $\Delta z = \Delta x$ seems to be the most suitable choice to achieve a uniform sampling in the space-depth domain (x, z). Table 1 shows a comparison of the proposed technique against a parabolic approximation [2] of the two-dimensional wave equation (2) which, in a new frame of coordinates, is modified into:

$$\frac{\partial^2 u(x,t,z)}{\partial z\,\partial t} = \frac{\partial^2 u(x,t,z)}{\partial x^2} \tag{25}$$

The results demonstrate the superior performance of the hyperbolic approximation using the all-pass phase filter previously discussed. Further improvements are possible by reducing the approximation domain in Fig. 8 to the shaded area since the zero phase constraints along the axes:

$$\phi(\omega_1, \omega_2) = 0 \qquad \text{for } (\omega_1 = 0 \cup \omega_2 = 0) \tag{26}$$

cannot be satisfied by any realizable recursive all-pass filter Fig. 10 shows the impulse response of the processing system in Fig. 4, using a cascade of four rotated phase filters. In this example all terms in cascade were identical and they have been designed using a unitary step $\Delta z = \Delta x$, a reduced bandwidth $B = 0.5\ \pi/\Delta'$ and a value $\varepsilon = 0.05\ \pi/\Delta'$ for the uncertainty region about the main axes. The resulting approximation error is $J = 0.077$ and it has to be compared to the value $J = 0.9438$ in Table 1. Table 2 shows the coefficients $\{d\}$ of the designed all-pass phase filter. By putting in cascade N identical phase filters the global approximation error J_N is related to the error J_Δ of the single term as follows:

$$J_N \approx N^2 J_\Delta \tag{27}$$

Of course we can improve this result by minimizing the approximation error of each term in cascade. This procedure has been used to design 10 phase filters and in Table 3 the computed error function J_i is compared against the estimated error (27).

6. EXAMPLE OF WAVE MIGRATION

The practical application of 2-D recursive filter for the solution of 2-D wave equation is demonstrated using the ideal seismic profile in Fig. 11 which represents the reflection intensity at time zero $u(x,o,z)$. The matrix size in this example is 128 samples in the x direction and 80 samples in the z direction. The wave propagation from the reflectors up to the geophones on the surface is simulated in the frequency domain. Following the same procedure of section 2, the 2-D wave equation is Fourier transformed in the variables (x,z) so that it becomes:

$$-(\omega_x^2 + \omega_z^2)\, U(\omega_x, t, \omega_z) = \frac{1}{c^2} \frac{\partial^2 U(\omega_x, t, \omega_z)}{\partial t^2} \tag{28}$$

and the solution of this differential equation is:

$$U(\omega_x, t, \omega_z) = U(\omega_x, o, \omega_z) \exp\left\{ jct\sqrt{\omega_x^2 + \omega_z^2} \right\} \tag{29}$$

Dependence on the depth variable is avoided by averaging:

$$U(\omega_x, t, o) = \int U(\omega_x, t, \omega_z)\, d\omega_z \tag{30}$$

and the space-time section is recovered for each value t, by computing the 1-D inverse Fourier transform:

$$u(x, t, o) = \mathcal{F}^{-1}\left[U(\omega_x, t, o) \right] \tag{31}$$

Since the operations are performed via FFT, a periodical pattern as in Fig 11 is recommended. The simulated record on the surface is reproduced in Fig 12 in which it is possible to recognize the interfering effect of the different time-delays on the surface. The processing scheme of Fig. 4 has been used to migrate this sinthetic seismic section. A single phase filter with the coefficients of Table 2 has been used in a cascade of N = 80 terms. The input data u(x,t,o) are processed backwards along the time direction to produce the output array $u(x,t,\Delta z)$, and so on. For each output in the cascade, a single record of the intensity profile is computed, as $u(x,o,m\Delta z)$. Each recursive filtering requires the storage of 4 rows of 128 elements in a working memory $\{v\}$, 17 multiplications and 26 additions per output sample. The recursive rotated phase filter is implemented as follows:

$$v(n_1,n_2) = x(n_1,n_2) - \sum_{\substack{m_1=0 \\ m_1+m_2\neq 0 \\ m_1\pm m_2 \text{ even}}}^{M} \sum_{m_2=0}^{M} d(m_1,m_2)\, v\left(n_1-\frac{m_1+m_2}{2}, n_2-\frac{m_2-m_1}{2}\right) \tag{32}$$

$$y(n_1,n_2) = \sum_{\substack{m_1=0 \\ m_1\pm m_2 \text{ even}}}^{M} \sum_{m_2=0}^{M} d(M-m_1, M-m_2)\, v\left(n_1-\frac{m_1+m_2}{2}, n_2-\frac{m_2-m_1}{2}\right)$$

where $\{x\}$ is the input and $\{y\}$ is the output of each section in cascade. The index n_1 is related to the time variable and n_2 is related to the space variable; (m_1, m_2) are defined in the rotated domain of Fig. 7. The result of wave migration is referred in Fig 13 where most of the original profile has been correctly recovered. Distorsions appearing for increasing values of the depth variable z are determined by the cascading process as illustrated in Table 3. Moreover a strong bandwidth reduction has been introduced for normalization purposes since $\Delta t = \Delta x/c$. An interesting comparison can be carried out between these results and the corresponding FFT realization. In this example it would be necessary to store and compute a (128x128) FFT of the input (N = 14 multiplications per sample), to perform the frequency product (10) and the inverse transform (128x128) for each of the terms in cascade, $z = n\Delta z$. The proposed methods is then slightly more efficient than FFT in terms of computation but it is strongly superior in terms of data storage in the working memory.

7. 3-D EXTENSION

The wave propagation is a 3-D process and it would be necessary to perform wave migration in a 3-D space. The same discussion outlined in Section 2 can easily be extended to three dimensions and the migrated section U can be computed from the surface record u(x,y,t,o) as:

$$U(\omega_x, \omega_y, \omega_t, z_o) = U(\omega_x, \omega_y, \omega_t, 0) \exp\left\{-j z_o \sqrt{\frac{\omega_t^2}{c^2} - (\omega_x^2 + \omega_y^2)}\right\} \tag{33}$$

Again 3-D Fourier transform are quite expensive and an efficent recursive processing method should be highly desirable. In equation (33) the circular symmetric constraint in the variables (x,y) is clearly visible and it suggests an easy way to obtain the coefficients of the 3-D recursive filter by transformation.

Due to the symmetrical characteristics of the 2-D polynomial (19) the frequency response can be written as:

$$D(\omega_t, \omega_x) = 1 + \sum_{n_1=1}^{n} \left\{ d(n_1, 0) + \sum_{n_2=1}^{n} d(n_1, n_2)\left(e^{-j\omega_x n_2} + e^{j\omega_x n_2}\right)\right\} e^{-j\omega_t n_1}$$

$$= 1 + \sum_{n_1=1}^{n} \left\{ \sum_{n_2=0}^{n} f(n_1, n_2) \cos \omega_x n_2 \right\} e^{-j\omega_t n_1} \tag{34}$$

where:

$$f(n_1, n_2) = \begin{cases} d(n_1, 0) & n_2 = 0 \\ 2\, d(n_1, n_2) & n_2 > 0 \end{cases}$$

Using the Chebyshev polynomials $\left\{T_n[x]\right\}$, the function $\cos \omega_x n_2$ can be expressed as a polynomial of degree n_2 in $\cos \omega_x$ [14].

$$\cos \omega_x n_2 = T_{n_2}\left[\cos \omega_x\right] \tag{35}$$

and

$$D(\omega_t, \omega_x) = 1 + \sum_{n_1=1}^{n} \left\{ \sum_{n_2=1}^{n} f(n_1, n_2)\, T_{n_2}\left[\cos \omega_x\right] \right\} e^{-j\omega_t n_1} \tag{36}$$

Then a 2-D frequency function F (ω_x, ω_y) can be substituted for $\cos\omega_x$. According to the input specifications (33) the McClellan circularly symmetric transformation [14] seems to be a reasonable choice:

$$\cos\omega_x = F(\omega_x, \omega_y) = \frac{1}{2}\left(\cos\omega_x \cos\omega_y + \cos\omega_x + \cos\omega_y - 1\right)$$

and a 3-D polynomial can be recovered $d_3(n_1\ n_2, n_3)$ starting from the 2-D coefficients $d(n_1, n_2)$ designed for wave migration (as in Table 2). In the recursive realization of a 3-D symmetric filter [15] of degree M it is necessary to store, in a working memory, a set of M 2-D arrays of the input data so that the memory requirements is heavily increased. By transformation of the rotated phase filter of Table 2, M = 4 and the number of multiplications per sample in the 3-D case would be N = 36.

ACKNOWLEDGMENTS

The author wishes to thank prof. Fabio Rocca for his suggestions and helpful discussion,and Mrs. Wanda Panelli for typing the manuscript.

APPENDIX

The phase function

$$\phi(\omega_x, \omega_t) = \pm z_0 \sqrt{\omega_t^2/c^2 - \omega_x^2} \qquad z_0 \geq 0 \tag{A1}$$

is defined in the frequency domain $S_1 \equiv \left\{(\omega_x, \omega_t): \ |\omega_t| > c|\omega_x|\ ,\ \omega_t \geq 0\right\}$

Its inverse Fourier transform is evaluated as

$$\begin{aligned} I(x,t) &= \frac{1}{4\pi^2}\iint_{S_1} e^{j\phi(\omega_x,\omega_t)}\, e^{j(\omega_x x + \omega_t t)}\, d\omega_x\, d\omega_t \\ &= \frac{1}{4\pi^2}\iint_{S_1} e^{j\left[\omega_x x + \omega_t t \pm z_0\sqrt{\omega_t^2/c^2 - \omega_x^2}\right]}\, d\omega_x\, d\omega_t \end{aligned} \tag{A2}$$

According to the stationary phase approach [2] , this integral has dominant values at pairs (x, t) which satisfy the equations

$$\left.\begin{aligned} \frac{\partial \phi(\omega_x, \omega_t)}{\partial \omega_x} + x &= 0 \\ \frac{\partial \phi(\omega_x, \omega_t)}{\partial \omega_t} + t &= 0 \end{aligned}\right. \Rightarrow \quad \begin{aligned} x &= \pm \frac{z_0 \omega_x}{\sqrt{\omega_t^2/c^2 - \omega_x^2}} \\ t &= \mp \frac{z_0 \omega_t}{c^2 \sqrt{\omega_t^2/c^2 - \omega_x^2}} \end{aligned} \tag{A3}$$

By using the variable transformation in polar coordinates

$$\begin{aligned} \omega_t / c &= \rho \cos \theta \\ \omega_x &= \rho \sin \theta \end{aligned} \tag{A4}$$

the phase function (A1) becomes

$$\phi(\rho, \theta) = \pm z_0 \rho \sqrt{\cos 2\theta} \qquad \cos 2\theta > 0 \tag{A5}$$

and (A3) is modified into

$$\begin{aligned} x &= \pm z_0 \sin\theta / \sqrt{\cos 2\theta} \\ t &= \mp z_0 \cos\theta / c \sqrt{\cos 2\theta} \end{aligned} \tag{A6}$$

which means that the integral (A2) has dominant values along the hyperbole of equation

$$c^2 t^2 - x^2 = z_0^2 \tag{A7}$$

When the phase function (A1) is taken with the positive sign, the integral (A2) is defined in the negative half-plane ($t < 0$) and vice versa for the negative sign choice, since

$$z_0 \cos\theta / c \sqrt{\cos 2\theta} > 0 \tag{A8}$$

in the frequency domain S_1.

REFERENCES

[1] O. Kulhanek, "Introduction to Digital Filtering in Geophysics", Elsevier Scientific Publishing Co., Amsterdam 1976.

[2] G. Bolondi, F. Rocca, S. Savelli, "A frequency domain approach to two-dimensional migration" 39th Meeting of the Europ Ass. of Exploration Geophysicists, Zagreb, Yugoslavia, June 1977.

[3] R.H. Stolt, "Migration by Fourier Transform", Geophysics, vol 43, n° 1, pp 23-48, Feb 1978.

[4] G. Garibotto, "2-D Recursive Phase Filters for the Solution of Two-Dimensional Wave Equations", IEEE Trans. on Acoust. Speech and Signal Process., Vol. ASSP-27 n° 4, pp 367-373, Aug. 1979.

[5] M.P. Ekstrom, R.E. Twogood, J.W. Woods, "Two-Dimensional Recursive Filter Design - A Spectral Factorization Approach", IEEE Trans. on Acoust. Speech and Signal Process. vol. ASSP-28, n° 1 pp 16-26, Feb. 1980.

[6] J.H. Lodge, M.M. Fahmy, "An Optimization Technique for the Design of Half-Plane 2-D Recursive Filters", IEEE Trans. on Circuits and Systems, Vol. CAS-27 n° 8 pp 721-724, Aug. 1980.

[7] D.B. Harris, "Design of 2-D Rational Digital Filters", Proceedings of the ICASSP' 81, pp 696-699.

[8] G. Garibotto, R. Molpen, "A new Approach to Half-Plane Recursive Filter Design", IEEE Trans. on Acoust. Speech and Signal Process. Vol. ASSP-29, n° 1 pp 111-115, Feb. 1981.

[9] G. Linnenberg, "Investigations on a class of Two-Dimensional LC Lattice Filters", Proceedings of the ECCTD '81, The Hague, The Netherlands, pp 425-432, Aug. 1981

[10] S. Aly, M.M. Fahmy, "Design of Two-Dimensional Recursive Digital Filters with Specified and Group Delay Characteristics", IEEE Trans. on Circuits and Systems, Vol. CAS-25 n° 11, pp 908-916, Nov. 1978.

[11] D.B. Harris, "Design of Stable Allpass Filters", Proc. of ICASSP '79 pp 813-817, Apr. 1979.

[12] G. Garibotto, "A New Stability Test for 2-D Digital Filters", Proceedings of the EUSIPCO '80, pp 413-416, Sept. 1980

[13] G. Garibotto, G. Ghisio, "2-D Recursive Filter Stabilization", Proc. of the ECCTD '81, The Hague, The Netherlands, Aug. 1981

[14] J.H. McCellan, D.S.K. Chan, "A 2-D FIR Filter Structure derived from the Chebyshev Recursion", IEEE Trans. on Circuits and Systems, vol. CAS-24 n° 7, pp 372-378, July 1977.

[15] G. Garibotto, G. Piretta, "Three-Dimensional Recursive Filtering", Proc. of the ICASSP '82, Paris, May 1982

$\Delta z = n\Delta x$ n	J_1	J_2
3	23.881825	40.801232
2	10.7817753	10.386983
1	2.634223	0.943889
1/2	0.651473	0.211116
1/3	0.288890	0.989966E-01
1/4	0.162369	0.567464E-01
1/5	0.103876	0.370008E-01
1/7	0.529808E-01	0.192972E-01
1/10	0.259560E-01	0.961479E-02
1/1000	0.259516E-05	0.999010E-06

Table 1 - Wave equation approximation error using the ideal parabolic operator (J_1) and the proposed rotated phase filter (J_2) for different values of Δz and N = 100 samples in the frequency domain.

$m_1 \downarrow$ / $m_2 \rightarrow$				
1.	0.	-0.413	0.	0.10248
0.	0.13449	0.	-0.12962	0.
-0.413	0.	0.3422	0.	-0.17742
0.	-0.12962	0.	0.32812	0.
0.10248	0.	-0.17742	0.	-0.02856

		$n_1 \downarrow$ / $n_2 \rightarrow$		
		1.		
	-0.413	0.13449	-0.413	
0.10248	-0.12962	0.3422	-0.12962	0.10248
	-0.17742	0.32812	-0.17742	
		-0.02856		

Table 2. Coefficients of the 2-D Phase Filter used in the text: (m_1, m_2) are the rotated indexes in the firts quadrant; (n_1, n_2) are the time-space indexes.

	J	E
z = Δ	0.077	0.077
= 2Δ	0.17	0.308
= 3Δ	0.27	0.693
= 4Δ	0.43	1.232
= 5Δ	0.67	1.925
= 6Δ	0.95	2.772
= 7Δ	1.26	3.773
= 8Δ	1.62	4.928
= 9Δ	2.01	6.237
= 10Δ	2.43	7.7

Table 3. Approximation error for 10 filters in cascade; J_i is the computed error, $E_i = i^2 J_1$ is the error estimate by putting the same filter in cascade.

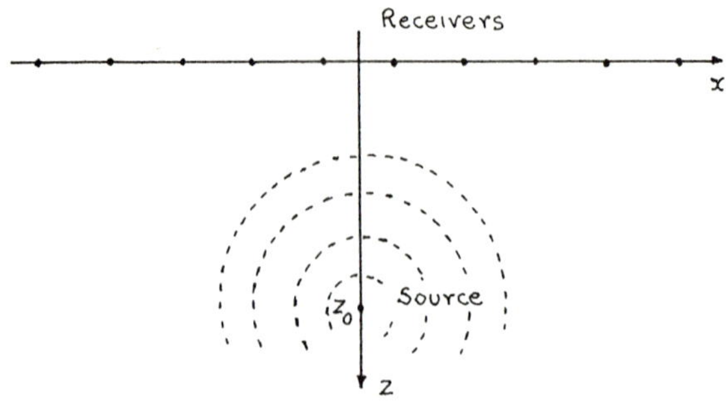

Fig. 1. Two-dimensional model of the reference system

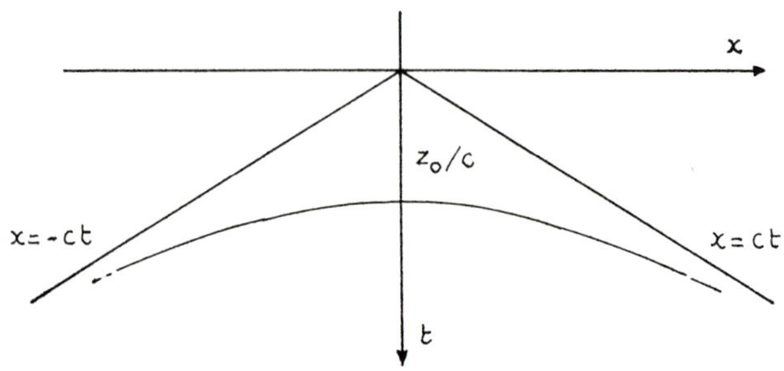

Fig. 2. Locus of the impulse response of the ideal phase filter

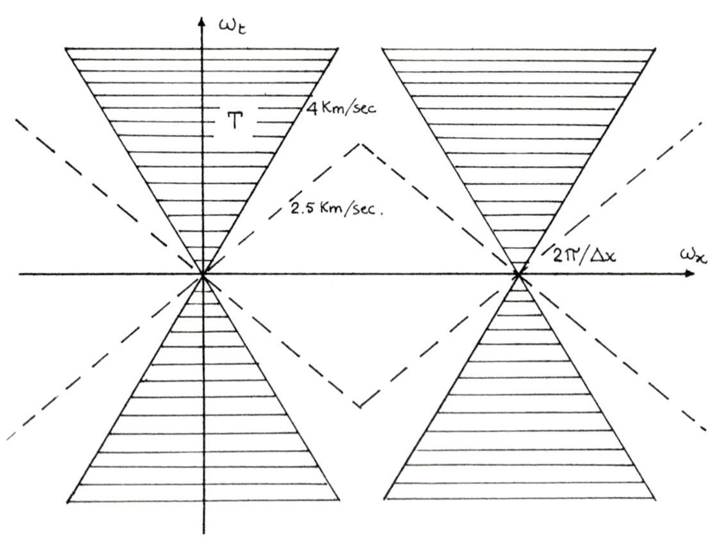

Fig. 3. Seismic signals in the frequency-wavenumber domain

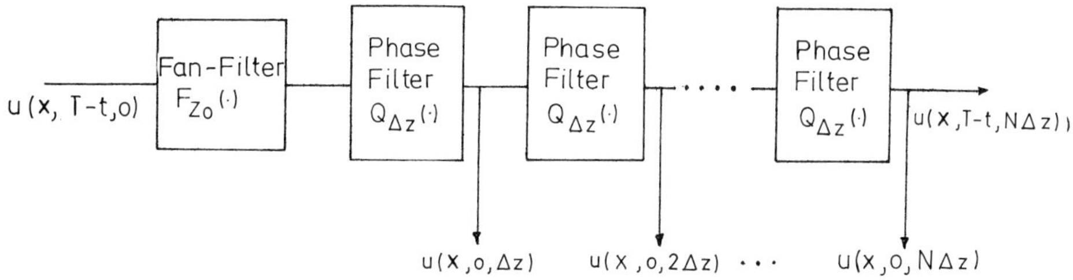

Fig. 4. Processing system for the solution of the 2-D Wave Equation

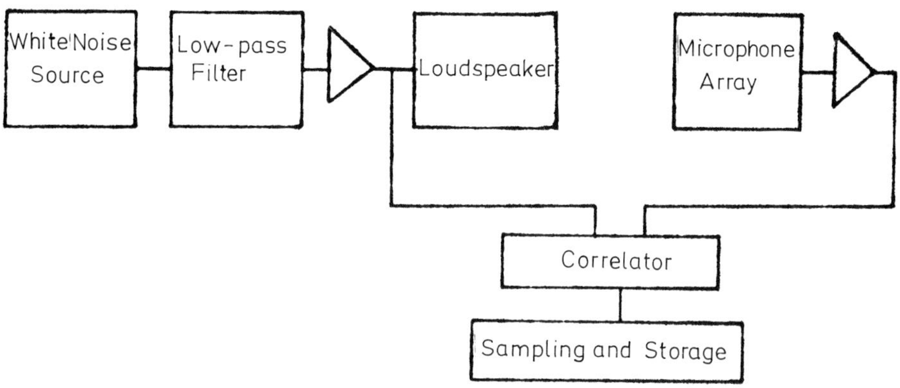

Fig. 5. Block diagram of the recording process for the analysis of loudspeakers

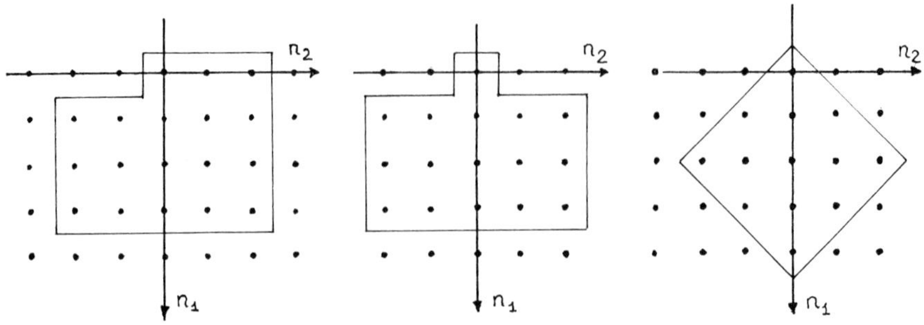

Fig. 6. Mask of coefficients: a) non symmetric half-plane, b) symmetric half-plane, c) rotated first-quadrant

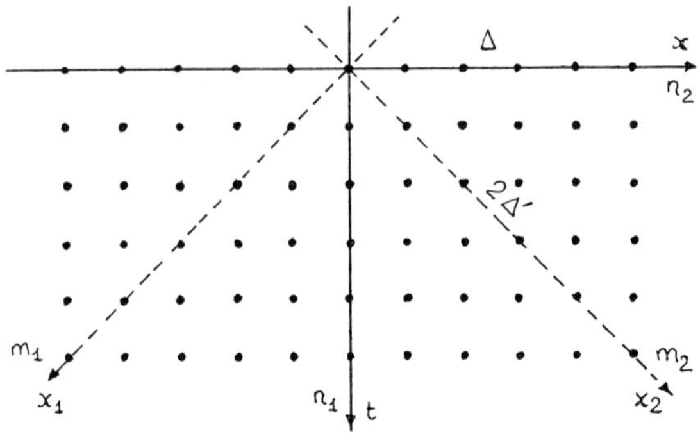

Fig. 7. Sampling grid in the rotated domain

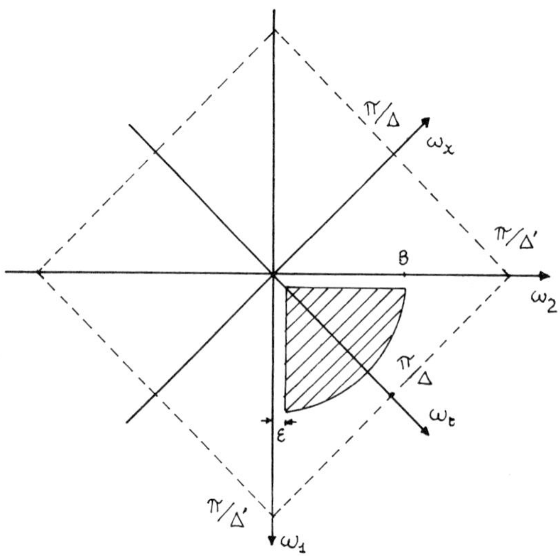

Fig. 8. Rotated frequency domain

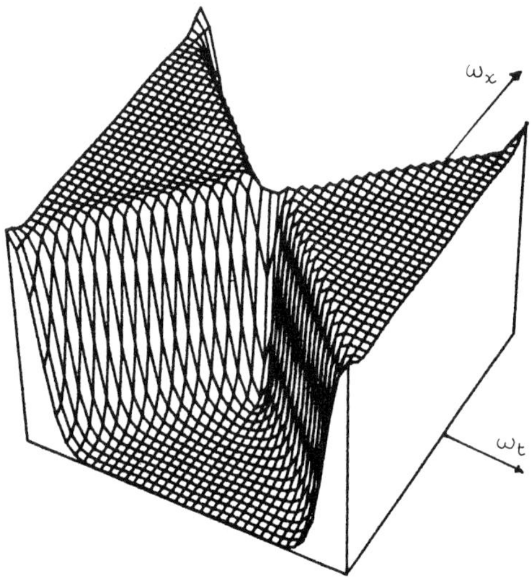

Fig. 9. Amplitude response of the rotated fan-filter

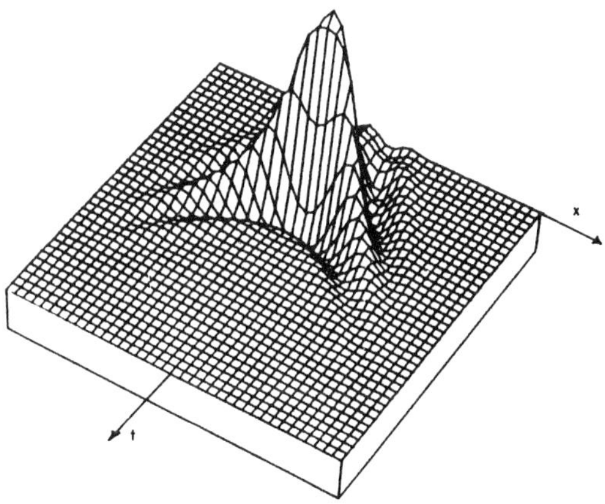

Fig. 10. Impulse Response of the 2-D phase filter; N = 4 sections in cascade

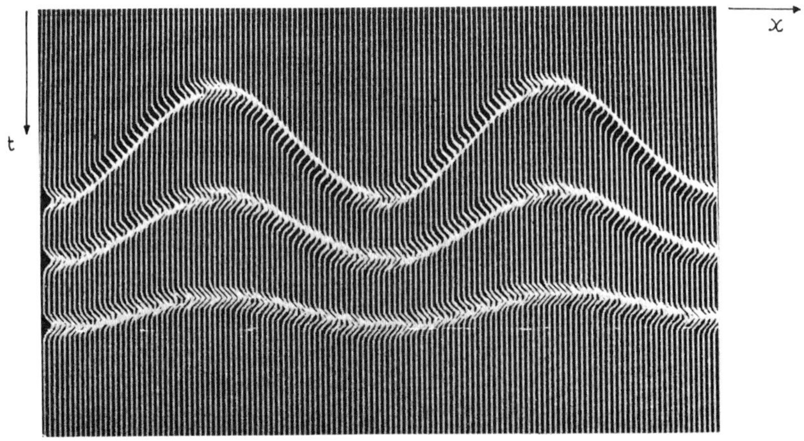

Fig. 11. Two-dimensional earth model

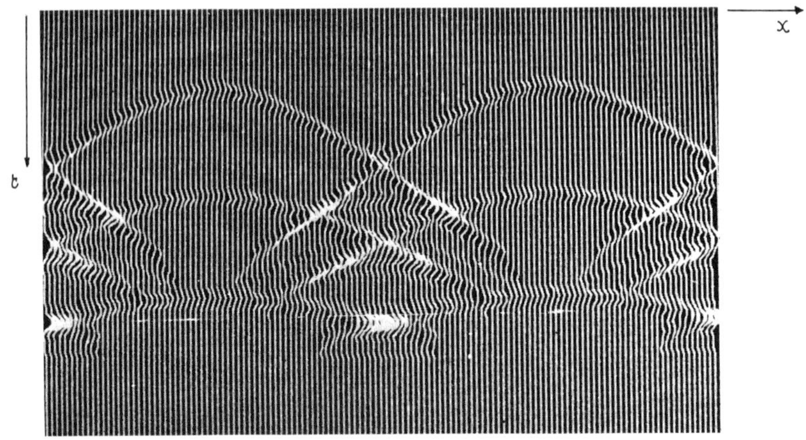

Fig. 12. Synthetic seismic section

Fig. 13. Wave migration via 2-D recursive filtering

SAMPLING THEORY AND WAVE PROPAGATION

J. MORLET
O.R.I.C. (ELF AQUITAINE)
370 bis, avenue Napoléon Bonaparte
92500 RUEIL MALMAISON - FRANCE

ABSTRACT

In the seismic reflection method, the seismic signal does vary in amplitude, shape, frequency and phase versus the propagation time.

Besides various artefacts due to the recording and processing methods, these variations reflect the frequency dependence of wave propagation, due to the following phenomena :

- absorption in heterogeneous media,
- scattering in multilayered and heterogeneous media.

From theoretical studies, it appears that, in high resolution seismic, the propagation of high frequency waves may be very sensitive to 1D scattering phenomena due to intrabed multiples. Such phenomena induce velocity dispersion, resulting, after standard processing, in attenuation of high frequencies.

On the other hand, backscattered waves may carry useful informations related to thicknesses of thin layers and layering of thin layered series. Extracting these informations from the seismic records gives us a tool to increase the seismic resolution.

In this aim, we need to develop multifrequency-band processing methods, involving multichannel interferometric methods in adjacent narrow frequency bands.

Such methods involve a time-frequency representation of the seismic trace, in other words "Instantaneous Frequency Spectra".

In this paper, we describe a practical 2D time-frequency sampling method for the seismic traces, suitable for optimization of the information/bit ratio. This sampling method is based on a new 2D transform, developed as an extension of the time-frequency Gabor expansion for complex signals.

In this new expansion, the 2D sampling grid is regular when using as the two coordinates the cycles (or unwrapped phases) and octaves rather than the times and frequencies. We may then call it the "Cycle-Octave" expansion.

NATO ASI Series, Vol. F1
Issues in Acoustic Signal/Image Processing and Recognition
Edited by C. H. Chen

Using this expansion, we may quantify the complex amplitude as a function of time and frequency (i.e. instantaneous energy and phase for any frequency).

Furthermore, using the "Uncertainty Principle", we may quantify the resolution of the method.

Finally, after theoretical considerations based on Quantum Mechanics, it appears that the practical value for the dimensions of each elementary cell in the sampling grid must be :

- 2 sampling points per cycle,
- 4 sampling points per octave,

an in each cell of the sampling grid, the sampled value represents a complex amplitude.

Practical methods to compute both direct and inverse transforms are given.

To avoid problems of dynamic range in field recording, it will be useful to develop new recorders for direct 2D sampling of the seismic traces, using the "Cycle-Octave" expansion.

SAMPLING THEORY AND WAVE PROPAGATION

INTRODUCTION

First introduced in the early 60's, the digital methods for recording and processing the seismic reflection data were based, from the beginning, on the following theories :

- Time series analysis,
- Statistics (Multichannel data enhancement methods),
- Wave propagation.

And after a strong and continous development during the last 20 years, the seismic reflection method now appears to be the best tool for oil exploration.

Due to a significant increase of drilling costs, especially in off-shore operations, and to a worlwide need for increasing recovery ratios in oil fields, we now have to solve new types of problems : those of oil field development.

The basic problem may be defined as follows :
Using seismic waves, we must obtain a good 3D representation of the producing reservoirs, especially :

- boundaries reservoirs/embedding rocks,
- oil/water boundaries,

and more generally a 3D representation of the distribution of physical properties of rocks and enclosed fluids in the neighborhood of the reservoirs.

That is the aim of the new developments of the seismic method, working now in two directions :

1) high resolution seismic,
2) 3D seismic reflection.

Nevertheless, the success of the seismic reflection methods in oil exploration does not imply that they will provide the best tool for oil field development.

In this paper, we will therefore discuss the fundamentals of wave propagation, then of recording and processing methods, and finally describe new tools to increase the resolution of the seismic reflection method.

These new tools are made to increase our knowledge of any oil field, by a specific method of survey close to the methods of "Non Destructive Testing". More precisely, the waves will be fully used to carry the information related to the oil field.

Furthermore, to measure the effective enhancement of the method, we will quantify the resolution using the "Uncertainty Principle".

We will then point out the following points :

1) The attenuation of the signal does not automatically involve absorption phenomena.

2) Multilayered media induce scattering phenomena in propagating waves. These phenomena are no more negligible when high resolution is needed.

3) Multichannel seismic processing methods appear to be the best tool for signal to noise ratio enhancement. Such methods practically operate an integration of the Wave Equation, simulating therefore interferometric methods. Consequently, the resolution is limited, as in Optics, by diffraction patterns.

4) The classical assumption on the "random distribution" of the layer thicknesses, made when using the standard methods of deconvolution in single channel processing is no more valid in high resolution seismic.

5) To handle backscattered informations coming from thin bed layering, related to an oil field, the standard sampling methods do not give the best way to preserve the useful information.

6) Time-Frequency Sampling Methods will improve the representation of the high resolution data, at recording and processing steps.

I - Wave propagation and signal attenuation

In the seismic reflection method, the seismic signal does vary in amplitude, shape, frequency and phase versus the propagation time. These variations reflect the frequency dependence of wave propagation.

The studies on "true amplitudes" and signal attenuation represent a very difficult area in seismic processing for oil exploration.

When signal attenuation is observed in digital seismic records, at least 4 different reasons may be involved besides the spherical divergence :

1) Recording methods,
2) Processing methods,
3) Scattering phenomena,
4) True absorption, involving energetic transformations.

We may summarize the effect of these 4 points on signal attenuation as follows :

1) Comparing the Fig. 1 and 2 gives us a very simple example of high frequency

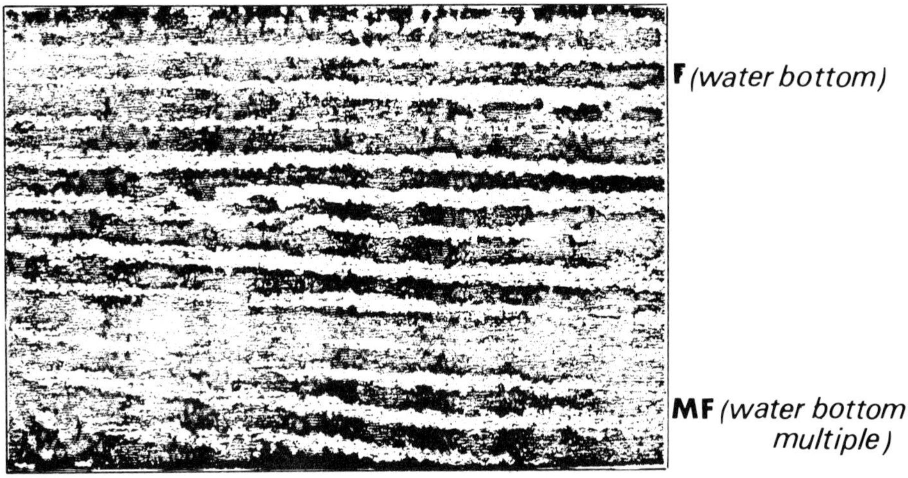

Fig.1. SHALLOW PART OF A CONVENTIONAL SEISMIC SECTION

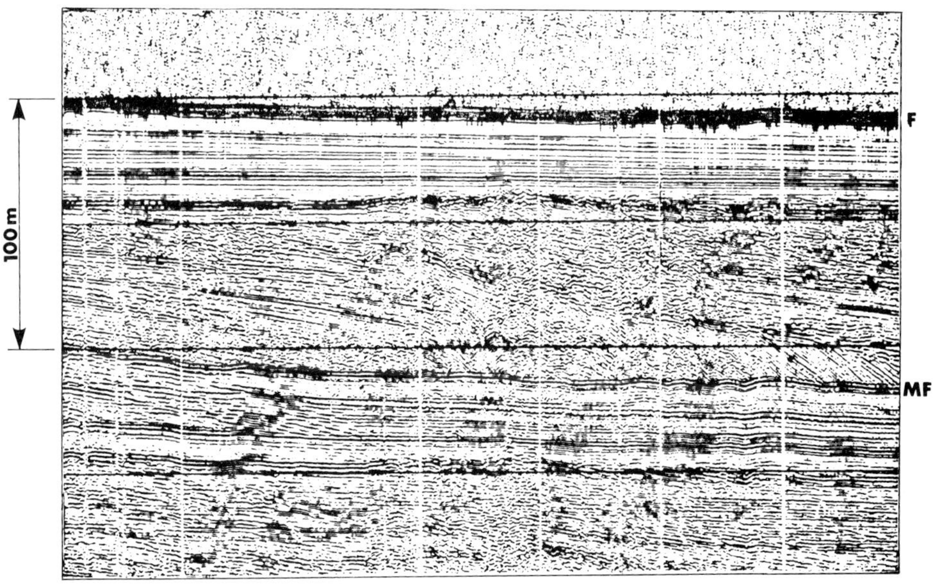

Fig.2. CORRESPONDING HIGH RESOLUTION SEISMIC IMAGE

attenuation due to the characteristics of the recording equipment. They represent the same section of the earth, using the same scales. Fig.1 was recorded and processed with standard parameters used for exploration. Fig.2 corresponds to single channel high resolution seismic, made for shallow gas field detection.

Numerous authors studied the effect of space distribution of the recording elements (i.e. antenna responses) on the filtering of high frequencies. We will mention here two recent papers presented at the S.E.G. Annual International Meeting in Houston in 1980 by K. Larner and N. Anstey.

2) Concerning the processing methods, we know the power of multichannel methods based on interferences and on the space-time coherence of the desired information.

To preserve the high frequency information, they need very accurate measurement of the 3D location of the elements of the recording pattern.

3) Scattering phenomena appear in heterogeneous media, at any scale of granularity in 3D. An interesting case is the case of multilayered media, inducing 1D scattering, as we will study later in this paper.

4) Let us come now to the true absorption in heterogeneous media. From the studies developed at the Stanford University by A. Nur and altera in the S.R.P. Project, such phenomena, for relative displacements lower than 10^{-6}, are significant in the specific case of porous rocks including gas and liquids.

To conclude this section, we will point out that the true absorption of the waves is very rare and practically restricted to the following cases :

- Source neighborhood,
- Unconsolidated sediments,
- Gas fields.

II - Wave propagation in Multilayered Media

To increase the resolution of the seismic reflection, we developed various single channel inversion methods resulting in pulse shortening.

These methods, well known under the general name of deconvolution, are powerful to suppress the water layer multiples.

On the other hand, when we try to use them to suppress intrabed multiples, we assume a random distribution of the layering.

But in quasicyclic series, that are commonly related to oil and coal fields, this last hypothesis is no more valid.

An alternate way is then to study the wave propagation in quasicyclic series.

From these studies, presented at the S.E.G. Annual International Meeting in Houston in 1980, and published in "GEOPHYSICS" in February 1982, we way explain the main part of the variations of the seismic signal and of the observed attenuations by 1D scattering due to intrabed multiples.

We may summarize the results of these works as follows :

The scattering of the signal
- depends on the ratio signal wavelength/layer thickness,
- is therefore frequency dependent,
- is significant in cyclic series,
- increases with impedance contrast.

These results were obtained by two different methods :

1.) Simulation by synthetic seismograms

We computed therefore a set of synthetic Vertical Seismic Profiles (i.e. amplitude of the wave versus propagation time and depth) using the Goupillaud-Kunetz algorithm for 1 D synthetic seismograms (Fig.3).

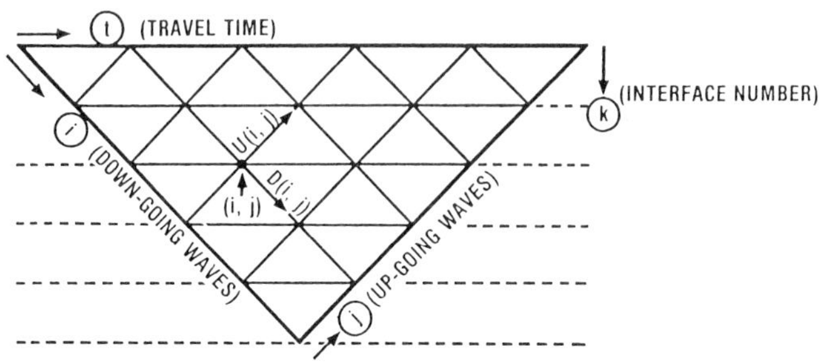

$$\begin{cases} U(i,j) = U(i,j-1) \times [1 + R(k)] + D(i-1,j) \times R(k) \\ D(i,j) = U(i,j-1) \times [-R(k)] + D(i-1,j) \times [1 - R(k)] \\ R(k) = \text{Reflection Coefficient at the } k^{th} \text{ interface} \end{cases}$$

Fig.3. 1D. PROPAGATION DIAGRAM (synthetic seismogram with multiples)

For the signal, we used simple zero-phase wavelets made of a Gaussian envelope modulated by a pure monochromatic cosine wave. The corresponding complex wavelet is a Gabor wavelet (Fig.4). As shown in the lower part of the figure, the Fourier transform of such wavelets is a Gaussian function.

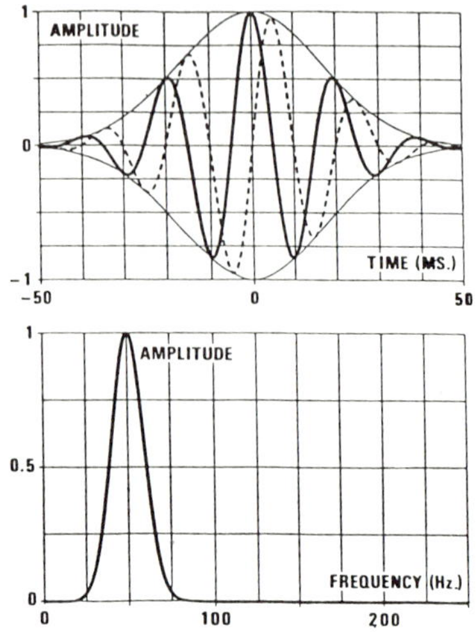

Fig.4. GABOR'S COMPLEX WAVELET

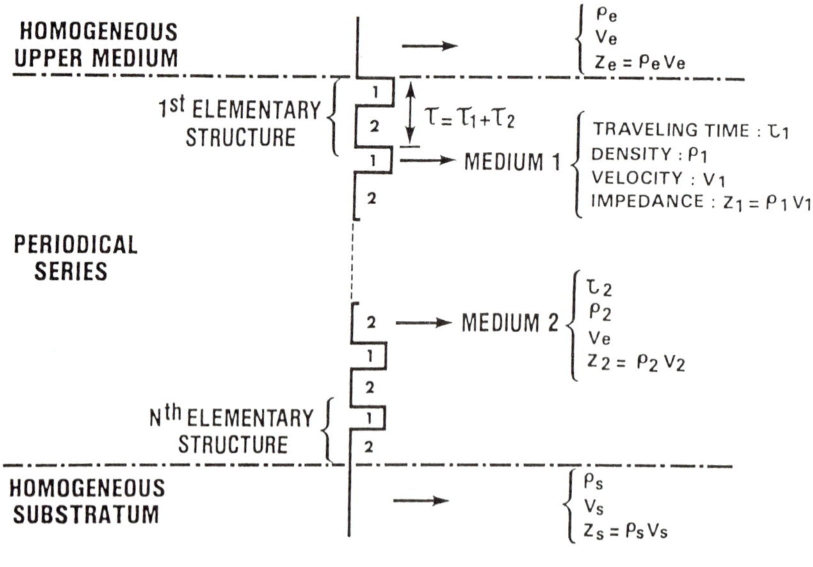

Fig.5. PERIODIC MODEL

In the simulation, we used such signals with constant shape for various frequencies, (i.e. constant number of peaks and troughs in the time representation of the signal).

For sake of simplicity, and to clearly illustrate the fundamentals of scattering phenomena, we used periodic models for the layering, as shown in Fig.5. In this figure, the physical parameters are given versus the one way propagation time, as used in the Goupillaud-Kunetz algorithm. In this case, the composite periodic medium is made up of two components.

Three different cases appear, depending on the ratio T/τ, i.e. : wavelength of the signal/period of the cyclic series.

We will present here the corresponding V.S.P. representations for the propagation of the wavelets, for an impedance ratio of 10.

- Low frequencies

In Fig.6, for low frequency wavelets, with a ratio T/τ of 16, the periodic medium is fully transparent. But, due to intrabeb multiples, the direct wave is evanescent and the effective wave is a delayed wave. Practically, this delayed wave is made of coherent interferences of intrabed multiples.

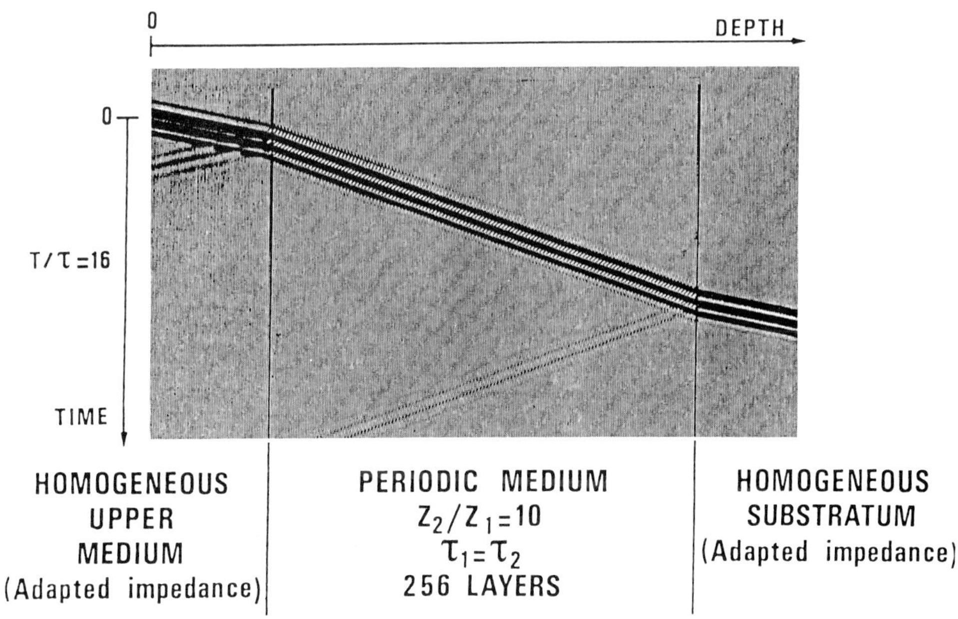

Fig.6. LOW FREQUENCIES : FULL TRANSPARENCY

- High frequencies

In Fig.7, for high frequency wavelets, with ratios T/τ of 2.83 and 5.66, the periodic medium is a superreflector : all the energy is backscattered by the first few layers at the top of the periodic medium. Such phenomena remain effective even for lower impedance contrasts and are well known in Thin Layer Optics.

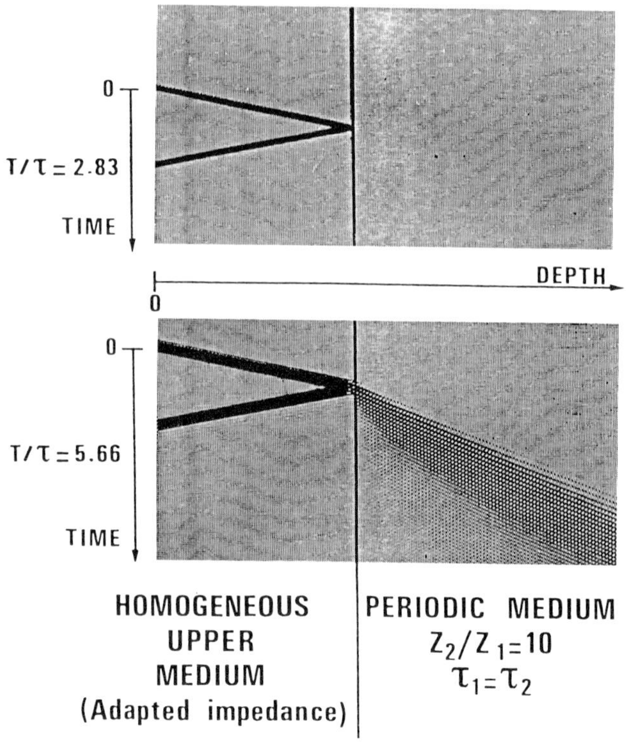

Fig.7. HIGH FREQUENCIES : SUPERREFLECTIVITY

- Intermediate frequencies

In Fig.8, for wavelets of intermediate frequencies, with a ratio T/τ of 8, the signal is dispersed inside the periodic medium and lower frequencies propagate faster than higher frequencies. Then the effective velocity is frequency dependent.

2) Analytical studies

These studies on scattering in periodic structures are well known in Solid State Physics as the "Theory of Bloch Waves".

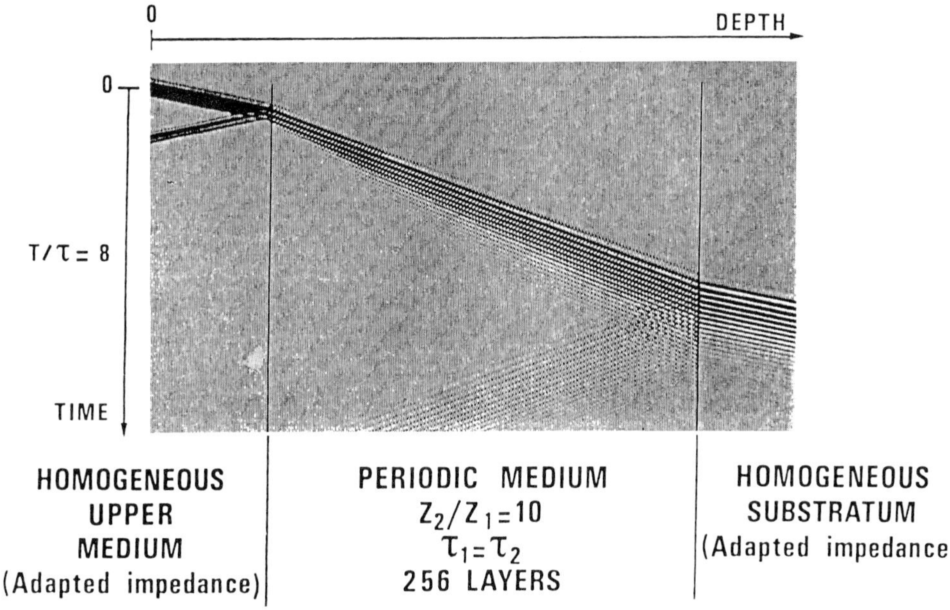

Fig.8. INTERMEDIATE FREQUENCIES : VELOCITY DISPERSION

Developed in the frequency domain by numerous authors for various applications, and leading to the introduction of the notion of "forbidden bands" for wave transmission, they agree perfectly with the Fourier transforms of the results obtained using the Goupillaud-Kunetz algorithm in the time domain.

The above results, for periodic media, may be extented to quasicyclic series, and more generally to 3D propagation in layered and heterogeneous media (cf. GEOPHYSICS, Feb. 1982).

Furthermore, as the signal varies slowly when propagating through multilayered media, we may call it a Quasi-signal, referring to the definitions of Quasi-pulses or Quasi-particles used in Quantum Mechanics.

And, as the structure of the diagram of the Goupillaud-Hunetz algorithm is similar to the structure of Feynmann diagrams, it may be possible to use them to model the wave propagation in layered media, introducing therefore some interaction of the wave with the elementary layers of the series.

From an other point of view, we may consider that, as the signal is progressively dispersed when propagating in layered media, the signal is not absorbed, but its "entropy" increases.

And, as the frequency dependence of the velocity is very complex in the layering related to real sedimentary series, this frequency dependence appears as a "velocity scattering" around a mean velocity.

Neglecting this velocity scattering in standard multichannel processing methods results in an attenuation of high frequencies on the final seismic section.

III - Wave Equation and Seismic Processing

In seismic processing, we use to build the wave fronts by constructive interferences of elementary wavelets related to a large number of elementary paths in the space-time domain.

In such multichannel methods (for example in Volocity Analysis, in Common Depth Point Stack and more explicitely in Wave Equation Migration), we simulate therefore the interferometric methods of Physical Optics.

These methods of image reconstruction are very powerful for signal to noise ratio enhancement, form study of the local space-time coherence of the reflected information.

Except for their large bandwidth, they are similar to holography.

Consequently, for the same reasons as in Optics, to secure the physical meaning of amplitudes and related energies obtained by such interferences, we must consider the amplitudes as complex functions.

The above considerations lead us to introduce complex amplitudes in the Wave Equation, defining therefore a complex wave function F.

Consequently, the velocity V becomes a complex value.

An the new Wave Equation is similar to a Schrödinger Equation (Fig.9).

$$\frac{\partial^2 F}{\partial z^2} = \frac{1}{V^2} \frac{\partial^2 F}{\partial t^2} \quad \text{(with F complex)}$$

Fig.9. WAVE EQUATION IN HETEROGENEOUS MEDIA

Furthermore, when simulating the integration of the Wave Equation by interferences in multichannel processing methods, we need to use complex amplitudes rather than real amplitudes.

From an other point of view, coming back to the meaning of the complex amplitudes, we may consider that :

- In single channel data, instantaneous amplitudes may be directly related to instantaneous energy flow of seismic waves.
- On the other hand, in final representations of multichannel processed data (i.e. in reconstructed images), instantaneous amplitudes may be directly related to the local coherence of the instantaneous information.

Furthermore these amplitudes are frequency dependent, as is the local space-time coherence of the layering in the seismic image.

And, as in Optics, when reconstructing a 3D image of the structures and of the layering using the Wave Equation Migration method, the resolution is limited by diffraction patterns.

IV - Seismic Method and Information Theory

Some of the seismic processing methods refer to standard methods developed in Communication Theory.

Practically, such methods apply quite well to Seismic Exploration.

But their first aim is to recover the signal transmitted by the source, then to suppress the noise and distorsions induced by the propagation medium (Fig.10).

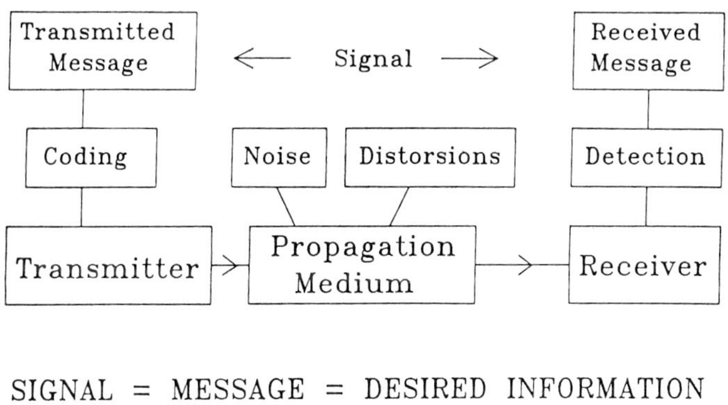

Fig.10. FLOW CHART FOR SIGNAL HANDLING (from Communication Theory)

Such models apply very well to standard SONAR and RADAR methods, when a single reflection is desired.

On the other hand, in high resolution seismic, as in "Non Destructive Testing" methods, the required information stays in the propagation medium.

Consequently, our first aim is to extract from the recorded data the information concerning local heterogeneities and physical characteristics of the propagation medium (Fig.11).

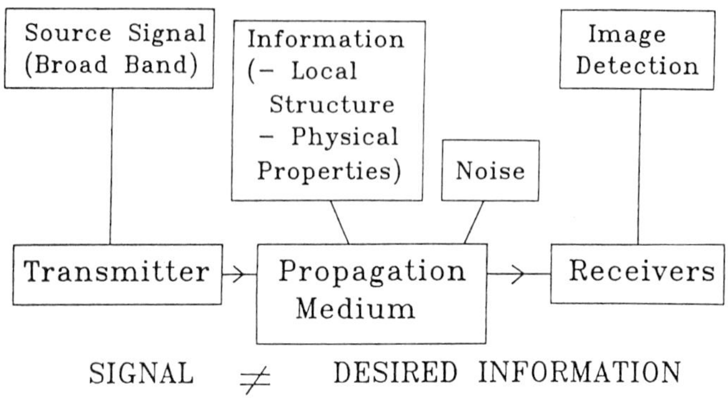

Fig.11. FLOW CHART FOR INFORMATION HANDLING (for Non Destructive Testing Methods by Backscattering or Transillumination)

The corresponding models must be appropriate.

More precisely, we may distinguish two types of applications of the seismic methods (Fig.12 & 13).

1/ Exploration seismic

In this case, we are searching for interfaces, and these interfaces produce reflections of the seismic signal.

The "Elementary Information" is then a Dirac pulse, which requires the use of waves having band spectra of infinite broadness.

Unfortunately, that is physically impossible.

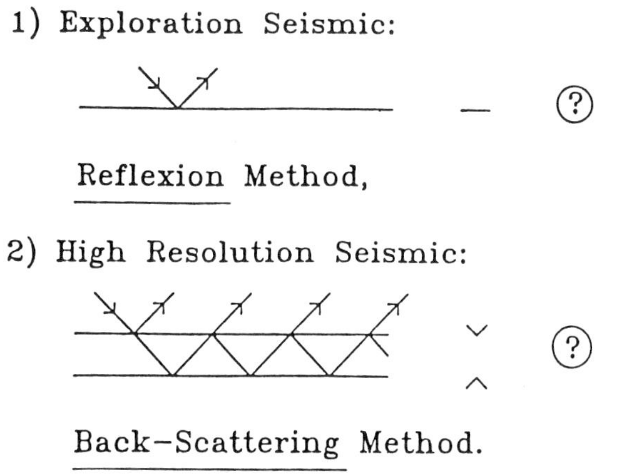

Fig.12. EXPLORATION SEISMIC AND HIGH RESOLUTION SEISMIC : Two types of problems

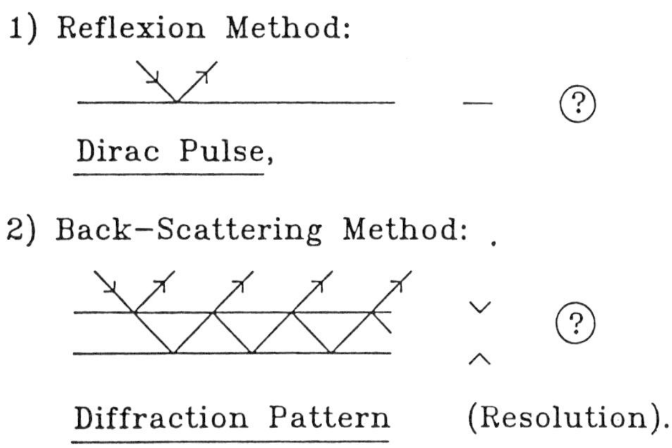

Fig.13. EXPLORATION SEISMIC AND HIGH RESOLUTION SEISMIC : Two types of elementary informations

2/ High resolution seismic

In field development problems, we are searching for layer thicknesses and their variations. To detect layers of various thicknesses, we may try to separate and interpret the corresponding back-scattering phenomena, and the resolution will be then limited by the dimensions of the corresponding diffraction patterns.

The seismic resolution may be defined therefore using diffraction patterns of various frequencies, obtained by interferences in the multichannel processing methods.

And the elementary information is then a diffraction pattern.

Finally, as wave propagation is frequency dependent, we must carry out multichannel processing methods separately and successively in adjacent narrow frequency bands.

Then, after Back-scattering Pattern Analysis, we may synthetise the results.

In some way, this analysis, based on intrabed multiples or on signal/layer coupling, would be similar to the standard Spectral Analysis. It would permit to attain, at least roughly, the physical properties of the corresponding layers.

For economic considerations, the above analysis must be restricted to the producing reservoirs and the embedding layers.

Consequently, to represent the seismic data, and their interferences in such processing methods, "Instantaneous Frequency Spectra" appear to be a powerful tool.

A 2 dimensional Time-Frequency sampling method is then appropriate.

V - Time-Frequency Expansions

To represent pseudoperiodic signals, D. Gabor first proposed in 1945 a 2D Time-Frequency decomposition.

This expansion is based on :

1/ A regular sampling grid in the time-frequency domain, as shown in Fig.14.

2/ A set of basic complex wavelets of various frequencies.

Each elementary wavelet is defined as the product of a monochromatic complex sine wave by a Gaussian envelope, as shown in Fig.15 (real part in straight line, imaginary part in dashed line).

In the frequency domain, such wavelets are represented by Gaussian functions.

The dimensions of the wavelets must be fitted to the dimensions of the elementary cells of the sampling grid.

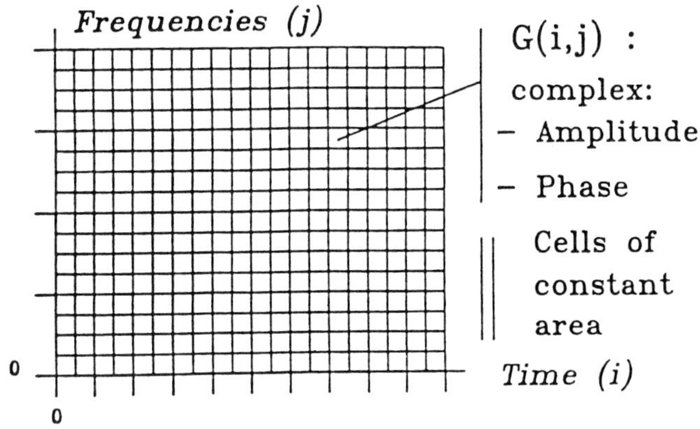

Fig.14. GABOR EXPANSION:Time-Frequency 2D Decomposition
(Sampling grid proposed by D. GABOR)

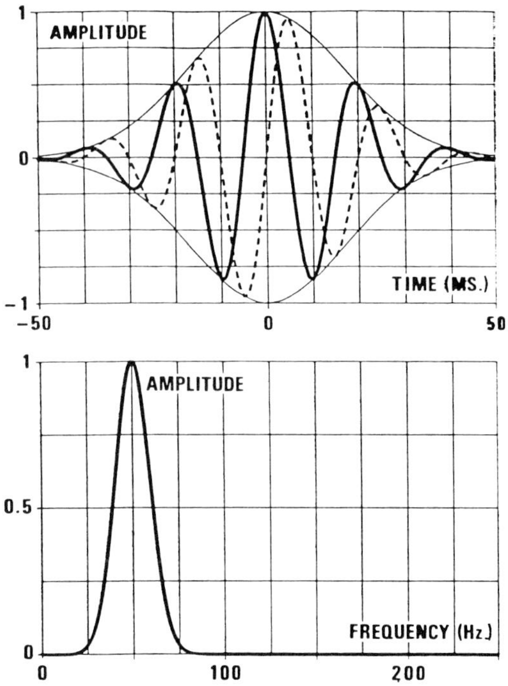

Fig.15. GABOR'S COMPLEX WAVELET

Furthermore, the principle of the Gabor Expansion is to decompose the signal on the set of basic wavelets above defined.

The result of this decomposition, in any cell of the sampling grid, is a complex number G (i,j) and the complex function of two variables G (i,j) represents the Gabor Expansion.

But this expansion does not permit a good local definition of the phase, especially for high frequencies.

VI - Cycle-Octave Expansion

1/ General considerations

We developed therefore a new type of expansion, and we call it the "Cycle-Octave Expansion", as later explained.

In this expansion, we modified the characteristics of the basic wavelets, as summarized in Fig.16, defining the basic wavelets accordingly to the formula of Fig.17.

- Constant shape ratio,
- Suitable for phase preservation,
- Close to the elementary information related to the diffraction pattern of a layer, obtained by interferences on backscattered data.

Fig.16. CYCLE-OCTAVE EXPANSION Complex Basic Wavelets Characteristics

$$g(T_j, t) = e^{-(t/T_j)^2 . \mathrm{Ln}2} . e^{2\pi i t/T_j}$$

where:

T_j = Mean Period of the Wavelet.

Fig.17. CYCLE-OCTAVE EXPANSION: General Formula for the Complex Basic Wavelets (Constant shape ratio)

In the time domain, the basic wavelets deduce from each other by simple dilation of the time scale : they have a constant shape.

Then for any discrete frequency of the sampling grid, we use a basic wavelet close to a diffraction pattern.

Fig.18 gives an example of such basic wavelet.

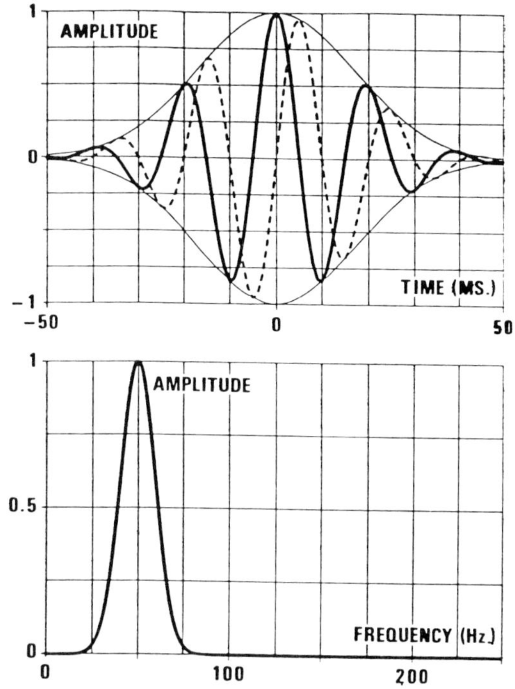

Fig.18. CYCLE-OCTAVE EXPANSION;Elementary Basic Wavelet

Consequently, in the frequency domain, the ratio :

wavelet bandwidth/mean frequency

is constant.

It follows that, if we use a logarithmic representation (i.e. octaves) in the frequency domain, the basic wavelets deduce from each other by a translation (Fig.19).

It seems worthy to notice that, in such a logarithmic representation of the frequency domain, the difficulties encountered with the physically meaningless

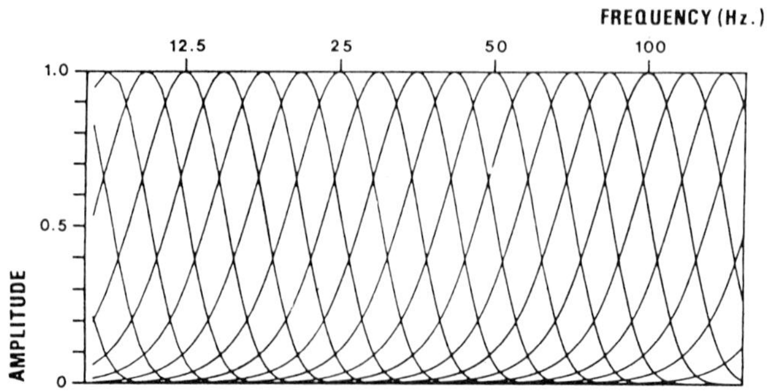

Fig.19. CYCLE-OCTAVE EXPANSION: Set of Complex Basic Wavelets (Frequency Domain : Octaves)

amplitudes associated to the zero frequency disappear.

But changing the characteristics of the basic wavelets involves a change in the distribution of the elementary cells in the sampling grid, as it appears in Fig.20.

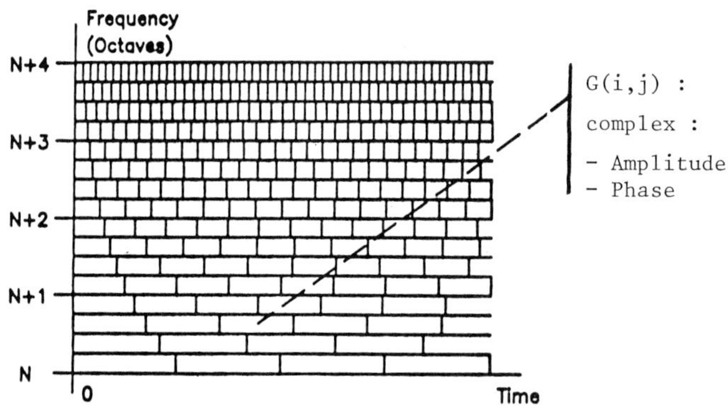

Fig.20. CYCLE-OCTAVE EXPANSION : Sampling grid characteristics (Time-Octave Coordinates)

Indeed, we need to keep the dimensions of the cells fitted to those of the basic wavelets.

In conformity with the "Uncertainty Principle", the cells of the grid have a constant area, when measured in the time-frequency domain.

But in the time-octave representation shown in Fig.20, it appears clearly that this condition involves a frequency dependent sampling rate in the time domain, with higher definition (i.e. higher information density) for higher frequencies.

2/ Computing Method

To come back to a regular sampling grid, we may replace the time scale by a scale graduated in cycles or periods, or better to say, in unwrapped phase (Fig.21).

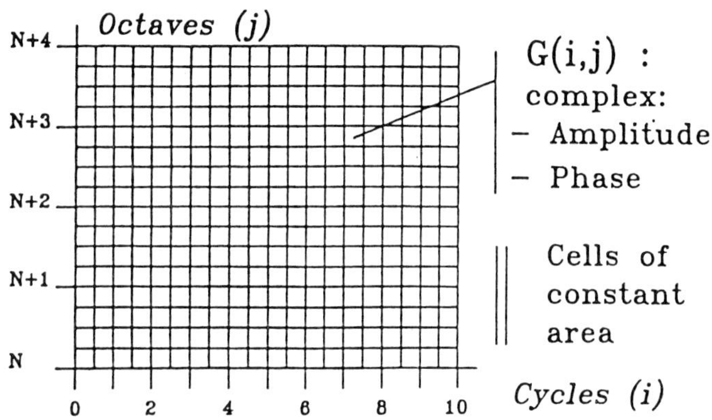

Fig.21. CYCLE OCTAVE EXPANSION (Cell dimensions fitted to wavelet dimensions)

Fig.22 gives the formula used to compute the complex value G (i,j) for each cell in the sampling grid, by complex crosscorrelation of the signal by the complex conjugate of the basic wavelet related to the cell (in other words this is an inner product).

The sampled values G (i,j) represent the Cycle-Octave Expansion of the signal s(t), in other words of the seismic trace.

Fig.23 gives the method to recover the signal from its Expansion by inverse transform.

– For each cell C(i,j) of the grid:

$$\begin{cases} T_j = T_0 . 2^{-j/4} \\ t_i = i.T_j/2 \end{cases}$$

where $1/T_0 =$ lower frequency

– Compute the crosscorrelations of the trace by the complex basic wavelets:

$$G(i,j) = \frac{2 \int^{t} s(t).g^{*}(T_j, t-t_i).dt}{\int^{t} g(T_j, t).g^{*}(T_j, t).dt}$$

Fig.22. CYCLE OCTAVE TRANSFORM : Formula for the direct transform

– 1) Compute G(t,j) from G(i,j) by Lagrange Interpolation in the time domain using a fine sampling rate (Amplitude and Phase)

– 2) Use the simplified formula

$$s(t) = K. \sum_j G(t,j)$$

where K = 0.25...

Fig.23. CYCLE OCTAVE INVERSE TRANSFORM (Simplified computing method)

All these operations have a mathematical support involving well lnown considerations of functional analysis and using representations of analytic functions in Hilbert and Hardy spaces.

New developments related to the Cycle-Octave Expansion were developed during the last few months by A. Grossmann in the Centre de Physique Théorique at the C.N.R.S. in Marseille Luminy.

Finally, we ask the following question :
What is the best choice for the cell area in the sampling grid ?

Two different ways are possible :
1/ From theoretical studies, made by different authors, to obtain an orthonormal decomposition, the cell area in the Time-Frequency domain must be unity (Fig.24).

- Using as sampling rates:
  - $\Delta t'$ in the time domain,
  - $\Delta f'$ in the frequency domain,
- We must have:

$$\Delta t' \,.\, \Delta f' = 1$$

(Cf. Bargmann, Janssen, Balian,..)

Fig.24. CONDITION REQUIRED FOR ORTHONORMAL DECOMPOSITION OF A TIME-FREQUENCY 2D TRANSFORM

2/ If we will attain the "Ultimate resolution", we must fit the cell dimensions to the wavelet dimensions, and the granularity of the grid is obtained therefore by juxtaposition of diffracation patterns.

Using standard deviations to define the dimensions of the basic wavelets, in both time and frequency domains, as usual in Quantum Mechanics, the cell area must be $1/4\pi$, involving therefore an oversampling ratio higher than 12 (Fig.25).

Experimentally, the second way and the fine sampling grid appear to be better for

- Using as standard deviations:
  - Δt in the time domain,
  - Δf in the frequency domain,
- We have:

$$\boxed{\Delta t \,.\, \Delta f = \frac{1}{4\pi}}$$ (Uncertainty Principle)

- Thus high resolution involves Oversampling (ratio $= 4\pi$)

Fig.25. CONDITION REQUIRED TO FIT THE CELL DIMENSIONS TO THE WAVELET DIMENSIONS

practical applications.

Practically, we may compare the characteristics of the two different sampling grids as follows :

The sparse sampling grid (Fig.26), with practical sampling rates of 1 point/octave and 1 point/1.5 cycle is fitted to an orthonormal expansion. But it assumes an infinite bandwidth.

When used in bandlimited and noisy environment, it induces end-effects and unstability.

- Sparse sampling density | 1 point/octave
- Fitted to orthonormality | 1 point/1.5 cycle condition.
- Much more economic (ratio # 1/12)
- But infinite frequency bandwidth is assumed.
- Unsuitable, due to end-effects and unstability in noise environment.

Fig.26. CYCLE-OCTAVE EXPANSION : Sparse sampling grid characteristics

On the other hand, the fine sampling grid (Fig.27), with practical sampling rates of 4 points/octave and 2 points/cycle is more suitable and robust in noise environment.

- Fine sampling density — 4 points/octave
- Fitted to maxima — 2 points/cycle
  possible Resolution
  (from Uncertainty Principle)
- Suitable for phase preservation.
- In spite of theoretical oversampling well adapted to real data.
- Robust in noise environment.

Fig.27. CYCLE OCTAVE EXPANSION : Fine sampling grid characteristics

It corresponds to a juxtaposition of diffraction patterns.

Furthermore the oversampling or overcompletness of the decomposition related to this fine sampling grid allows a correct interpolation between the sampled values in the grid.

Notice that we encounter no difficulties when unwrapping the phases, which is needed for phase preservation in the reconstruction problem.

An example of computation of direct then inverse transform is presented in Fig.28. The input signal was the impulse response of a digital recorder.

The relative accuracy obtained after a two way transform of such signal is 1/1000, leading to a perfect optical superposition of input and recovered signals.

The Fig.29 summarizes the specifications of the parameters used to obtain this accuracy :

Bandwidth : 5 octaves

Sampling rates of the grid : 2 points/cycle, 4 points/octave.

And to compute the crosscorrelations of the signal by the successive basic wavelets,

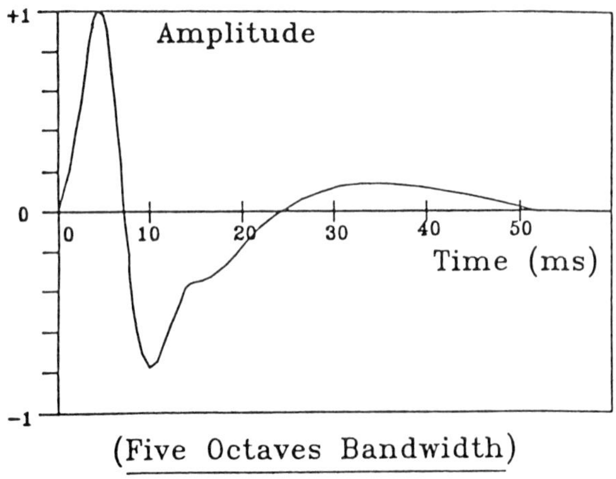

Fig.28. TEST SIGNAL FOR THE CYCLE OCTAVE EXPANSION
(Input response of an actual seismic recorder)

Relative accuracy: 10^{-3}

for:

- 1) Computing bandwidth:

Input signal	Bandwidth
Basic Wavelet	3 Octaves
Recorder Response	5 Octaves

- 2) Sampling rate (Time)
  - Real Data : 8 points/Cycle
  - Complex Data: 2 points/Cycle

Fig.29. PRACTICAL ACCURACY OF THE CYCLE OCTAVE EXPANSION
(after a 2 way transform of the signals)

we used 8 points per cycle.

The last question is the following :
How many bits do we need to record the complex value associated to any cell in the sampling grid ?

It appears that, to preserve the information on the instantaneous phase and to be able to unwrap this phase, including the neighborhood of singular points, 4 bits (i.e. 1/16 of cycle) are enough for the phase.

On the other hand, for the amplitude, 12 bits would certainly be sufficient if we refer to the common use in standard seismic recording. But, if we intend to work with a large number of channels, it is probably unnecessary to record the amplitude (cf. Sign Bit Recording methods). In this last option, only 4 bits per cell are needed.

CONCLUSION

The Cycle-Octave Expansion defined in this paper will be a powerful tool to quantify and increase the resolution of the Seismic Method.

It permits to develop new processing methods using parallel interferometric processing in adjacent narrow frequency bands.

Such processing methods are made to recover the backscattered information by Backscattering Pattern Analysis.

But, as clearly stated by N. Anstey and K. Larner in two recent papers, to process broad band spectra on seismic data, we need to record valuable data, especially for high frequency information.

Actual technology, using either digital or hybrid circuits is available to develop 2D Time-Frequency Sampling Methods for real time field recording, with separate gain for any frequency band using the Cycle-Octave Expansion.

ACKNOWLEDGEMENTS

I wish to thank the ELF AQUITAINE Group for permission to publish this material, and also MM. B. DELAPALME, J. BERTHELOT, P. ALBA and my colleagues in O.R.I.C. and S.N.E.A.(P) for their effectual support.

I would especially express my gratitude to R. BALIAN of the Service de Physique Théorique at the C.E.A. (Saclay) and A. GROSSMANN of the Centre de Physique Théorique at the C.N.R.S. (Marseille-Luminy) for their help concerning the developments related to Theoretical Physics.

REFERENCES

- Anstey N.A., 1980, Resolution, bandwidth and money : 50th Annual International S.E.G. Meeting (Houston, Tx, Nov. 1980).

- Bacry H., Grossmann A., Zak J., 1975, Proof of completness of lattice states in the kq-representation : Physical Review, B12, Series 3, P. 1118 - 1120.

- Balian R., 1981, Un principe d'incertitude fort en théorie du signal ou en mécanique quantique : Compte rendu à l'Académie des Sciences, Paris.

- Bargmann V., Butera P., Girardello L., and Klauder J.R., 1971, On the completness of the coherent states : Reports on Mathematical Physics, vol.2, Nov. 6, P. 221 - 228.

- Bastiaans M.J., 1980, Gabor's Expansion of a signal into Gaussian elementary signals : Proceedings of I.E.E.E., vol. 68, N° 4, p. 538 - 539.

- Gabor D., 1946, Theory of communication : Journal of I.E.E.E., vol. 93, p. 429 - 441.

- Jansen A.J.E.M., 1980, Gabor representation of generalized functions : Journal of Math. Anal. Appl. (in press).

- Larner K. and Dragoset B., 1980, Data enhancement from a 500 channel streamer : 50th Annual International S.E.G. Meeting (Houston, Tx, Nov. 1980).

- Morlet J., 1975, Seismic tomorrow : Interferometry and quantum mechanics : 45th Annual International S.E.G. Meeting (Denver, Co, Oct. 1975).

- Morlet J., Arens G., Fourgeau E. and Giard D., 1980, Signal filtering and velocity dispersion through multilayered media : 50th Annual International S.E.G. Meeting (Houston, Tx, Nov. 1980).

- Morlet J., Arens G., Fourgeau E. and Giard D., Feb. 1982, Wave propagation and sampling theory, Parts I and II, Geophysics, vol. 47, p. 203 - 236.

- Morlet J., Sampling theory and wave propagation, 51th Annual International S.E.G. Meeting (Los Angeles, Ca, Oct. 1981).

- Morlet J., Sampling theory and wave propagation, 44th E.A.E.G. Meeting (Cannes, France, June 1982).

- Shannon C., 1949, The mathematical theory of communication : University of Illinois Press.

IMPLEMENTATION OF IMAGE ENHANCEMENT AND LEAST SQUARES FILTERING TECHNIQUES FOR ACOUSTIC DATA

Dr. T. S. Durrani
Department of Electronic and Electrical Engineering
University of Strathclyde
Glasgow G1 Scotland UK

ABSTRACT:

Two complementary problems in image processing are addressed in this paper, these are concerned with the restoration of blurred and noise corrupted images; and the enhancement of resolution when images are collected over limited apertures, as in many acoustic applications.

Kalman filtering techniques are discussed and shown to facilitate desmearing of space variant and invariant blurs, as well as to suppress noise. 1-D smearing effects are tackled within a Gauss-Bernoulli frame work for data observed over a circulant aperture.

Several two dimensional techniques are included for the enhancement of spectral resolution. These are based on extensions of 2-D MEM spectral estimation schemes. Further, a set of ARMA type models are considered for aperture extension when associated sensor arrays have limited coverage.

1. INTRODUCTION

An image may be considered as the representation of a two dimensional field which is sampled by an array of sensors at a fixed set of locations. If the wave-field originates from optical radiation, the image represents a conventional visual picture, and each pixel intensity may be considered as the output from an optical sensor located at that point. If the sensors comprise an array of antennas sampling electromagnetic radiation, the image is essentially a snapshot of the incident wave-field. If the radiation is acoustic, the stored image may be either a hologram, when mixed with a reference field, or for passive sensors this could be a plot of (field) intensity from a reflecting surface; at seismic frequencies this could be a seismogram from a reflecting geologic interface; at sonic frequencies - a snapshot, as in a time-bearing plot if beam forming techniques are employed.

This paper addresses two complementary problems that arise when data is available, in image form, from a set of sensors arranged in a 1-D or 2-D configuration. The first is concerned with the problem of image restoration and enhancement when the data represents a blurred or smeared version of the original image corrupted by noise; and the second is concerned with the requirement for improving spatial frequency resolution when the data is sampled over a finite aperture.

The first problem originates from a variety of reasons - lack of focussing or collimating facility; image (sensor) or object motion while data is being collected;

NATO ASI Series, Vol. F1
Issues in Acoustic Signal/Image Processing and Recognition
Edited by C. H. Chen

transmission effects - such as refractive index perturbations for optical fields; diffractive, refractive and multipath effects in acoustics, or in seismology, passive sonar and ultrasonics; and due to the finite spatial resolution offered by an array of sensors. Attendant noise may arise from the environment such as isotropic acoustic noise, sensor noise, i.e. noise in the associated hardware, and in transmission and digitisation of data.

Digital techniques for improving the quality of an image are by now well established, and methods such as fixed filter restoration or Wiener inverse filters for removing image blur are well known [1,2]. This paper presents an overview of techniques for restoring images which are affected by non-uniform (spatially varying) blurs, due to tilt or motion. These effects cannot be accommodated within the frame work of Wiener filtering. Modified Kalman filtering methods are presented which exploit sampling symmetries to minimise computing effort, and for the case of images obtained from linear arrays, it is shown that the data may be considered as affected by a one dimensional smearing process, and the image desmeared using Gauss-Bernoulli models [3].

The finite aperture limitations are caused by the size of the array of sensors; if the data is collected by the spatial movement of a single sensor by mechanical means, then the finite time restriction for data collection, and associated finite mechanical traverse, leads to a limited aperture size, as in acoustic holography. Here the requirement is for extrapolating the available support in order to achieve improved resolution in a transformed domain other than the domain of measurement, such as the frequency domain; to separate out closely occurring peaks or point sources, as in acoustic imaging.

Two approaches are considered for improving the available resolution. The first relies on extending the finite support by using 2-D Maximum Entropy (MEM) techniques, and the second employs an ARMA model for the available data to extrapolate beyond the given aperture [5].

2. IMAGE DESMEARING USING A KALMAN FILTER APPROACH

The discretised model for an image which has been smeared and corrupted by additive noise may be expressed as

$$y_{k,s} = \sum_{u,v}\sum c(k,u;s,v)x_{u,v} + \eta_{k,s} \tag{1}$$

where $\{c(k,u;s,v)\}$ is the spatial convolution representing a known blur. For spatially invariant blur this reduces to $\{c(k-u;s-v)\}$ (point spread function). $\{\eta_{k,s}\}$ is an additive white noise process, and $\{x_{u,v}\}$ is the intensity of the 'object' which is to be extracted from the observations $\{y_{k,s},\ k,s = 0,1,2 \ldots N\}$.

The conventional frequency domain approach is valid for space invariant point spread functions only, and leads to a two dimensional Wiener estimator (or inverse filter) which yields the optimum frequency spectrum $\{X(\Omega_1,\Omega_2)\}$ of the desired object as

$$\hat{X}(\Omega_1,\Omega_2) = \left[|C(\Omega_1,\Omega_2)|^2 + \sigma^2\right]^{-1} C^*(\Omega_1,\Omega_2) \left[Y(\Omega_1,\Omega_2)\right] \tag{2}$$

where

$C(\Omega_1,\Omega_2)$ = 2-D Fourier transform of $\{c(k,s)\}$

$Y(\Omega_1,\Omega_2)$ = 2-D Fourier transform of the observed image $\{y_{k,s}\}$

σ^2 = variance of additive noise

$*$ = complex conjugate

The filter is obviously invalid for the space-varying smearing effect, as Eq (2) represents a global operation on the complete image, and $\{C(\Omega_1,\Omega_2)\}$ cannot be defined if the smearing varies from region to region within the image.

An alternative is to consider a space domain approach. By re-ordering the pixel addresses, Eq (1) may be recast as

$$\text{vec } \underline{Y} = C \text{ vec } \underline{X} + \underline{\eta} \tag{3}$$

where $y_{k,s}$ is equal to the k+(N-1)s element in vec $\underline{Y}$, and $\underline{Y}$ is an N^2 vector for an (NxN) pixel image. For an (NxN) object, vec $\underline{X}$ is (N^2x1), and C is now a banded matrix of size (N^2xN^2), which includes the smearing effects over the complete domain of the object.

A linear least squares estimator may be set up to extract vec $\underline{X}$ from observation vec $\underline{Y}$. Thus

$$\text{vec } \hat{\underline{X}} = D \text{ vec } \underline{Y}$$

where the least squares filter D may be obtained by minimising the mean square error criterion

$$J = \text{Min } || \underline{X} - \hat{\underline{X}} ||^2 \tag{4}$$

This leads to the solution

$$D = R_x C^T \left[C R_x C^T + \sigma^2 I\right]^{-1} \tag{5}$$

where R_x = covariance matrix for the object E $\left[X X^T\right]$, I = identity matrix. In the absence of apriori R_x, which is the usual case, this reduces to

$$D = C^T \left[C C^T + \sigma^2 I\right]^{-1} \tag{6}$$

the well known least squares solution.

The method has the advantage of accommodating all types of smearing effects, as well as allowing desmearing of a complete image as a once-off. However, it suffers from a major computational disadvantage in that the solutions in Eq (5) or (6) require the inversion of an (N^2xN^2) matrix plus excessively large data storage, e.g. a (10x10) pixel image involves manipulation of matrices of order (100x100). Thus in practice while Eq (6) yields an optimal least squares desmearing filter, it is rarely used due to its computational complexity.

An alternative approach is to consider a sequential estimator which operates on small regions of the image [4], and performs deconvolution (desmearing) before moving on to adjoining regions. Such an estimator offers substantial reduction in computing effort, and as it operates over localised regions, it can be easily adapted or changed to take into account spatially varying blurs.

Fig 1 represents the case of a point spread function (PSF) covering a 5x5 pixel square $\{c_i, i = 1,2 \ldots 25\}$. Converting the image into a vector, an impulse at location n in the object plane leads to a banded vector, as shown in Fig 1, where the serrated image represents the point spread function for an NxN pixel image.

In general for a PSF defined over a (2m+1)x(2m+1) grid the state space representation for the object may be expressed as

$$\begin{bmatrix} x_1 \\ \cdot \\ \cdot \\ \cdot \\ \cdot \\ \cdot \\ \hline x_{2m(N+1)} \end{bmatrix}_{k+1} = \left[\begin{array}{c|c} \underline{0} & I \\ \hline 0 & \underline{0}^T \end{array}\right] \begin{bmatrix} x_1 \\ x_2 \\ \\ \hline x_{2m(N+1)} \end{bmatrix}_k + \begin{bmatrix} \underline{0} \\ \hline 1 \end{bmatrix} u_{k+2m(N+1)} \tag{7}$$

or

$$\underline{X}_{k+1} = A\,\underline{X}_k + B\,u_{k+2m(N+1)} \tag{8}$$

The image arises via a convolution of the object, as

$$y_k = [\overbrace{c_1 \ldots c_5 00 \ldots 0}^{N} \quad c_6 \ldots c_{10} 0 \ldots 0 \quad \ldots \ldots \quad c_{20} \ldots c_{25}] \begin{bmatrix} x_1 \\ x_2 \\ \cdot \\ \cdot \\ \cdot \\ x_{2m(N+1)} \end{bmatrix}_k + \eta_k$$

$$\text{or} \quad y_k = \underline{C}^T \underline{X}_k + \eta_k \tag{9}$$

where $\{y_k\}$ and $\{u_k\}$ are elements of 1-dimensional sequences comprising the image $\{\text{vec } \underline{Y}\}$ are object $\{\text{vec } \underline{X}\}$. $\underline{X}$ is a vector which stores $2m(N+1)$ values of the object matrix, and represents only those values of the object which contribute to the image point y_k, due to the smearing process. Note the convolution, which is manifest in the smearing process, is contained in the vector $\underline{C}$, while η_k is the additive noise effecting the image pixel at location k.

Eqs (8) and (9) form the basic model which can be applied to a Kalman filter for estimating the state vector $\underline{X}_k$ from observation $\{y_k\}$. Note that the vector $\underline{X}_k$ is a dummy vector which stores the elements of the object field prior to the convolution in the measurement equation (9). With the mean image level removed, the image may be desmeared (restored) by passing it through the Kalman equation,

$$\hat{\underline{X}}_{k+1} = A\,\hat{\underline{X}}_k + \underline{G}_{k+1}\,[y_{k+1} - \underline{C}^T \underline{A}\,\hat{\underline{X}}_k] \tag{10}$$

$$k = 0,1,2 \ldots$$

where

$$\underline{G}_{k+1} = \text{Kalman gain}$$

$$= P_{k+1}\, \underline{C}^T \left[\underline{C}^T P_k\, \underline{C} + \sigma^2\right]^{-1} \quad (11)$$

with

$$P_{k+1} = A\, Q_k\, A^T + \underline{B}\, \nu(k)\, \underline{B}^T \quad (12)$$

$$Q_{k+1} = \left[I - \underline{G}_{k+1}\, \underline{C}^T\right] P_{k+1} \quad (13)$$

where $\nu(k) = E\left[u^2_{k+2m(n+1)}\right]$ (= power of object field); $\sigma^2_n = E\left[\eta^2_k\right]$, E denotes expectation, Q_k and P_k are the error covariance and the projected error covariance matrix respectively. The algorithm may be initialised by setting P_o as a large diagonal matrix, and $\underline{X}_o$ as a null vector.

Fig 2 illustrates the application of the Kalman filter for image restoration. Note that the object model and smearing effect are included within the filter. The Kalman gain vector $\underline{G}$ is updated with each new pixel of observation.

The following points are worth noting. The computational burden of the algorithm is largely due to matrix multiplication, as the model in (8) ensures only a scalar inversion (division) in (11). Further, the exact solution of the Riccati equations (11-13) involve matrix operation on very long vectors. However the computing effort can be further reduced by considering the fact that direct information concerning the point to be deconvolved is only available where non zero elements of the smearing vector $\underline{C}$ exist. The Kalman gain corresponding to all zero elements of $\underline{C}$ can be constrained to be zero. After some manipulation the Riccati equations can be simplified such that the matrix size is only (2m+1)x(2m+1) leading to considerable computational saving. Note that for the (5x5) smearing example. a Kalman filter with 25 element gain vector will effect the deconvolution. Also, deconvolution of spatially varying blurs can be easily incorporated in the Kalman filter framework by setting the smearing vector $\underline{C}$ as a space-varying vector $\underline{C}_k$ which varies from location to location. Note it is this facility of the Kalman filter, which traditionally can be applied to time varying estimation problems, that can be exploited to achieve image restoration of spatially varying blur effects.

In practice, for a spatially invariant PSF and constant noise and object variances, the Kalman gains rapidly converge to the steady state values, and the Kalman solution reduces to the Wiener solution.

The Kalman filter will provide minimum mean square error (MMSE) estimates of the object field provided the following conditions hold:

i) η_k is an uncorrelated random field,

ii) the object is uncorrelated,

iii) the smearing effect as portrayed by $\underline{C}$ is known.

The additive noise condition is generally acceptable, as noise arising from the sources indicated in Section 1, such as sensor noise or environmental noise is usually broadband and can be regarded as white. In practice condition (ii) is rarely satisfied.

An extensive study of the correlation of incoherent optical sources, has been made [1], and an acceptable model for the correlations is a simple two dimensional first order system

$$R_{k+1,j+1} = a_1 R_{k,j+1} + a_2 R_{k+1,j} - a_1 a_2 R_{k,j} \tag{14}$$

where $R_{k,j}$ is the correlation value for spatial displacements of $(k\Delta s, j\Delta s)$, and Δs is the sampling interval. For a causal model for the data, this correlation may be incorporated within the system model of Eq (8) by modifying it to

$$\begin{bmatrix} x_1 \\ \cdot \\ \cdot \\ \cdot \\ \cdot \\ \hline x_{2m(N+1)} \end{bmatrix}_{k+1} = \left[\begin{array}{c|c} \underline{0} & I \\ \hline a_1 \cdot a_2 & a_2 \; 0 \; 0 \; \dots \; 0 \; a_1 \end{array}\right] \begin{bmatrix} x_1 \\ x_2 \\ \cdot \\ \cdot \\ \cdot \\ \hline x_{2m(N+1)} \end{bmatrix}_k + \begin{bmatrix} \underline{0} \\ \hline 1 \end{bmatrix} u_{k+2m(N+1)} \tag{15}$$

$$y_k = \underline{C}^T \underline{X}_k + \eta_k$$

a_1, a_2 are usually chosen in the range $0.7 \leqslant |a_{1,2}| \leqslant 0.9$. The filter is relatively insensitive to the choice of a_1, a_2, suffice it to say that it allows deconvolution, though, in theory, the implementation is sub-optimal.

The evaluation of the point spread function may be achieved either by the use of calibration targets, or known incident fields. In cases where calibration is not possible, either point locations, or smeared out image edges may be used to evaluate $\underline{C}$ by observing points in close vicinity of either. An alternative approach involves modelling $\underline{C}$ in the frequency domain by observing the zeros of the frequency response, or using extended Kalman filtering. For further details see [6,15].

3. TECHNIQUES FOR ONE-DIMENSIONAL DESMEARING

The previous section was concerned with techniques for restoring images smeared by arbitrary point spread functions. In certain applications the smearing effect is in one dimension only. For instance when incident radiation is observed via a linear array of sensors the received data may be considered as a smeared version of the desired image. The smearing effect being a manifestation of the array response. Thus if the array look direction is θ, the output at any time instant may be expressed as

$$y_t(\theta) = \int A(\theta;\phi)\, x_t(\phi)\, d\phi + \eta_t(\theta) \qquad \theta \in (-\pi,\pi) \tag{16}$$

where

$x_t(\phi)$ = Incident radiation appearing on the array from direction ϕ at time t

$A(\theta;\phi)$ = Array response in direction θ when incident radiation at bearing ϕ.

The discretised version of Eq (16) is given in Eq (17),

$$y_t(\theta) = \sum_{\phi} \tilde{A}(\theta;\phi)\, x_t(\phi) + \eta_t(\theta) \tag{17}$$

which may be written in matrix notation for equispaced directional steering as

$$\underline{y}_t = A\, \underline{x}_t + \underline{\eta}_t \tag{18}$$

Note that elements in the matrix A are components of the array response $A(\theta;\phi)$. Further, for most arrays $\tilde{A}(\theta;\phi) = \tilde{A}(\theta-\phi)$, i.e. a function of difference between the look direction and source direction.

Successive records of vector $\underline{y}_t$ for different time instants lead to a two dimensional record (image), which is smeared in the θ dimension by the array response. To enhance such images, it is important to restore the effect of the array smearing, i.e. it is necessary to estimate $\underline{x}_t$ from observation vector $\underline{y}_t$.

The least squares estimate for $\underline{x}_t$ is

$$\hat{\underline{x}}_t = (A^T A)^{-1} A^T \underline{y}_t \tag{19}$$

Experimental studies on data sets for desmearing representative observations have been unsatisfactory [3]. This is due to the fact that the matrix A is highly ill-conditioned, and the least squares estimate $\underline{x}_t$ is very unstable, particularly when the array response has high sidelobes.

A technique for desmearing such images when the source locations are unknown has been developed in [3] as an extension of the 'most likely replacement' method of Mendel and Kormylo [7]. Here an iterative scheme is proposed based on minimising the residual sum of squares for each source location until all sources have been identified. This is achieved when the associated likelihood measure does not increase with the addition or subtraction of a source at any single location. It is a computationally expensive algorithm, and work continues towards achieving a speedier search technique. Extension of the algorithm to two dimensional smearing would allow tracking of source movements.

4. IMPROVEMENT IN IMAGE RESOLUTION USING 2-D MEM TECHNIQUES

This section is concerned with a discussion of techniques for improving the frequency resolution offered by an image defined over a finite support. Conventionally, frequency resolution is inversely proportional to the size of the image in each dimension, thus to increase the resolution by a factor of N in each dimension, the original support must increase by a factor of $\sim N^2$. If conventional DFT techniques are employed a very large support is required to yield a fine frequency resolution.

Maximum entropy (MEM) techniques in one dimensional spectral analysis have been known to yield better frequency resolution than conventional DFT techniques, for identical data lengths. Thus recent efforts in improving spatial frequency resolution for image data are directed towards exploiting 2-D MEM techniques to increase frequency resolution without the need for a large original support. A number of algorithms have been recently proposed to achieve this:

- Data extrapolation by linear prediction [8]
- 2D Linear prediction - Pseudo-Inverse method [9]
- 2D Quarter plane causal autoregressive method [10]
- 2D MEM Spectral estimator [11].

A comparative assessment of these algorithms is given in [12]. Frost [8] was the first to propose a method for extrapolating the observed data beyond the observation window which employs 2-D least squares prediction filters. The original 2-D support (observation window) is extended by passing it through the prediction filters. The 2-D spectrum of the extended support yields higher resolution than the conventional 2-D DFT.

For an original image (data support) defined over a rectangular grid of $N_1 x N_2$, four rectangular $P_1 x P_2$ ($P_1 < N_1$, $P_2 < N_2$) prediction error filters are designed from the original support. These are illustrated in Fig 3, including the samples they predict. The filter coefficients are derived by finding the least squares solution to a $2(N_1-P_1)(N_2-P_2)$ system of equations in $P_1 P_2$ unknowns.

$$X_o \underline{a} = \underline{x}_p \tag{20}$$

where X_o = $(N_1-P_1)(N_2-P_2) x P_1 P_2$ matrix of support observations where the rows of X_o are vector $\underline{X}$ as defined below

$\underline{a}$ = $(P_1 P_2 x1)$ vector of prediction coefficients

$\underline{x}_p$ = $(N_1-P_1)(N_2-P_2)x1$ vector of observations representing the predicted values.

Each row in (20) represents a prediction equation given by

$$\hat{x} = \underline{X}^T \underline{a}$$

with $\underline{a}$ the prediction coefficient vector, and $\underline{X}$ a $(P_1 P_2 x1)$ vector of observations covered by the filter, and $\hat{x}$ is the predicted value (scalar).

The least squares solution to eq (20) is

$$\underline{a} = (X_o^T X_o)^{-1} X_o^T \underline{x}_p \tag{21}$$

The four prediction filters are determined by solving eq (21) for different orientations of the support. Each filter is then used to extrapolate the data in the corresponding quadrant (see Fig 3). The original support is expanded one layer at a time, and the procedure continued until the enlarged support is as large as desired. The 2-D FFT is then applied to the enlarged support to yield higher frequency resolution.

Kumaresan and Tufts [9] have proposed a method which generates two separate quarter plane prediction error filters and then combines these to produce high resolution spectral estimates. The filter and data support for each filter A and B are given in Fig 4. The coefficients for the forward and backward prediction filters are reversed versions of each other. These coefficients ($\underline{a}$ and $\underline{b}$) are determined by solving two equations in (N_1N_2-1) unknowns. These are

$$X_1 \underline{a} = -\underline{x}_a \quad ; \quad X_2 \underline{b} = -\underline{x}_b \tag{22}$$

where X_1, X_2 = $2x(N_1N_2-1)$ matrices of support data

$\underline{a}, \underline{b}$ = $(N_1N_2-1)x1$ vector of coefficients

$\underline{x}_a, \underline{x}_b$ = (2x1) vector of prediction values (these are the corner samples of the original support.

The pseudo-inverse solution to the above equation yields the prediction filter coefficients

$$\underline{a} = -X_1^T \left[X_1 X_1^T\right]^{-1} \underline{x}_a$$

$$\underline{b} = -X_2^T \left[X_2 X_2^T\right]^{-1} \underline{x}_b \tag{23}$$

The high resolution spectral estimate is then determined by using the 2-D spectrum definition

$$S(\omega_1,\omega_2) = \frac{1}{|A(\omega_1,\omega_2)|^2 + |B(\omega_1,\omega_2)|^2} \tag{24}$$

where

$$A(\omega_1,\omega_2) = \sum_{k,p=0}^{N_1-1} \sum^{N_2-1} a_{kp} \exp^{-j(k\omega_1+p\omega_2)} \qquad a_{oo} = 1$$

and similarly for B. $\{a_{kp}\}$ are ordered values of the predictor weights.

The technique uses the complete image (support) and it offers high resolution, along with little computing burden as it involves inversion of only a (2x2) matrix.

Cadzow [10] employs a weighted least squares criterion for evaluating a single quarter-plane forward linear prediction filter from which autoregressive spectral estimates can be derived.

A (P_1P_2) order AR predictor is defined as

$$\hat{x}(k,s) = \sum_{\substack{p=q=0 \\ p\neq q}}^{(P_1-1)} \sum^{(P_2-1)} a_{pq}\, x(k-p,s-q) \tag{25}$$

with prediction error

$$e(k,s) = x(k,s) - \hat{x}(k,s)$$

and the cost criterion is

$$J(\underline{a}) = \sum_{n_1=P_1}^{N_1} \sum_{n_2=P_2}^{N_2} w(n_1,n_2)\,|e(n_1,n_2)|^2 \tag{26}$$

where Cadzow proposes a weighting sequence

$$w(n_1,n_2) = (N_1-n_1+1)\,(N_2-n_2+1) \tag{27}$$

Turning the prediction weight sequence into a (P_1P_2x1) vector, and minimising the weighted least squares criterion of eq (26) leads to

$$\underline{a} = -(X_o^T W X_o)^{-1} X_o^T W \underline{x}_p \tag{28}$$

where X_o and $\underline{x}_p$ are as defined in eq (20), and W is the equivalent weighting matrix from eq (27). Note that for W = I, the weighted predictor coefficients reduce to those given in eq (21).

The associated AR spectral estimator for the coefficients $\{a_{p,q}\}$ is given by

$$S(\omega_1,\omega_2) = \frac{\sigma^2}{|A(\omega_1,\omega_2)|^2} \tag{29}$$

where

$$A(\omega_1,\omega_2) = \sum_{k,p=0}^{(P_1-1)} \sum^{(P_2-1)} a_{kp} \exp^{-j(k\omega_1+p\omega_2)} , \quad a_{oo} = 1$$

and $\sigma^2 = J_{min}(\underline{a})$ the minimum value for the cost function in eq (28). The above quarter plane approach ensures that the autoregressive estimator is causal.

Lim and Malik [11] have proposed an iterative scheme for evaluating a 2-D MEM spectral estimator which ensures that the resulting spectrum when Fourier transformed matches the available image correlation. The approach is based on optimising the following cost function

$$J = \iint d\omega_1 d\omega_2 \{\log S(\omega_1,\omega_2) - \sum_{p,q}\sum \lambda(p,q) [S(\omega_1,\omega_2)e^{j(\omega_1 p+\omega_2 q)} - R(p,q)]\} \tag{30}$$

where the Lagrange multiplier sequence $\{\lambda(p,q)\}$ allows inclusion of the constraints that the available autocorrelations are Fourier transformed values of the 2-D spectrum. Using a variational technique a solution to the above constrained optimisation criterion is

$$S(\omega_1,\omega_2) = \frac{1}{\sum_{p,q}\sum \lambda(p,q) \exp j(\omega_1 p+\omega_2 q)} \tag{31}$$

The $\{\lambda(p,q)\}$ may be evaluated by using the constraint condition

$$R(p,q) = DFT^{-1} \{S(\omega_1,\omega_2)\} \tag{32}$$

Lim and Malik propose an iterative scheme for computing $\{\lambda(p,q)\}$. This involves starting with an initial estimate of $\{\lambda(p,q)\}$, the corresponding correlation sequence is evaluated, this is modified making a correction based on the available R(p,q), and a new coefficient sequence $\{\lambda(p,q)\}$ is formed. The process is repeated till convergence occurs. Apriori requirement is the computing of the autocorrelation function $\{R(p,q)\}$ from the available image support.

Mathematically, the algorithm is a simplified version of the ART algorithm commonly employed in digital image reconstruction [13], and can be stated as

$$[\lambda(p,q)]_{n+1} = [\lambda(p,q)]_n + \gamma[\hat{R}(p,q)-R(p,q)] \tag{33}$$

where subscript n denotes nth iteration.

$\hat{R}(p,q)$ is obtained as in eq (32) from the nth spectral estimate, while R(p,q) is the apriori correlation sequence obtained from the data. γ is a step length parameter, and has a range $0<\gamma<2$ (see [13]). Note that the model in eq (31) is restricted to an autoregressive representation of the image. Again, as an MEM spectral estimator is employed, high resolution is possible in the frequency domain even for a comparatively small data support. It is worth stating that of all the schemes discussed in this section, the last is by far the most computationally demanding.

It is interesting to observe that if in the cost function of eq (32) $\{\log S(\omega_1,\omega_2)\}$ was replaced by $|S(\omega_1,\omega_2)|^2$, then the associated 2-D spectral density estimate would be the conventional 2D periodogram (i.e. a 2D DFT of the autocorrelation of the available support).

5. RESOLUTION ENHANCEMENT FOR FINITE APERTURE ACOUSTIC IMAGING

In acoustic imaging, as in acoustic holography, a requirement exists for enhancing the lateral resolution by extrapolating the hologram beyond the available aperture. In most practical applications the aperture size is limited due to the number of points at which the hologram plane can be sampled either by a finite sensor array, or because of physical restrictions such as the available scanning region. Usually the aperture size, defined in terms of wavelengths, is much smaller in acoustic holography than in optical holography.

In this section two approaches are discussed for aperture extension; the first method is based on the use of discrete prolate spheroidal wave functions [14] which allow both interpolation and extrapolation; and the second method employs an ARMA model for the hologram [5], when the object is considered as a collection of finite radiating sources. Both approaches are based on a least squares modelling of the available data. (For an up to date review of earlier work in this area refer to [5]).

Fig 5 illustrates a simplified version of a holographic imaging system. The hologram plane is separated by a known distance D from the object plane, the maximum scale of the object is not greater than $S = 2X_o$, where S is much smaller than D, such that the distance d from any point x in the object plane to a detecting point ξ in the hologram plane can be approximated by

$$\begin{aligned} d &= \left[D^2+\xi^2+x^2-2x\sin\theta\,(D^2+\xi^2)^{\frac{1}{2}}\right]^{\frac{1}{2}} \\ &\simeq (D^2+\xi^2)^{\frac{1}{2}} - x\sin\theta \end{aligned}$$

such that the hologram signal $H(\xi,k)$ at the detecting point ξ, for a wave-number, k, and wavelength λ may be expressed as

$$\begin{aligned} H(\xi,k) &= \beta(\xi,k)\int_{-X_o}^{X_o} a(x)\exp(jk\,x\sin\theta)\,dx \qquad (34) \\ &\triangleq \beta(\xi,k)\cdot\bar{H}(\xi,k) \end{aligned}$$

where a(x) is the spatial distribution of the object considered to be radiating acoustic energy coherently, and

$$\beta(\xi,k) = \left[1/\lambda(D^2+\xi^2)^{\frac{1}{2}}\right] \exp -jk\ (D^2+\xi^2)$$

According to the modified hologram $\bar{H}(\xi,k)$ is a Fourier transform of the spatial distribution of the object. If the hologram is sampled at unit intervals in spatial frequency, say Δ, with

$$\Delta = k\left[\sin\theta_{m+1} - \sin\theta_m\right] / (2\ X_o)\ , \quad (m = 0, \pm1, \pm2 \ldots) \tag{35}$$

where $\theta_o = 0$, then

$$\bar{H}(m\Delta,k) = \sum_{p=-N}^{N} A(pf_o)\ \sin\left(\frac{m\Delta}{f_o} - p\pi\right)/\left(\frac{m\Delta}{f_o} - p\pi\right) \tag{36}$$

where $A(f)$ is the Fourier transform of $a(x)$, considered band-limited to $|f| < N f_o = N/(2\ X_o)$, and f_o is the fundamental frequency of the hologram.

To establish a least squares estimate of the hologram is an extended area, the sampling interval Δ may be considered uniform over the aperture and equal to a function of the fundamental such that $\Delta = \pi f_o/K \triangleq \pi W$ (by definition). Then eq (36) may be recast as

$$\bar{H}(m\Delta;k) = \sum_{p=-N}^{N} A(pf_o)\ \frac{\sin \pi W(m-pK)}{\pi(m-pK)}\ , \quad m = 0, \pm1, \ldots \pm M \tag{37}$$

Note that only a small number of sampled values (= 2M+1) representing a few complete wavelengths of the aperture are considered to have been observed.

Fig 6 represents the relationships between the sampling points of the observed (modified) hologram $\bar{H}$, the true hologram $\{A(pf_o),\ p = 0, \pm1, \pm2 \ldots \pm M\}$, and the extended hologram $\{A(pf_o),\ M < |p| < N\}$. Note the sampled values $\{\bar{H}\}$ do not necessarily coincide with the desired values $\{A\}$ even within the available aperture. Thus to obtain $\{A(pf_o)\}$ from measurements $\{\bar{H}\}$ involves both interpolation within the aperture, and extrapolation outside the aperture.

Considering (2M+1) values of $\bar{H}$ in vector form

$$\underline{H}_M = \left[\bar{H}(-M\Delta;k) \ldots \bar{H}(M\Delta;k)\right] \tag{38}$$

and similarly the (2N+1) values of the true and extrapolated hologram,

$$\underline{A}_N = \left[A(-Nf_o) \ldots A(+Nf_o)\right] \tag{39}$$

Eq (37) may be recast as

$$\underline{H}_M = B\ \underline{A}_N \tag{40}$$

where the (ij)th element of matrix B is $\{\sin \pi W(i-jK)/\pi(i-jK)\}$. Note B is a (2M+1)x(2N+1) matrix. More generally due to measurement noise the observation model will be

$$\underline{H}_M = B\ \underline{A}_N + \underline{\eta}_M \tag{41}$$

A least squares solution to the above yields the desired values of the enhanced aperture as

$$\hat{\underline{A}}_N = B^T\left[B\ B^T\right]^{-1} \underline{H}_M \tag{42}$$

which is the pseudo-inverse solution to eq (40) or (41).

Note that if the vector $\underline{A}_N$ is restricted to only (2M+1) values, aperture extrapolation is still possible, since Δ is a fractional value of f_o. For this case

$$\underline{H}_M = B\,\underline{A}_M$$

where B is now a (2M+1)x(2M+1) matrix whose eigenvectors are discrete prolate spheroidal sequences (DPSS) [14], and a solution to the desired hologram vector $\underline{A}_M$, is given by

$$\underline{A}_M = \left[\sum_{p=-M}^{M} \frac{1}{\lambda_p} \underline{V}_p(W;2M+1)\, \underline{V}_p^T(W;2M+1)\right] \underline{H}_M \tag{43}$$

where λ_p = eigenvalue associated with the DPSS of order p - $\{\underline{V}_p(W;2M+1)\}$.

Note that the above solution holds even for the case of noisy data represented by (41). However in practice the method has been found to be sensitive to signal to noise ratios.

Abdel-Aal et al [5] have proposed a modelling scheme for achieving hologram aperture extension when the object is considered as a sequence of coherent point sources. Here adjacent values of the hologram are modelled as

$$H(m\Delta;k) = \chi_p(m)\, H(\overline{m-1}\Delta;k)$$

with

$$\chi_p(m) = \sum_{i=1}^{p} \alpha_i\, (m\Delta)^{i-1} \tag{44}$$

and $\{\alpha_i\}$ are M unknown coefficients to be determined from p observations within the restricted aperture. Note that $\chi_p(m)$ is a least squares polynomial form, where $(m\Delta = \xi)$ is the distance of the hologram sample $H(m\Delta;k)$ the centre of the aperture.

The above model has been found to be sensitive to noise, and so an ARMA model has been proposed:

$$H(m\Delta;k) = \left(\sum_{i=1}^{p} \alpha_i\, (m\Delta)^{i-1}\right) H(\overline{m-1}\Delta;k) + \sum_{j=1}^{q} c_j\, H(\overline{m-q+j-1}\Delta;k) \qquad m = 1, 2, 3 \ldots M \tag{45}$$

where p is the number of moving average (MA) terms corresponding to the space variant effects, and q is the number of autoregressive (AR) terms which cover space-invariant effects. $\{\alpha_i\}$ represent the pth order MA coefficients, while $\{c_j\}$ are the qth order AR coefficients.

By employing (p+q) observations of $\{H(m\Delta;k),\ m = 1,2 \ldots (p+q)\}$, the (p+q) coefficients of the ARMA model may be evaluated after solving the corresponding (p+q) equations from eq (45). Once the model coefficients $\{\alpha_i;c_j\}$ are obtained, these can then be used to predict hologram values outside the available aperture, according to the relationship

$$\hat{H}(n\Delta;k) = \left(\sum_{i=1}^{p} \alpha_i\, (n\Delta)^{i-1}\right) \hat{H}(\overline{n-1}\Delta;k) + \sum_{i=1}^{q} c_j\, \hat{H}(\overline{n-q+j-1}\Delta;k) \qquad n > (p+q) \tag{46}$$

Ref (5) includes several studies on point sources, and modifications to the MA model to achieve relative stability in the presence of noise. In general the inclusion of the AR model ensures much more stable performance than the model in eq (44).

6. DISCUSSION

Several aspects of image processing have been presented covering problems associated with image desmearing and techniques for improving image resolution.

In the first instance, Kalman filtering techniques are shown to allow desmearing of images effected by space variant or space invariant blurs. This is made possible by the use of a state space model for the image which is primarily valid for a randomised image data sequence. However the filter is robust enough (though sub-optimal) to allow desmearing even when this condition is not met. The requirement of apriori information regarding the smearing function and the additive noise variance cannot be relaxed.

Fig 7 shows a smeared image representing motion blur. Fig 8 is the restored output after passing data through a Kalman filter as discussed in section 2. Fig 9 shows an image effected by space variant blur, and Fig 10 represents the restored image after Kalman filtering. Figs10-13 show the 2-D response of the filter for different signal to noise ratios when effecting desmearing and noise suppression. Note in Fig 10 , when the S/N ratio is high, the filter essentially acts as a 2-D high pass filter, compensating for the low pass smearing effect. As the S/N ratio increases, the filter still acts as a high pass filter compensating for the smearing effect, however at higher frequencies, where the noise break through occurs, the filter starts suppressing the higher frequencies, hence suppressing the noise component. For very poor S/N ratios, the filter primarily acts as a low pass filter ensuring noise suppression, while effecting little desmearing.

When linear smearing affects the image, as, for instance, the smearing associated with a broad main lobe of an array response and related side-lobes, the image may be restored by employing an inversion approach based on Gauss-Bernoulli models for incident data from point sources.

To improve frequency resolution MEM techniques are discussed using prediction filters. Though the methods discussed are computationally expensive, the spectrum achieved as reciprocals of the squared magnitudes of the filter response offer significantly better resolution than conventional 2D FFT methods. Similar comments hold for the hologram extension technique discussed in section 5.

7. REFERENCES

[1] W.K. PRATT, 'Digital Image Processing', Wiley-Interscience Publ., NY, 1978.

[2] R.C. GONZALEZ and P. WINTZ, 'Digital Image Processing', Addison-Wesley, Reading, Mass, 1977.

[3] M. BISHOP and T.S. DURRANI, 'An estimator for image desmearing using a Bernoulli-Gaussian model', Proc. IEEE ICASSP 82, Paris, May 3-5, Vol.2,pp1154-8.

[4] T.S. DURRANI et al, 'Signal processing' (text), to be published, Addison Wesley, Reading, Mass, 1983.

[5] R. ABDEL-AAL, C.J. MACLEOD and T.S. DURRANI, 'Expansion of acoustic hologram apertures using ARMA modelling techniques', Proc. IEE Conf. "Acoustical Imaging", Savoy Place, London, 19-22 July 1982.

[6] K.R. CASTLEMAN, 'Digital signal processing', Prentice-Hall, NJ, 1979.

[7] J.M. MENDEL and J. KORMYLO, 'New fast optimal white noise estimators for deconvolution', Trans IEEE Geoscience Electronics, Vol.GE-15, pp.32-40, 1977.

[8] O.L. FROST Jr, 'High resolution 2-D spectral analysis at low SNR', Proc. IEEE ICASSP 80, Denver Co., pp.580-3, 1980.

[9] R. KUMARESAN and D.W. TUFTS, 'A two dimensional technique for frequency waveno. estimation', Proc IEEE, Vol.69, pp.1515-6, 1981.

[10] J.A. CADZOW and K. OGINO, 'Two dimensional spectral estimation', Trans IEEE, ASSP-29, pp.396-401, 1981.

[11] N.A. MALIK and J.S. LIM, 'Implementation of a new algorithm for 2-D maximum entropy spectral estimation', Proc. UK Inst. of Acoustics (Underwater Acoustics Group) Conf. 'Spectral analysis and its use in underwater acoustics', Imperial College, London, 29-30 April 1982, p.11.1-8.

[12] E.J. FREY, 'Two dimensional spectral estimation', MSc Thesis, University of Colorado, Boulder, USA, 1982.

[13] T.S. DURRANI and C. GOUTIS, 'Optimisation techniques for digital image reconstruction from their projections', IEE Proc. Vol.127, Pt.E, pp.161-169, 1980.

[14] D. SLEPIAN, 'Prolate spheroidal wave functions, Fourier analysis and uncertainty - V: the discrete case, BSTJ, Vol.57, pp.1371-1430, 1978.

[15] S.S. DIKSHIT, 'A recursive Kalman window approach to image reconstruction', Trans IEEE, ASSP-30, pp.125-140, 1982.

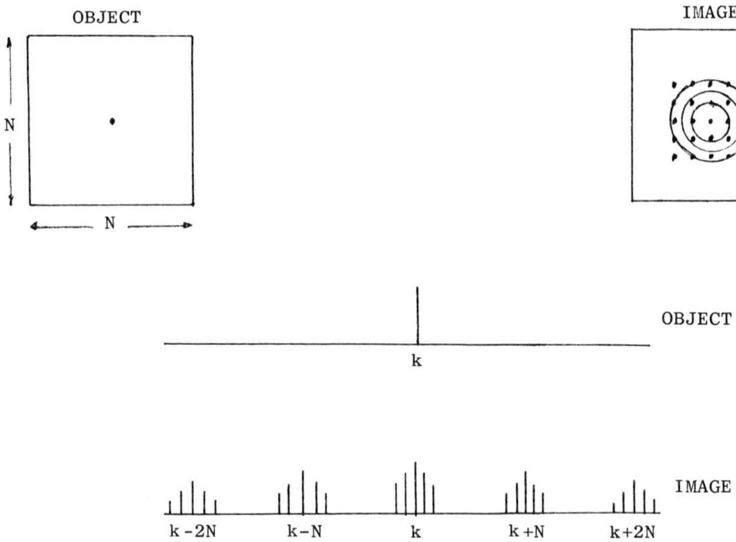

Fig.1. Point spread function covering a 5x5 pixel square

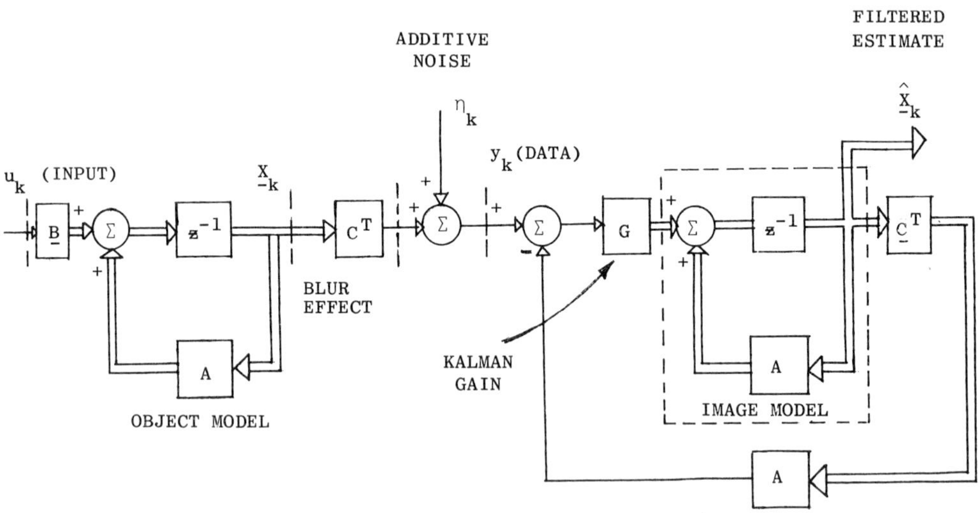

Fig.2. Kalman filter implementation of sequential algorithm for image desmearing

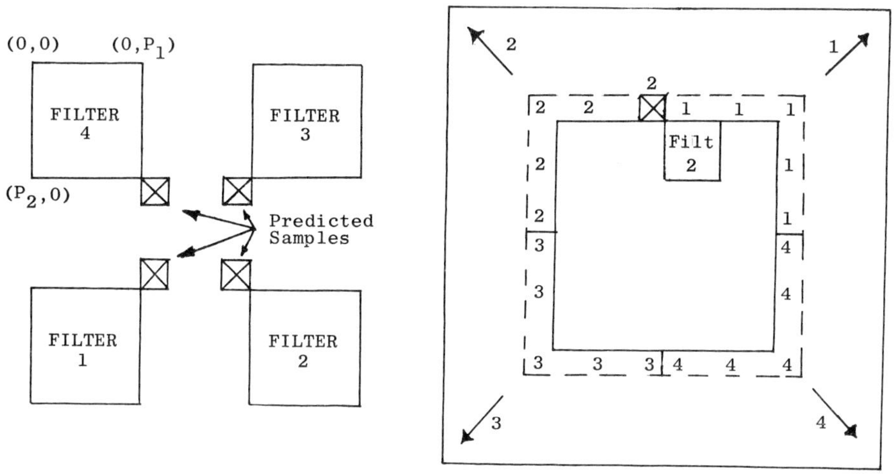

Fig. 3. Two dimensional prediction filters
(one layer at a time is extrapolated)

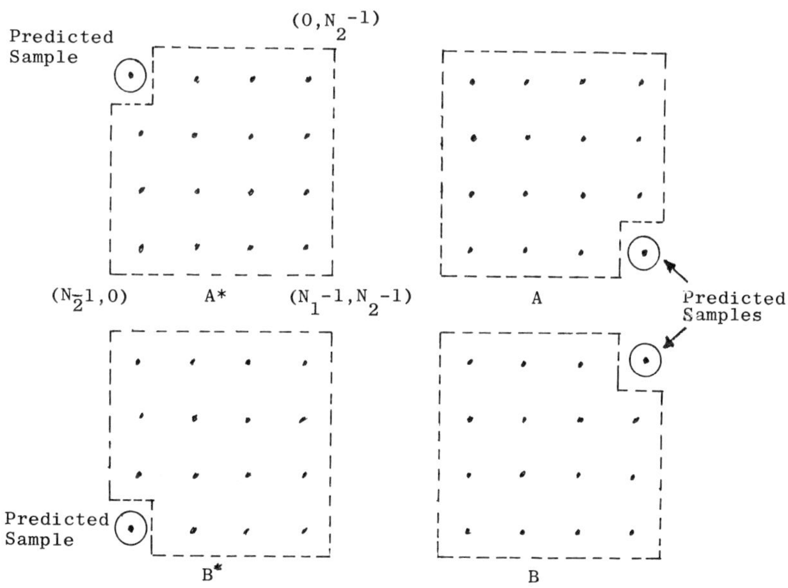

Fig. 4. Filter and data support for 2-D forward and backward linear prediction

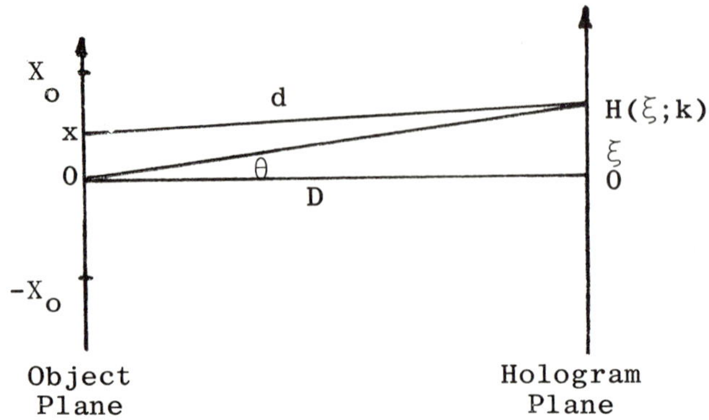

Fig. 5. Geometry for acquisition of hologram signal

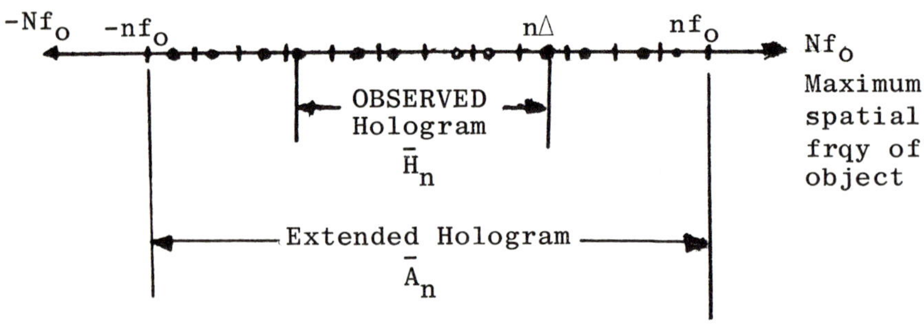

Fig. 6. Schematic showing relationship between hologram vector $\bar{A}_n$ in extended area, and the observed hologram $\bar{H}_n$

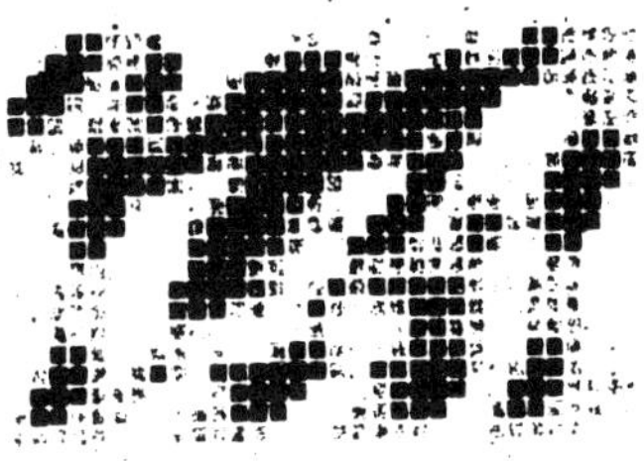

Fig. 7. Motion blur simulation (30db SNR)

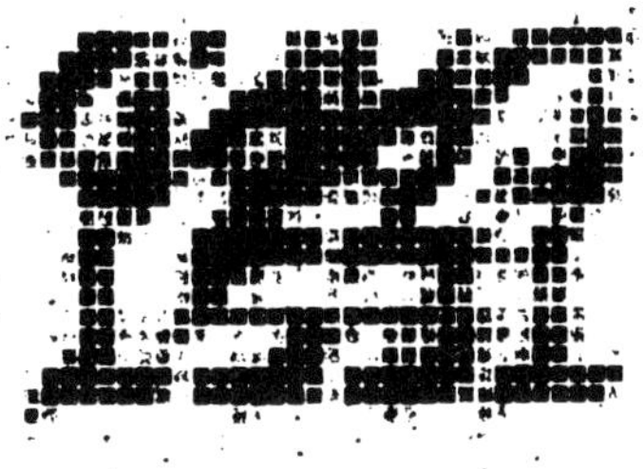

Fig. 8. Restoration of Fig 7 using 2-D filter

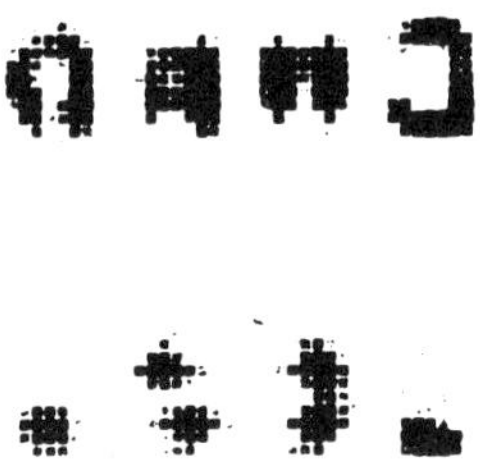

Fig. 9. Spatially varying psf (40dB SNR)

Fig. 10. Restoration of Fig 9

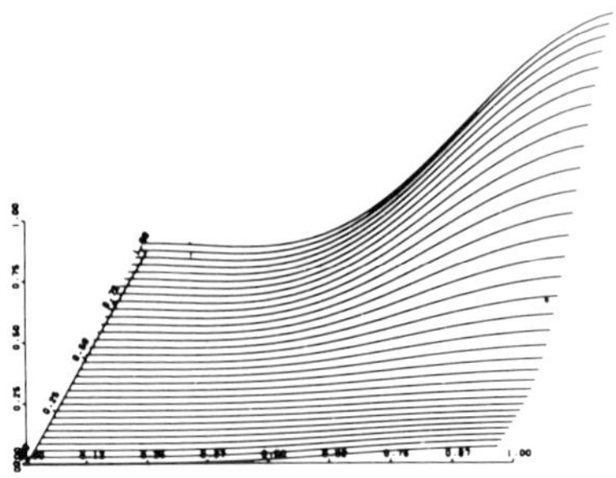

Fig. 11. Frequency response of 2D restoration filter (50dB SNR)

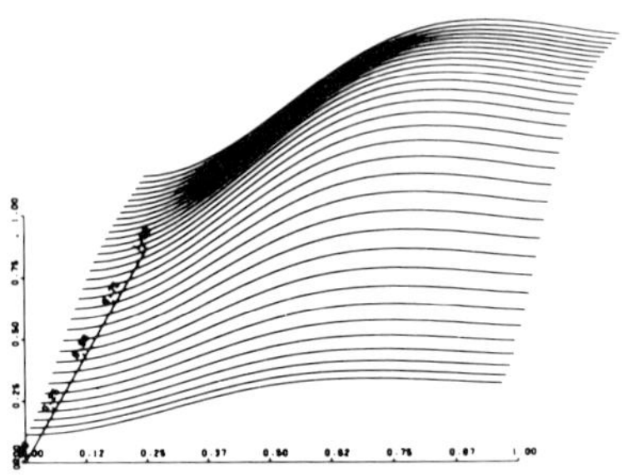

Fig. 12. Frequency response for 25dB SNR

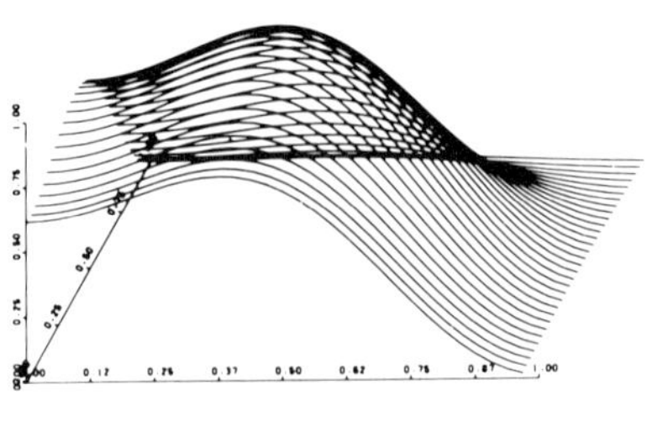

Fig. 13. Frequency response for 12dB SNR

DIGITAL PROCESSING OF TIME VARYING IMAGES

V. Cappellini and A. Del Bimbo
Istituto di Elettronica, University of
Florence and IROE -C.N.R.
Florence, Italy

Abstract

Digital processing techniques of time varying images are briefly reviewed. A reliable procedure for analysis of time varying images and recognition of image patterns is in detail presented. The procedure consists of the following main steps: subsequent observed images are digitized; a pre-processing is performed to reduce the noise and enhance useful patterns; pattern edges are extracted; a scanning of the pattern contours is performed; FFT of contour points is evaluated; the obtained contour FFT is compared with pre-stored data corresponding to patterns to be recognized. The good efficiency of the proposed technique is outlined. Experimental results obtained in processing different real world varying images (in particular given by remote sensing detectors, TV or infrared cameras) are shown, confirming the good performance of the above technique for recognizing several different patterns in observed images.

NATO ASI Series, Vol. F1
Issues in Acoustic Signal/Image Processing and Recognition
Edited by C. H. Chen

1. Introduction

Digital processing of time varying images (or dynamic scene analysis) is becoming of increasing interest as theoretical and experimental research area [1] [2] [3] .

Three main cases of varying images can be considered:

a) rigid objects moving in the space (spatial image sequences or frames);

b) phenomena changing in time in a constant space region, e.g. the radio-isotopes counting in nuclear medicine (temporal image sequences);

c) non rigid objects moving in the space (spatio-temporal image sequences).

Two basic problems are encountered in the investigation of the systematic variations between consecutive images in a sequence for object recognition and tracking:

a) how to decompose an image into "segments", significant to descrive the image (gray level, shape, texture...) , all or part of which are remaining invariant from one image frame to the next one;

b) how to find a transformation which uses the smallest number of parameters to describe the observed changes along an image sequence.

From the practical point of view, during the last decade, significant advances in digital sensors, storage and processor technologies facilitated the sampling and analysis of image sequences as above outlined. Problems however remain to be solved: data obtained through common sensing devices (as TV cameras, infrared detectors and laser scanners) usually give an observer centered representation, while the interactive systems that specify the objects most naturally involve object centered representations; computing time to perform all the digital processing steps in the recognition and tracking procedure can be to high for real-time applications. For the last problem solution fast parallel processors or special purpose processors can be used. The applications of time varying image processing are fast expanding: tracking in satellite images (to derive in particular wind velocity),underwater detector images processing (array, sonar),ultrasound image processing of heart, nuclear medicine image analysis, traffic monitoring, missile target tracking.

As a general observation (the state of the art) we can observe that more algorithmic approaches and studies are required for image sequence analysis by using digital processing, pattern recognition and artificial intelligence methodologies to obtain a more general formulation of problems and solution techniques.

In the following some procedures for recognition of moving objects are firstly considered and after a specific technique,actually implemented on a minicomputer system, is presented.

2. Some procedures for recognition of moving objects

Systematic displacements of some image components from frame to frame can be taken as a strong hint to look for the image of a moving object. This assumption opens a way to extract objects shilhouettes from a sequence of frames without any detailed a-priori knowledge about size, shape and texture of the stationary part of the scene. Application of this process leads to a quick and reliable identification of nonstationary image components in a sequence of frames. There have been several studies which deal with moving pictures, e.g. measuring gray level differences between consecutive frames and separating non stationary from stationary components [4] , or finding and tracking prominent gray level features from the moving pictures [5] . Although these methods are efficient in simple scenes, it seems to be difficult to apply them for more complex scenes which are common in real applications. Furthermore this problem was studied under the assumption that objects show high contrast against background, so there was a little problem in detecting boundaries of objects [6] . Further an approach was recently developed which seems to overcome some of the above mentioned difficulties [7] . An advantage in automatic analysis of moving images compared with static images is that we can use interframe information in addition to spatial difference of the gray level. This key idea is the starting point to solve various problems in analysis of moving pictures and an efficient system was recently presented just using this method [8] . In the system described in the following, this idea is further developed to eliminate some unpleasant effects, e.g. produced by light reflexion on metal surface. Really the previous algorithm requires the hypothesis that objects are darker than the background, which may be accepted only in the applications where objects are known. This assumption was removed using an edge detection processing before moving object-detection and, as first step, a non-linear smoothing of original sequence to reduce noise effects. After separating non stationary components, each object is isolated and the Fourier transform of boundary points is evaluated to get a shape description for recognition purpose. In the tracking step, a model of the first frame is generated and used in a matching processing to identify the same object in the subsequent frames.

3. Hardware configuration

Fig. 1 shows a block diagram of the used hardware system.

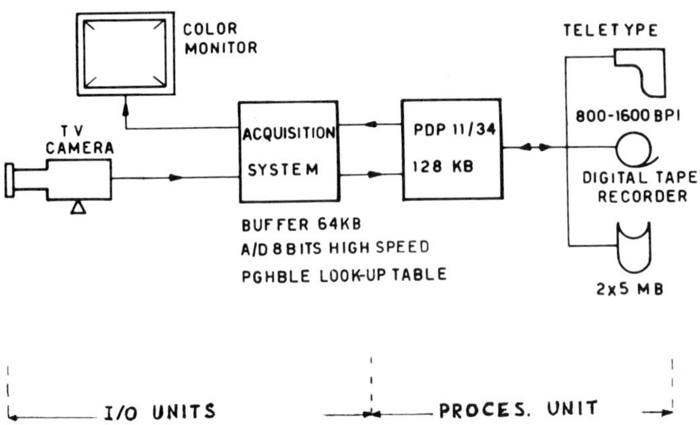

Fig. 1. Block diagram of the used hardware system

The computer is a PDP 11/34, a 16 bits machine with 128 K bytes of memory, two disks (5 + 5 M bytes), a magnetic tape (800-1600 BPI) and a printer terminal. A TESAK interfacing unit is provided, containing an external memory of 64 K bytes to store the pictures acquired by a black and white TV camera. The same unit performs an 8 bits analog-to-digital conversion at 30 M bytes/s rate. Programmable look up tables are also available to obtain display color monitoring. Once a frame has been sampled, a digitized picture stored in the external memory is sent to the system and recorded in the magnetic tape or in a disk. Any sequence of picture data stored in the magnetic tape or in a disk can be displayed as a dynamic picture on the color display, which is very useful to examine and evaluate the intermediate processing results.

4. Software description

In the processing sequence of frames,the following steps are performed:

1 - Learning phase, including the acquisition and the modelling of the objects which the system has to recognize (off-line processing).

2 - Tracking of moving objects, which requires also the following sub-steps:

- 2.1 - Preprocessing of input sequence by a non-linear smoothing algorithm.
- 2.2 - Edge detection of each frame by the gradient method.
- 2.3 - Moving objects detection: non-stationary components of the input sequence are separated from stationary ones which constitute the background of the image.
- 2.4 - Post-processing.
- 2.5 - Segmentation of the single frame (with silhouettes of the moving objects in a two levels code) to get single object.
- 2.6 - Object modelling - This algorithm takes into consideration only the shape of the silhouttes of the detected objects (interior part is ignored). This process involves the FFT of the boundary.

3 - Object recognition - It requires a matching processing of the spectra to evaluate the "distance" between the shape of detected object and model shapes stored in the memory during the learning phase.

5. Detailed description of each step

In the following, a more detailed description of each step processing is given.

1 - In this phase spectra are obtained and then they are stored in a disk.

2 - The tracking requires evaluation of a model which describes kinetic state and some features of the shape for each object.As the descriptors of geometrical characteristics, the inertial invariants of the boundary curve and its length were chosed. Kinetic state included centroid position and rate. These are used to predict the position of the objects in the next frame. A matching algorithm is used for coupling the objects detected in the subsequent frames with those ones detected before. This matching is based on the comparison of the geometrical features above described, which are updated in each frame as well as the kinetic state of each tracked object. If the differences of the shape description are in absolute value less than a fixed threshold and if the estimated centroid position has a distance from real position less than a fixed tolerance, the object is considered matched and the model is updated. We suppose that no occlusion occours during the tracking and that there are small variations of the object rate. This processing is useful also to get a sequential estimate of the spectrum module of each object. The following substeps are also performed in each frame:

- 2.1 - This processing is a variation of the smoothing described in [9] , which helps to eliminate the noise in the image [10] . In this processing, the intensity value at each point is replaced by average of itself and its neighbourhood, except those which have gray level differences greater than a fixed threshold in absolute value. For the point P_0 (Fig. 2), its updated value is given by

$$P'_o = \frac{1}{n} \sum_{P_i \in S} P_i \qquad (1)$$

where $S = \{P_i : | P_i - P_o | \leq k \}$ and $i = 0, 1, 2 \ldots 8$.

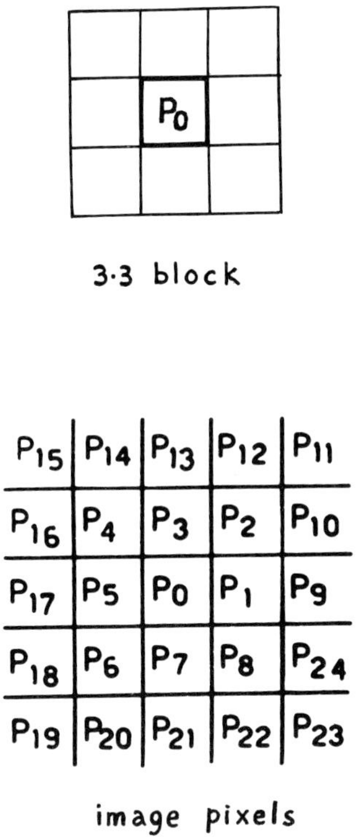

Fig. 2. Pixels of an image to be processed

This procedure has the following advantages:

a) near a boundary whose contrast is greater than k, it doesn't include in the average evaluation any point around the edge. This allows a smoothing of the points on either side of the boundary without any damage for the boundary as common linear smoothing would do.
b) In an homogeneous region, noise, such that its differences are less than K, is smoothed.
c) A textured region, whose elements have differences in intensity less than K,

will be smoothed into a homogeneous region.
By appropriately choosing the threshold value, the noise in the image can be reduced without any loss of significant informations and this processing can be iterated to obtain better results. It is sometime useful to use masks larger than 3 · 3 and reduce the number of iterations (two iterations of a 9 · 9 operator produce similar results to eight iterations of a 3 · 3 operator with the same threshold). After experimental testing of this conditional average, it was subjectively concluded that two iterations of a 3 · 3 operator are enough to obtain good results.

2.2 - An approximate Sobel gradient operator is used for detecting edges between regions with different gray level. The edge magnitude was obtained by applying gradient method at every pixel of frame sequence:

$$\underline{G}(x,y) = \nabla I\,(x,y) \tag{2}$$

$$G(x,y)_{P=P_o} = \left(\frac{\partial I}{\partial x}\right)_{P=P_o} = P_4 + 2P_5 + P_6 - P_8 - 2P_1 - P_2 \tag{3}$$

$$G_y(x,y)_{P=P_o} = \left(\frac{\partial I}{\partial y}\right)_{P=P_o} = P_2 + 2P_3 + P_4 - P_6 - 2P_7 - P_8 \tag{4}$$

The module is approximated as follows:

$$\|\,\underline{G}(x,y)\,\| = \left|\,G_x^2\,(x,y) + G_y^2\,(x,y)\,\right|^{1/2} \cong |G_x(x,y)| + |G_y(x,y)| \tag{5}$$

(see Fig. 3)

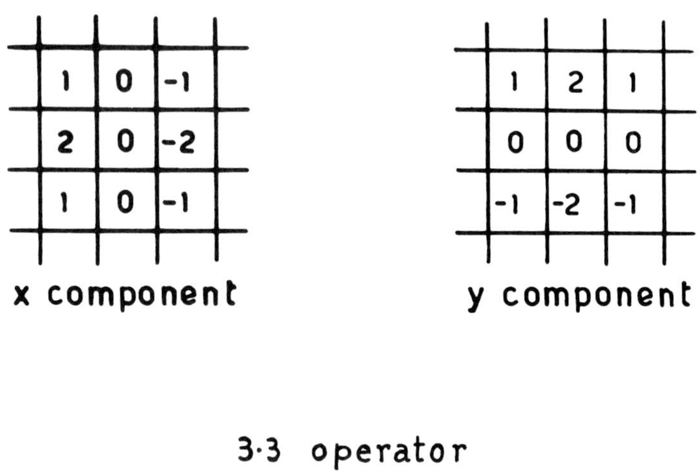

Fig. 3. Definition of the approximate Sobel gradient operator: x component and y component

2.3 -Moving objects detection is performed using a point to point subtraction between two pictures of K-th and (k + n)-th frame:

$$D(i,j,k,k+n) = F(i,j,k) - F(i,j,k+n) \tag{6}$$

a thresholding of D(i,j,k,k+n) is used to get a two levels image as follows:

$$B(i,j,k,k+n) = 1 \qquad \text{if } D(i,j,k,k+n) > T \tag{7}$$

$$B(i,j,k,k+n) = 0 \qquad \text{elsewhere}$$

As a final result, the shapes of the objects are obtained by the union of previous binary images

$$U(i,j,k) = \bigcup_{n \in N} B(i,j,k,k+n) \qquad k \in N \qquad n = \{1,2.3,\dots n_{max}\} \tag{8}$$

where k is the number of the frame in which objects are detected.

2.4 - Reducing of noise is often necessary to get better results. Small areas are eliminated and small holes are filled and also irregularities of boundary curve are reduced. This simple but efficient processing uses a 3 · 3 operator and is based on statistical decision criterion. We have to decide, for each pixel, if it would change its level or not, and the decision depends on its originary level and those ones of the pixels which surround it. Larger masks are not advisable due to the related computing difficulties. The algorithm is defined as in the following:

$$\text{if } P_o = 0 \quad \begin{cases} P'_o = 0 & \text{if } \sum_{i=1}^{8} P_i < 5 \\ P'_o = 1 & \text{otherwise} \end{cases} \tag{9}$$

$$\text{if } P_o = 1 \quad \begin{cases} P'_o = 0 & \text{if } \exists\, i \in \{1,2, \dots 8\} : P_i = P_j = 0 \\ & \text{where } j = i + 1 \text{ if } i < 8 \\ & \text{and } j = 1 \text{ if } i = 8 \\ P'_o = 1 & \text{otherwise} \end{cases} \tag{10}$$

where P'_o is the updated value of P_o and P_i i = 1,2, ...8 are the pixels surrounding it as illustrated in Fig. 2.

2.5 - Binary image segmentation was developed using tracking of boundary points. To detect a boundary point a scanning of the input image is performed. Subsequent points are obtained using a counter-clockwise boundary tracking according to an 8-points chain coding scheme shown in Fig. 4, starting from the first detected point up to return to the starting one. Thus a closed curve is obtained, enclosing a single object, and this surrounded object can be separated from the other ones and also subtracted from the group. In this way the separation of each object,detected by iteration of this process, can be accomplished.

2.6. - A model which describes the shape of that object is to be obtained. We developed a simple and fast method based on circular scanning of the object boundary. The centroid of the curve, which enclose the object, is evaluated; the intersection points with the boundary of straight lines from the center (with a costant angular increment) are evaluated as shown in Fig. 5.

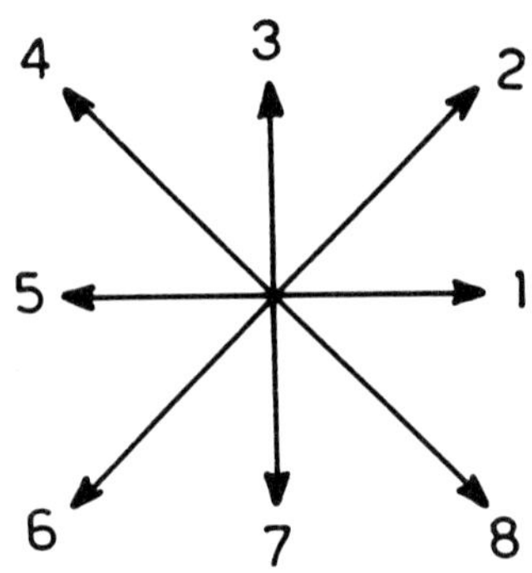

Fig. 4. 8-points chain coding scheme

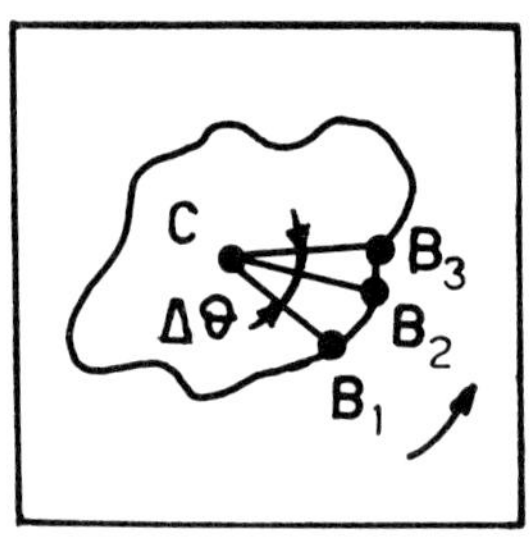

Fig. 5. Circular scanning of the object boundary

In this step, also samples of distance of boundary points from the centroid of the curve are obtained. A common standard FFT algorithm is used to evaluate the spectrum of the boundary function.

3 - Object recognition is based on the key idea of a distance in the frequency domain. Two objects with the same shape but different size are supposed equivalent. Distance between the object and the model is defined as follows:

$$\delta = \sum_{k=1}^{M} \left| |F_o(j\omega_k)| / |F_o(0)| - |F_m(j\omega_k)| / |F_m(0)| \right| \tag{11}$$

where $F_o\{(j\omega_k)\}$ are the samples of the object spectrum, $\{F_m(j\omega_k)\}$ are the ones of the model spectrum and M is a suitable positive integer number (by using 256 samples, the value M = 20 was selected). Each detected object is compared with each model stored in computer memory and, if the evaluated distance is less than a fixed threshold, the object is considered recognized. Only modules of spectra are used in the matching processing, so it is impossible to distinguish differences among the phases of the spectra, but this is really very useful because we get a rotational invariance. In fact if a rotation occours, it determines a shift of boundary function which implies a linear phase in its Fourier transform, but no module change.

6. Experimental results

A sequence of frames showing a spanner, a nut, a washer as in Fig. 6 (at left) has been acquired by the TV camera.

Moving objects are detected using the processing as described above in sub. step 2.3. The goodness of this method can be realized also in the comparison with the results of a previous moving objects detection method [7] . Following this way, in fact, we get indipendence from the background gray level constraints and - at least in part - from the number of required frames. Sampling of the object edges is obtained by the circular scanning method to get objects spectra. These spectra are used to classify the detected objects. Fig. 6 shows the tracking for subsequent positions of the detected objects in five consecutives frames. In the third frame the nut is covered by an undesirable object and is not matched; in the subsequent frame, being removed the undesirable object the nut is still matched and the tracking goes on to the end.

7. Conclusions

A simple but efficient software system is presented, which has been developed to get batch moving objects detection and tracking. It takes advantage from previous works on the same topic, getting however a greater simplicity.

It can be successfully used anywhere it is required to analyse sequences of frames to follow the shape and to recognize the trajectory of moving objects. Finally, by using a suitable fast image acquisition system and computer (parallel processing, special purpose processors), the presented system can be extended to real-time applications.

Acknowledgments

We acknowledge the useful cooperation of Drs. P. Borghesi, C. Conese and M. Poggi, Facoltà di Ingegneria, University of Florence.

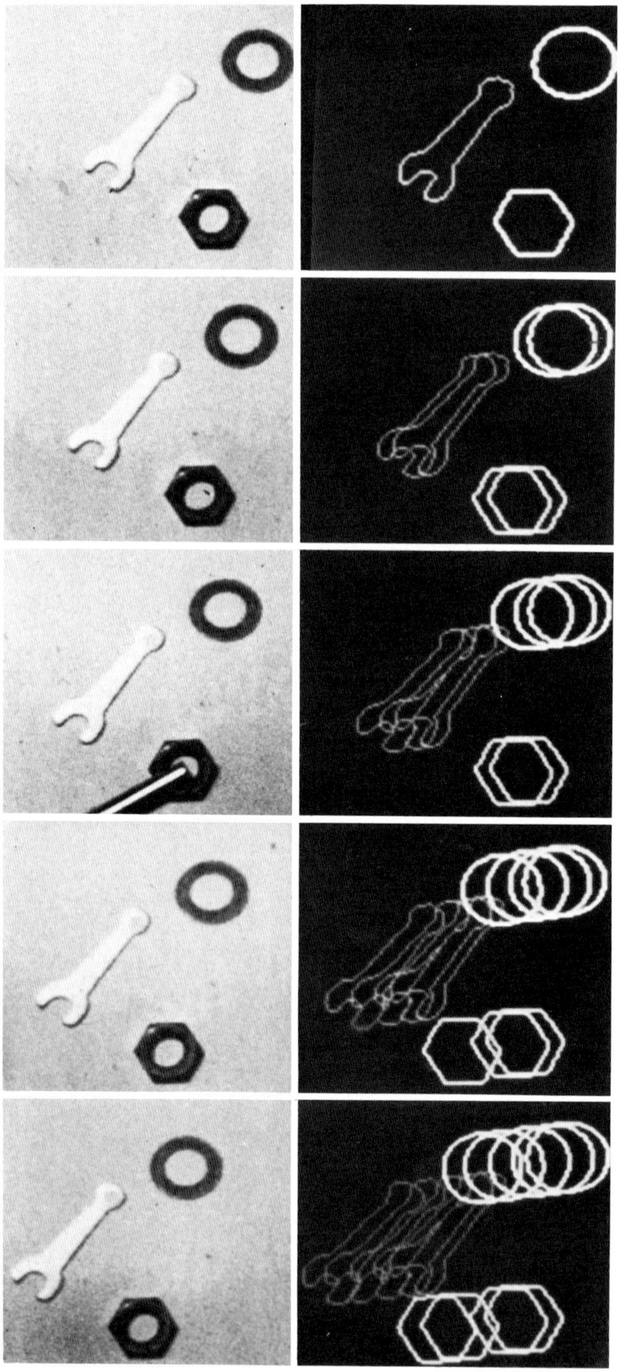

Fig. 6. A sequence of digitized five consecutive frames (at left) and object recognition and tracking (at right): in the third frame the nut is covered by an undesirable object which however is not affecting the recognition procedure

References

[1] T.S. Huang Ed.: Image Sequence Analysis, Springer Verlag, Berlin-New York, 1981.

[2] L.N. Kanal and A. Rosenfeld Ed.: Progress in Pattern Recognition, North Holland Publishing Comp., Amsterdam, 1981.

[3] H.H. Nagel, Image sequence analysis: What can we learn from applications?, Report Fachbereich Informatik, University of Hamburg, I1I-HH-M-79/80.

[4] J.L. Potter: Scene segmentation by velocity measurement obtained with a cross-shaped template, in Proc. 4th Int. Conf. Artificial Intell., (1975), 803-810.

[5] R.T. Chien and V.C. Jones: Acquisition of moving objects and hand-eye coordination, Proc. 4th Int. Conf. Artificial Intell., (1975), 737-741.

[6] W.K. Chow and J.K. Aggarwal: Computer analysis of planar curvilinear moving images, IEEE Trans. comp. vol. C-26, (1977), 179-185.

[7] M. Yachida, M. Asada and S. Tsuji: Automatic motion analysis system of moving objects from the records of natural processes, in 4th Int. Conf. Pattern Recognition, (1978), 726-730.

[8] W.B. Thompson: Combining motion and contrast for segmentation, IEEE Trans. on Pattern Analysis and Machine Intelligence, vol. PAMI, 2, n. 6, November 1980.

[9] A. Lev, S.W. Zucker and A. Rosenfeld: Iterative enhancement of noisy images, IEEE Trans. Syst. Man. Cybern. vol. SMC - 7, (1977).

[10] V. Cappellini, A.G. Constantinides, P. Emiliani: Digital Filters and Their Applications, Academic Press, London-New York, 1978.

THE TEMPLATE EDGE DETECTOR AND ITS PERFORMANCE

J. Kittler
Technology Division
SERC Rutherford Appleton Laboratory
Chilton, Didcot, OX11 0QX, UK

1. Introduction

The two main prerequisites of accurate and reliable automatic image analysis and interpretation are low noise and distortion free performance of the image forming device and accurate image segmentation.

Before image segmentation can even be attempted, it is essential that the image at the output of the sensing device faithfully represents the sensed scene. To achieve a faithful representation of the imaged scene, it is necessary to model the characteristics of the sensing device and its response. The propagation properties of the media between the scene and the sensor in relation to the physical phenomenon used for image forming can also play a role which should not be ignored in modelling. Additional factors include the geometry of the sensing array, the viewing angle, method of illumination and others.

Based on the models for each of these various aspects of the image forming process, the data at the output of the sensing device can be appropriately corrected for the respective distortions. The reduction or elimination of noise, and image enhancement by filtering, for instance, are also normally considered as integral parts of the problem of representation of the imaged scene.

Once an image corrected for the various forms of distortion is acquired, its automatic analysis would proceed with image segmentation. This step results in partitioning the image into homogeneous parts in which picture elements have similar properties. The aim of this process is to separate an object in the scene from the background or other objects that may be present.

There are various approaches to image segmentation depending on the pixel properties that characterise image segments. The simplest methods are grey level thresholding or colour slicing but these are often inadequate in practical applications. Other possibilities involve the detection of local features such as lines and shapes, or the global matching of an image with, for instance, the templates of objects it contains [1-3, 20].

Objects may also be discernable from the background or from each other by the prescence of rapid image intensity changes which mark object boundaries. A suitable

NATO ASI Series, Vol. F1
Issues in Acoustic Signal/Image Processing and Recognition
Edited by C. H. Chen

image transform that enables us to detect these abrupt changes of grey level values is the gradient operation. Various implementations of this differentiation process which take into account other factors such as object boundary continuity, noise, the inherent grey level discontinuity at the object boundary, and the boundary directionality, then lead to specific edge detection algorithms abundance of which has been proposed in the literature [4, 5].

The majority of edge detection algorithms fall into the category of local methods. A small window (mask, kernel) centred at a pixel is used to detect any local intensity variation which would be indicative of the presence of an edge at the pixel. As a typical example we can mention the KIRSCH operator [6] which is one of the first edge detectors suggested.

The HUECKEL operator [7] on the other hand is representative of regional methods where a step function giving rise to an ideal edge is fitted to the pixel intensities in a region of interest. In global methods [8], edge extraction is viewed as an image filtering problem. Another alternative is to detect edges sequentially using a line following algorithm [3]. Also approaches based on dynamic programming and relaxation methods have been considered for edge detection [9, 10].

The purpose of this paper is to consider in detail one of the local edge detector operators - the SOBEL edge detector [11, 12]. This particular edge detector is very important because of its robustness and ease of implementation. For these reasons it is very popular in the image processing community. Here we shall be concerned with its edge angle and edge magnitude accuracy. First the SOBEL detector edge magnitude and angle will be expressed as a function of the edge direction and subsequently analysed in Section 3. The analysis assumes a particular image acquisition system model which will be described in Section 2 as a prerequisite to our discussions.

In Section 4, a four mask edge detector will be considered and its accuracy studied. It will be shown that this template edge detector not only has a better magnitude accuracy than the SOBEL detector to which it has strong affinity, but it is also computationally simpler. Section 5 is devoted to discussion and conclusions.

2. The Image Acquisition Model

The relationship between the image at the output of the sensing device and the scene the sensor is imaging will, as a rule, affect the performance of any edge detector. The physical process of image acquisition should therefore be taken into account in the edge detector design and analysis.

Several researchers adopted this broader view of the problem in their edge detection studies and assumed various models between the scene and grey level values of its image or directly between the scene and the edge detection output. The notable work in this area is that of HERSKOWITZ and BINFORD [13] who studied the effect on the edges of noise and blurring introduced by the imaging device. O'GORMAN [14] discussed the discrepancy between the actual edge angle and the detected one obtained by fitting a plane to the luminance function in a square window. This discrepancy motivated BROOKS [15] to develop a 2 x 2 edge detector operator which produced an error free response (from the edge angle point of view) for an ideal step luminance function. Other interesting work in this context is that of GRIFFITH [16, 17].

In this paper we shall concentrate on the model considered by IANNINO and SHAPIRO [18] and ABDOU and PRATT [19] which is a quite realistic, although somewhat simplistic representation of the imaging process of standard television cameras. This model will now be introduced in more detail.

Let F (x,y) be the spatial luminance function characterizing the image source in the field of view of the sensor. We assume that the sensor samples the function F(x,y) at the resolution in the x and y directions of N_x and N_y points respectively. This sampling scheme imposes a grid of $N_x N_y$ picture elements on the viewed scene.

The actual grey level values of an image of the scene are obtained by a spatial integration of the function F(x,y) over some finite resolution cell defined by the photodector surface. We shall assume that the area of the cell is identical to the size of one picture element (pixel), that is it is defined by the pixel width and height Δx and Δy respectively, ie

$$\Delta x = \frac{X}{N_x} \qquad \Delta y = \frac{Y}{N_y} \tag{1}$$

where X and Y denote the range of the field of view in the x and y directions. Thus the grey level value at the (k,l)th pixel, G (k,l) is given as

$$G(k,l) = \int_{(k-1)\Delta x}^{k\Delta x} \int_{(\ell-1)\Delta y}^{\ell\Delta y} F(x,y)\,dx\,dy \tag{2}$$

$$k = 1,2,\ldots N_x$$

$$\ell = 1,2,\ldots N_y$$

For the above model of a practical luminance function sampling system, in the next section we shall look in detail at the SOBEL edge detector which is widely used in image processing applications. In the discussion it will be assumed that the effects due to aliasing and the cell overlap in practice are negligible.

3. The Analysis of the SOBEL Edge Detector

The basic SOBEL edge detection operator [11, 12] is defined in terms of two 3 x 3 masks which measure the edge signal values in the x and y directions respectively. Suppose that the 3 x 3 operator is centred at an arbitrary pixel (k,l) and let the pixel values in the window be denoted using the convention of Fig.1. Then the edge signals in the x and y directions at the pixel (k,l) are defined as

$$e_x = c + Kf + i - [a + Kd + g] \tag{3}$$

$$e_y = a + Kb + c - [g + Kh + i] \tag{4}$$

In practice the coefficient K is normally set to 2 but this particular value has an intuitive basis only. For the purpose of the analysis and subsequent discussion here we shall treat K as a general constant.

a	b	c
d	e	f
g	h	i

Figure 1

In terms of these x and y components the edge signal magnitude yielded by the SOBEL edge detector is given by

$$E = \sqrt{e_x^2 + e_y^2} \tag{5}$$

and the edge angle as

$$\alpha = \tan^{-1} \frac{e_y}{e_x} \tag{6}$$

Intuitively the SOBEL edge detector is very simple and reasonable and its performance therefore should be non-controversial. Nevertheless, it is pertinent to ask, whether an ideal edge in the scene imaged by a device that can be modelled as in Section 2, will be detected by the SOBEL operator with the correct magnitude and angle. In the remaining part of this section we shall address ourselves to this problem.

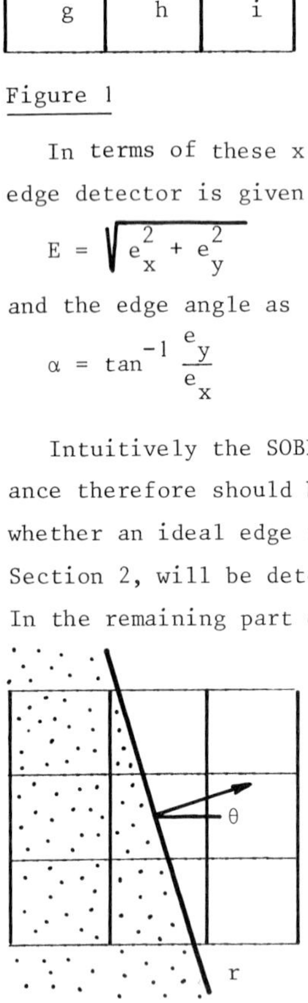

Figure 2

Let us consider the simplest possible case where a straight edge r of an object in the scene passes through the centre of a pixel at an angle θ as illustrated in Fig.2. We shall assume that the lighting conditions and the object are such that the luminance function F (x,y) in the 3 x 3 window centred at the pixel is given by

$$F(x,y) = \begin{cases} B & \forall\ (x,y) \text{ such that } r(x,y) \geq 0 \\ D & \forall\ (x,y) \text{ such that } r(x,y) < 0 \end{cases} \tag{7}$$

F (x,y) is therefore a two-dimensional step function which changes its value from D to B at the object boundary r. D is the background luminance and B is the object luminance.

According to the adopted model, at the output of the sensor, the grey level values of the pixels in the window in the interval $0 \leq \theta \leq \tan^{-1} 1/3$ will be

$\underline{0 \leq \theta \leq \tan^{-1} 1/3}$

$$\begin{aligned} a &= d = g = D \\ c &= f = i = B \\ b &= B\ (0.5 + \tan\theta) + D\ (0.5 - \tan\theta) \\ h &= 0.5\ (B + D) - (B - D)\tan\theta. \end{aligned} \tag{8}$$

Denoting

$$\eta = \frac{(3\tan\theta - 1)^2}{8\tan\theta} \tag{9}$$

and

$$\varepsilon = \frac{(1 - \tan\theta)^2}{8\tan\theta} \tag{10}$$

the window pixel intensities in the interval $\tan^{-1} 1/3 \leq \theta \leq \tan^{-1} 1$ can be written

$$\begin{aligned} d &= g = D \\ c &= f = B \\ a &= (1 - \eta)\ D + \eta B \\ b &= \varepsilon D + (1 - \varepsilon)\ B \\ h &= (1 - \varepsilon)\ D + \varepsilon B \\ i &= \eta D + (1 - \eta)\ B \end{aligned} \tag{11}$$

Owing to symmetry, it is not necessary to carry out the analysis for the edge angles θ in the range $45° < \theta < 360°$.

We shall now consider the SOBEL edge detector outputs for the above pixel grey level values. In the interval $0 \leq \theta \leq \tan^{-1} 1/3$ we have

$$\begin{aligned} \varepsilon_x &= (B - D)\ (2 + K) \\ \varepsilon_y &= 2K\ (B - D)\tan\theta \end{aligned} \tag{12}$$

Substituting (12) into (6) we find

$$\tan\alpha = \frac{2K\tan\theta}{2 + K} \tag{13}$$

As the detected edge angle α should equal the actual edge angle θ we have the following condition for K

$$\frac{2K}{2 + K} = 1 \tag{14}$$

which will be satisfied if K = 2. Thus by setting the constant K to 2 as originally suggested by SOBEL, the edge detection operator will correctly detect the true angle of the edge direction at the boundary between the object and the background.

Let us now consider the edge signal magnitude. If the resolution of the imaging device was infinite then the edge signal magnitude corresponding to an ideal luminance step function would be the Dirac delta function. However, the resolution of a practical sensor is finite and, therefore, the edge signal magnitude function will be a pulse of at least one pixel width and of finite amplitude. For the sensor considered here the actual pulse amplitude corresponding to the step function in (7) will be given by the difference between the grey level intensities B and D of the bright object and the dark background respectively, ie

$$E_a = B - D \tag{15}$$

(taking the resolution Δx and Δy to be unity).

By definition, the magnitude detected by the SOBEL edge operator is

$$E = \sqrt{(2 + K)^2 + (2K)^2 \tan^2 \theta}\ (B - D) \tag{16}$$

For K = 2 the magnitude becomes

$$E = 4 \sqrt{1 + \tan^2 \theta}\ E_a \tag{17}$$

From (17) it is apparent that in order to obtain the actual magnitude E_a, the measured signal E must be modified (divided) by the factor $4\sqrt{1 + \tan^2 \theta}$.

From (11), in the interval $\tan^{-1} 1/3 \leq \theta \leq \tan^{-1} 1$ the edge signal in the x and y directions is given as

$$\begin{aligned} e_x &= (B - D)\ (2 + K - 2\eta) \\ e_y &= (B - D)\ (2\eta + K - 2K\varepsilon) \end{aligned} \tag{18}$$

For the detected edge angle we now have

$$\tan \alpha = \frac{(3 \tan \theta - 1)^2 + 4K \tan \theta - K\ (1 - \tan \theta)^2}{(2 + K)\ 4 \tan \theta - (3 \tan\theta - 1)^2} \tag{19}$$

From the requirement that the detected angle α be equal to θ, ie

$$\tan \alpha = \tan \theta \tag{20}$$

we get the following condition on K

$$K = \frac{9 \tan^3 \theta - 5 \tan^2 \theta - 5 \tan\theta + 1}{5 \tan^2 \theta - 6 \tan \theta + 1} \tag{21}$$

From (21) it follows that, for $\theta > \tan^{-1} 1/3$, the optimal coefficient K in the sense of the requirement (20), is a function of the actual edge angle θ. When $\tan \theta = 1$ ($\theta = 45°$) using the l'Hospital rule it can be shown that

$$\lim_{\theta \to 45°} K = 3 \tag{22}$$

Thus the optimal value of the coefficient K varies (as a function of θ) between K = 2 and K = 3. As the actual edge angle θ is not known a priori and in any case it will vary within one scene, the optimal K cannot be predetermined.

To complete our analysis of the SOBEL edge detector for the case when the ideal edge passes through the centre of a pixel, we now derive the edge signal magnitude. From (5), (15) and (18) we have

$$E = E_a \sqrt{(2 + K - 2\eta)^2 + (2\eta + K - 2K\varepsilon)^2} \tag{23}$$

Setting K = 2 we can now analyse the function E/E_a. Taking the derivative of E/E_a with respect to the angle θ and setting it to zero, we find the condition for the function extrema, ie

$$130 \tan^4 \theta - 156 \tan^3 \theta + 28 \tan \theta - 2 = 0 \tag{24}$$

It is easy to verify that

$$\tan \theta = 1 \tag{25}$$

is one root of the equation. Dividing (24) by ($\tan \theta - 1$) we get

$$130 \tan^3 \theta - 26 \tan^2 \theta - 26 \tan \theta + 2 = 0 \tag{26}$$

Two roots of the equation in (26) lie outside the interval of interest for θ. The only remaining root of interest is

$$\tan \theta = 0.5251 \tag{27}$$

At these two extrema points the function E/E_a takes the following values respectively

$$\frac{E}{E_a} = 4.30567 \qquad \text{at } \tan \theta = 0.5251 \tag{28}$$

$$\frac{E}{E_a} = 4.2426 \qquad \text{at } \tan \theta = 1 \tag{29}$$

It should be recalled that the measured edge signal magnitude E is the aggregation of contributions from the various pixel grey level value differences in the 3 x 3 window, weighted by either 1 or K. Thus to obtain the normalized edge signal magnitude $\bar{E}$, the SOBEL detector magnitude output E must be scaled by a factor (2 + K), ie

$$\bar{E} = \frac{E}{2 + K} \tag{30}$$

In the particular case of K = 2, the scaling factor equals 4. In terms of the normalized edge signal $\bar{E}$ the actual edge magnitude E_a is from (28) and (29)

$$E_a = \bar{E} \cdot 0.929 \qquad \tan \theta = 0.5251 \tag{31}$$

$$E_a = \bar{E} \cdot 0.943 \qquad \tan \theta = 1 \tag{32}$$

Thus the error between the edge signal magnitude detected and the actual magnitude can be as much as 7% in the noise free case considered here.

We can conclude from the analysis that the standard SOBEL edge detector (K = 2) will give the correct edge angle only for θ from the interval $0 \leq \theta \leq \tan^{-1} 1/3$. At larger angles θ, the maximum error between the angle detected and the actual one will be of the order of 2° (under estimate). The magnitude will be correct only when the actual edge is vertical or horizontal. At any other angle the detected edge signal magnitude will be greater than the actual one but the error will not exceed 7%.

The SOBEL edge detector gives, therefore, a relatively good balanced performance. It cannot be overemphasised however, that the analysis from which this conclusion is drawn is based on a number of simplifying assumptions. In practice, additional factors will affect the detector performance. These include quantization errors, edge displacement from the centre of the pixel, noise and various artifacts.

In some situations it may be desirable to use an edge detector with a better magnitude and direction accuracy than that of the standard SOBEL detector. This requirement motivated the development of the iterative SOBEL edge detector [18]. While this edge detector corrects the angular error of the standard SOBEL operator, this is achieved only at the expense of degrading its magnitude accuracy [22]. Error-correcting formulae for both the magnitude and angle of the SOBEL edge detector are given in [22].

Not surprisingly, both the iterative and error correcting edge detectors are computationally more involved than the standard SOBEL edge detector. The computational complexity of a detector can be the decisive factor in its applicability. As a matter of fact, in many instances even the SOBEL detector cannot be used for computational reasons. A simplified method of calculating the edge magnitude is often adopted wherein the expression (5) is approximated by

$$E = |e_x| + |e_y|. \tag{33}$$

This method requires only two absolute value operations and one summation but the error on the edge signal magnitude can be almost as high as 50%.

In the following section we shall consider an alternative method of reducing the computational complexity of the SOBEL edge detector. In the method, the edge detection operator is defined in terms of four masks (templates) instead of only two. The use of this template edge detector has been suggested by ABDOU and PRATT [19] for applications where a coarse quantization of the detected edge directions is adequate. Here it will be shown that the template edge detector not only is computationally simple but also that its magnitude is more accurate than that of the SOBEL detector.

4. The Template Edge Detector and its Performance

In the case of the SOBEL operator the edge signal magnitude and angle are defined in terms of two 3 x 3 kernels outputing the edge signals e_x and e_y in the two orthogonal x and y directions. We shall now consider an edge detector defined by four 3 x 3 kernels detecting the edge signal components in four directions 45° apart. Thus in addition to the basic SOBEL masks giving the signals e_x and e_y, denoted in the following e_1 and e_3 respectively, we have two other masks which yield the signals e_2 and e_4, ie

$$e_2 = b + Kc + f - d - Kg - h \tag{34}$$

and

$$e_4 = Ka + b + d - f - h - Ki \tag{35}$$

As we are concerned with the ease of implementation of the detector, we shall fix K to 2 (integer value).

We shall now define the edge signal magnitude of the template edge detector as

$$E = \max_i \left\{ |e_i| \right\} \tag{36}$$

It is apparent that the calculation of the output of this edge detector does not involve any arithmetic operation on the edge signal components e_i apart from the absolute value operation and comparisons.

To study the magnitude accuracy of the template edge detector let us analyse the dependence of the edge signal components e_i, $\forall_i$, on angle θ. By analogy to Section 3, we shall derive the expressions for e_2 and e_4 for the intervals $0 \leq \theta \leq \tan^{-1} 1/3$ and $\tan^{-1} 1/3 \leq \theta \leq \tan^{-1} 1$ separately. Noting the pixel intensity values in (8) - (11) and using the notation in (15) we can write:

<u>$0 \leq \theta \leq \tan^{-1} 1/3$</u>

$$\begin{aligned} e_1 &= 4E_a \\ e_2 &= E_a\,[3 + 2\tan\theta] \\ e_3 &= 4E_a \tan\theta \\ e_4 &= E_a\,[2\tan\theta - 3] \end{aligned} \tag{37}$$

<u>$\tan^{-1} 1/3 \leq \theta \leq \tan^{-1} 1$</u>

$$\begin{aligned} e_1 &= E_a\,(4 - 2\eta) \\ e_2 &= E_a\,(4 - 2\varepsilon) \\ e_3 &= E_a\,(2\eta + 2 - 4\varepsilon) \\ e_4 &= E_a\,(4\eta - 2 - 2\varepsilon) \end{aligned} \tag{38}$$

where ε and η are defined in (10) and (9).

Owing to the use of the four templates rather than two, we now have a symmetry along four axes and, therefore, it is sufficient to investigate the angle dependence of the components e_1 and e_2 only without any loss of generality. From (37) it follows that after normalization the component e_1 will give the correct value of the edge signal magnitude in the whole interval $0 \leq \theta \leq \tan^{-1} 1/3$. It is easy to verify that in this interval neither of the components e_2, e_3 and e_4 assume an absolute value that would exceed $|e_1|$. We shall concentrate therefore on the interval $\tan^{-1} 1/3 \leq \theta \leq \tan^{-1}$

1.

First it should be noted that at $\theta = \tan^{-1} 1/3$ the (normalised) component e_1 still gives the correct value of the edge magnitude. When θ increases, the absolute value of the component e_1 decreases. The edge signal component e_2 on the other hand assumes an inferior value to the edge magnitude at $\theta = \tan^{-1} 1/3$ but this value monotonically increases with θ until $\theta = \tan^{-1} 1$. At this point the (normalised) component e_2 becomes identical with the true edge magnitude. Thus there is an angle θ in the interval of interest where

$$|e_1| = |e_2| \tag{39}$$

at which the error between the detected and true edge signal magnitude is maximum. Substituting from (38) into (39) it is easy to show that the point of maximum error is given by

$$\theta = \tan^{-1} 0.5 \tag{40}$$

At this point the normalised detected edge magnitude $\bar{E}$ is related to the true magnitude as

$$\frac{\bar{E}}{E_a} = 0.9845 \tag{41}$$

Again it would be easy to verify that in this interval the components e_3 and e_4 assume values which are inferior (in the absolute sense) to those of e_1 and e_2.

We can conclude that the error between the detected and true magnitudes is less than 1.6%. It is interesting to note that this error is more than four times better than that of the conventional SOBEL edge detector which, in addition, is computationally more complex.

Finally, a few remarks are in order concerning the edge angle output of the template edge detector. If we are interested only in a coarse 45° angle quantization of the edge directions, then this information is also provided directly by the edge signal components. We can then consider the detected edge direction to be consistent with the directionality of the template giving the maximum output, ie

$$\alpha = [1+i-2\,\mathrm{sgn}\, e_i]\frac{\pi}{4} \qquad \text{if } E = |e_i| \tag{42}$$

where $\mathrm{sgn}\, e_i$ is the sign function taking values plus or minus one depending on whether e_i is positive or negative.

In many situations however this coarse quantization of the edge angle is not adequate and the edge angle must be computed in the usual way as in (6), ie

$$\alpha = \tan^{-1} \frac{e_3}{e_1} \tag{43}$$

or using the error-correcting formula derived in [22].

5. Discussion and Conclusions

In the above analysis a number of simplifying assumptions has been made. To start with, we have assumed that the grey level values can be measured with infinite precision. Additional errors will be introduced when the light intensities are digitised with a finite number of bit accuracy. However, if the required edge detector output angular accuracy is of the order of 1° the effect of the quantization errors introduced by conventional 8 bit A - D convertors will be negligible.

Second, our model assumes that the edge passes through the centre of a pixel. With the exception of vertical, horizontal and diagonal edges, if the edge passes through the centre of one pixel, it will be displaced from the centre at adjacent pixels. The effect of edge displacement has been studied experimentally by IANNINO and SHAPIRO [18], and by ABDOU and PRATT [19].

Also the practical luminance function giving rise to edges will never approach the ideal step function. The effect of shadows and non-uniform lighting will also affect the performance of edge detectors. Moreover, in general the object boundaries will not be straight. Thus even from this point of view the adopted model is only a crude approximation of the real situation. A detailed analysis should, in addition, consider the effects of noise and, if relevant, the sensor cell overlap.

In conclusion, the reported results can be considered only as the first step towards a comprehensive analysis of the SOBEL and template edge detectors. Nevertheless, it is believed that the findings usefully contribute to the knowledge and understanding of the detectors. In particular it is interesting to note that the computationally simple template edge detector has a better magnitude accuracy than the standard SOBEL operator.

References

1. Rosenfeld, A., "Image pattern recognition", Proc. IEEE, Vol.69 (1981), pp 596-605.

2. Ullmann, J. R., "Aspects of visual automation", in Physical and biological processing of images, Springer Verlag, Berlin (1983).

3. Rosenfeld, A. and Kak, A. C., Digital picture processing, Academic Press, New York (1976).

4. Davis, L. S., "A survey of edge detection techniques", Computer Graphics and Image Processing, Vol.4 (1974), pp. 248-270.

5. Levialdi, S., "Finding the edge", in Digital image processing, J. C. Simon and R. M. Harralick eds. D. Reidel, Dordrecht (1981).

6. Kirsch, R. A., Cahn, L., Ray, C. and Urban, G. H., "Experiments in processing pictorial information with a digital computer", Proc. Eastern Joint Computer Conference (1957), pp. 221-229.

7. Hueckel, M. H., "An operator which locates edges in digital pictures", Journal A C M, Vol.18 (1971), pp. 113-125.

8. Modestino, J. W. and Fries, R. W., "Edge detection in noisy images using recursive digital filtering", Computer Graphics and Image Processing, Vol.6 (1977), pp. 409-433.

9. Lenkin, P., "The boundary trace transform", Computer Graphics and Image Processing, Vol.9 (1979), pp. 150-165.

10. Schachter, B. J., Lev, A., Zucker, S. W. and Rosenfeld, A., "An application of relaxation methods to edge reinforcement", IEEE Trans. Systems, Man, and Cybern., Vol.SMC - 7 (1977), pp. 813-816.

11. Tennenbaum J. M., Kay, A. C., Binford, T., Falk, G., Feldman, J., Grape, G., Paul, R., Pingle, K. and Sobel, I., "The Stanford hand-eye project", Proc. IJCAI (1969), pp. 521-526.

12. Duda, R. O. and Hart, P. E., Pattern classification and scene analysis Wiley, New York (1973).

13. Herskowitz, A. and Binford, T. O., "On boundary detection", A. I. Memo 183, MIT (1970).

14. O'Gorman, F. O., "Edge detection using Walsh functions", AISB Summer Conference (1976), pp. 195-206.

15. Brooks, M. J., "Rationalizing edge detectors", Computer Graphics and Image Processing, Vol.8 (1978), pp. 277-285.

16. Griffith, A. K., "Edge detection in simple scenes using a priori information", IEEE Trans. Computers, Vol.C - 22 (1973), pp. 371-381.

17. Griffith, A. K., "Mathematical models for automatic line detection", J. Assoc. Comput. Mach., Vol.20 (1973), pp. 62-80.

18. Iannino, A. and Shapiro, S. D., "An iterative generalization of the Sobel edge detection operator", Proc. IEEE Conf. Pattern Regognition and Image Processing (1979), pp. 130-137.

19. Abdou, I. E. and Pratt, W. K., "Quantitative design and evaluation of enhancement/thresholding edge detectors", Proc. IEEE, Vol.67 (1979), pp. 753-763.

20. Pratt. W. K., Digital image processing, Wiley, New York (1978).

21. Brady, M., "Artificial intelligence approaches to image understanding", in Pattern recognition theory and applications, J. Kittler at al eds., D. Reidel, Dordrecht (1982).

22. Kittler, J., "On the accuracy of the Sobel edge detector" (submitted for publication).

ON A SPATIAL CLUSTERING ALGORITHM FOR IMAGE ANALYSIS

C. H. Chen
Electrical Engineering Department
Southeastern Massachusetts University
North Dartmouth, MA 02747 U.S.A.

Abstract

A computationally efficient spatial clustering algorithm is presented for image segmentation and compression. The algorithm can automatically determine the minimum number of clusters and can also work on a specified number of clusters. Examples are given on the processing of Seasat images using the algorithm.

Introduction

Besides the use of acoustic sensors, remotely sensed images can provide essential information on the object extraction and tracking in the ocean environment. Seasat SAR images are a good example. Many automatic image analysis algorithms have been developed. Such algorithms are generally application dependent. For remote sensing images, cluster analysis is important as it reveals the structure of the data from which useful information can be derived. Conventional clustering methods do not preserve the spatial relations in a image. Spatial clustering for image analysis has been considered [1][2]. However feature extraction was not taken into account. Furthermore the computation involved is quite extensive. A more efficient spatial clustering algorithm is developed for minicomputer processing, that employs properly selected features. The clustered image shows various regions (segments) from which desired objects may be extracted. Furthermore considerable image data compression is accomplished with essentially no loss of information. Examples are based exclusively on Seasat images dealing with the ocean environment.

Algorithm for Spatial Clustering

The algorithm proceeds as follows:

(1) Form a feature set, for each pixel, consisting of local mean and gradient. Other features may also be used.

(2) For each 2x2 subarea, measure the mean vector and dispersion.

(3) Determine the critical dispersion, and calculate the merging distance d.

(4) Merge adjacent subareas with distance less than d tc form subregions. Calculate the mean vectors of subregions.

(5) Group these mean vectors into clusters using K-mean algorithm which converges to several cluster centers representing the mean vectors of regions.

For a given inter-region threshold distance, the algorithm can automatically adjust to an appropriate number of clusters. If the number of clusters is specified as in conventional cluster analysis, the algorithm will provide the specified number of clusters.

NATO ASI Series, Vol. F1
Issues in Acoustic Signal/Image Processing and Recognition
Edited by C. H. Chen

The image speckles in synthetic aperture are multiplicative in nature. Without removing such noise, the cluster results may still be very noisy. A simple preprocessing method is to use the Sigma filter suggested by J.S. Lee [3]. For each 5x5 (or 3x3) subarea the average gray level value is compared with the three-standard deviation (3σ) of the normalized image histogram (first order probability density). If the value is within 3σ then the center pixel is replaced by the average value. If the value exceeds 3σ, then it is an indication of edges or object boundary and the original gray level is retained. The procedure thus provides a compromise between noise filtering and edge preserving and adds only slight amount of computation to the clustering algorithm.

Computer Results

The original Seasat images are all of 256 gray levels. For convenience with minicomputer processing the digitized pictures are all reduced to 256x256 pixels even though the original images are much larger in size. The cluster algorithm is set such that a maximum of 7 clusters is selected. Figure 1 is for the scene of a ship off Chesapeake Bay. Figure 1a is the original. Figure 1b is the spatially clustered image. Figure 1c has the histograms of original (left) and clustered (right) images. Figure 1d uses the Sigma filter preprocessing with 3x3 subarea while Fig. 1e is the corresponding result with 5x5 subarea. Preprocessing with 5x5 subarea followed by spatial clustering appears to be the best. Figure2 is the scene of Anchorage, Alaska with Fig. 2a for original and Fig. 2b for 5x5 preprocessing followed by spatial clustering. Figure 3 is the scene of Nantucket Shoals with Fig. 3a for original, Fig. 3b for 5x5 median filtering and Fig. 3c for 5x5 preprocessing followed by spatial clustering. Computer results all show that the "natural" clusters of the original images are very much preserved while noises are considerably reduced, and at the same time the contrast is enhanced, with the use of the spatial clustering algorithm.

Acknowledgement

This work was suported by ONR Contract N00014-79-C-0492. The digital Seasat SAR images were kindly provided by Dr. Jong-Sen Lee of Naval Research Laboratory who also brought to our attention of his work on filtering of multiplicative noises.

References

1. R.M. Haralick and I. Dinstein, "A spatial clustering procedure for multi-image data", IEEE Trans. on Circuits and Systems, Vol. C-22, pp. 440-450, 1975.
2. Y. Fukada, "Spatial clustering procedures for region analysis", Pattern Recognition, Vol. 12, no. 6, pp. 395-403, 1980.
3. J.S. Lee, "The Sigma filter and its application to speckle smoothing of synthetic aperture radar image", presented at the ONR Signal Processing in the Ocean Environment Workshop held at U.S. Naval Academy, May 1982.

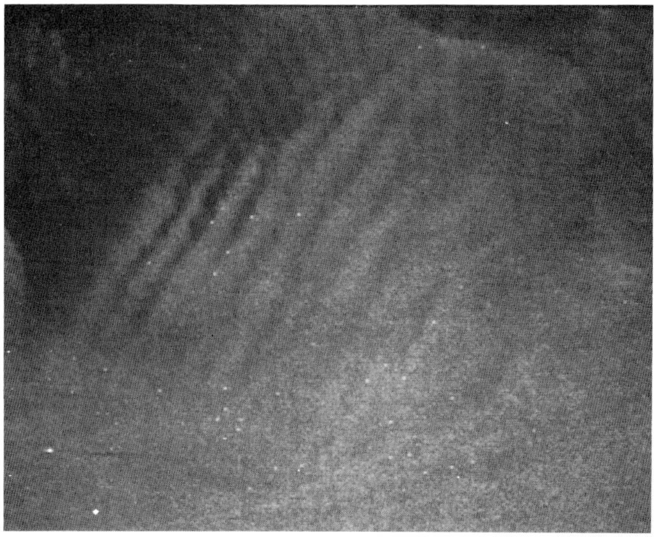

Fig. 1a

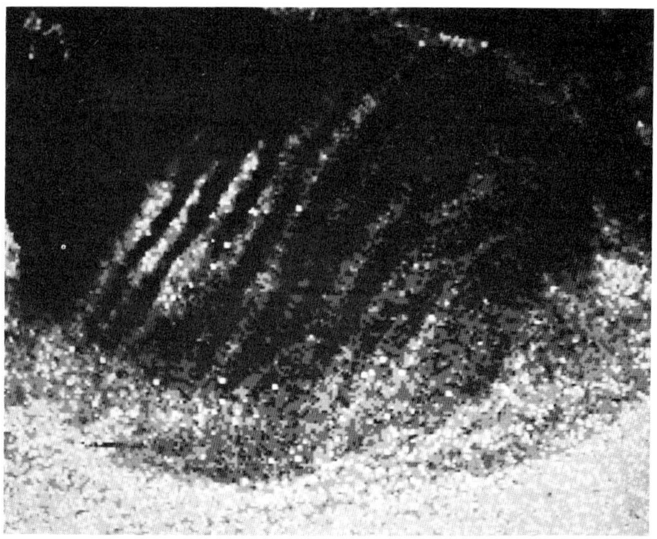

Fig. 1b

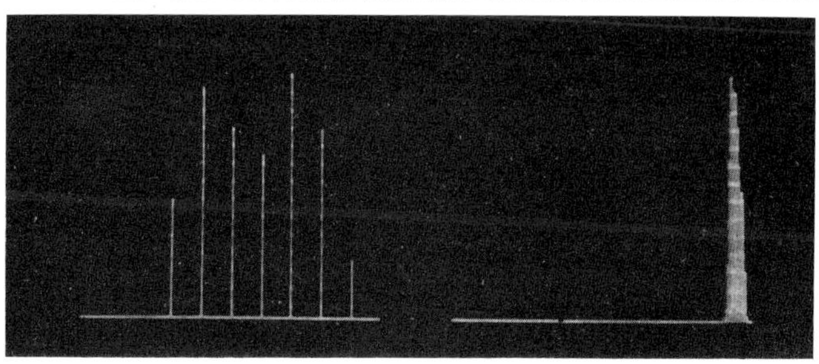

Fig. 1c

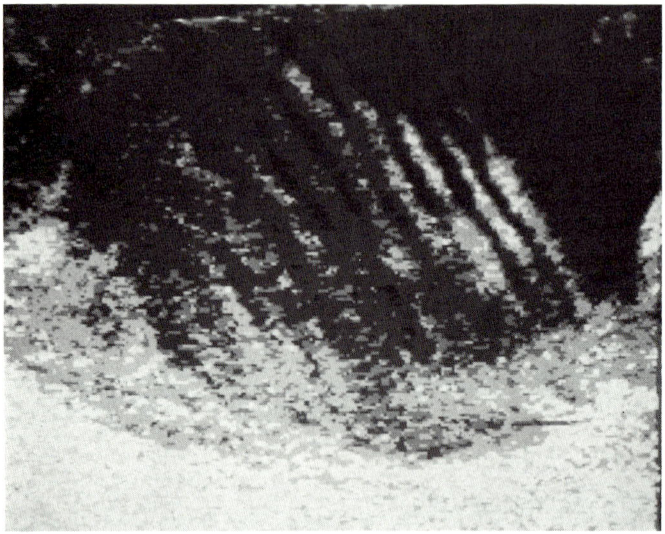

Fig. 1d

Fig. 1e

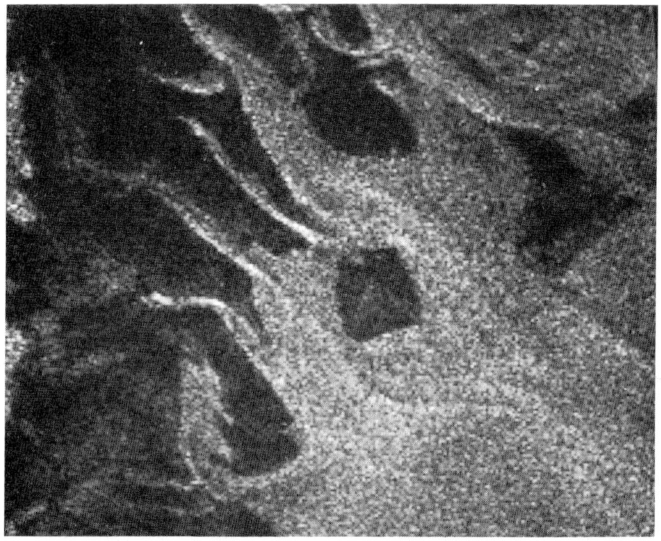

Fig. 2a

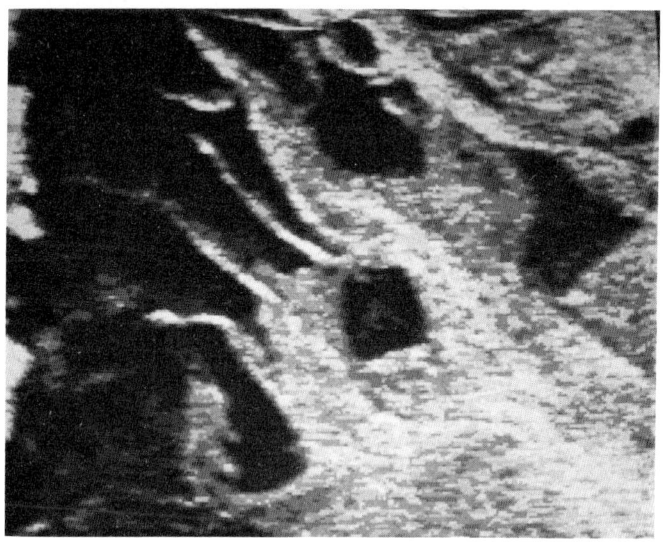

Fig. 2b

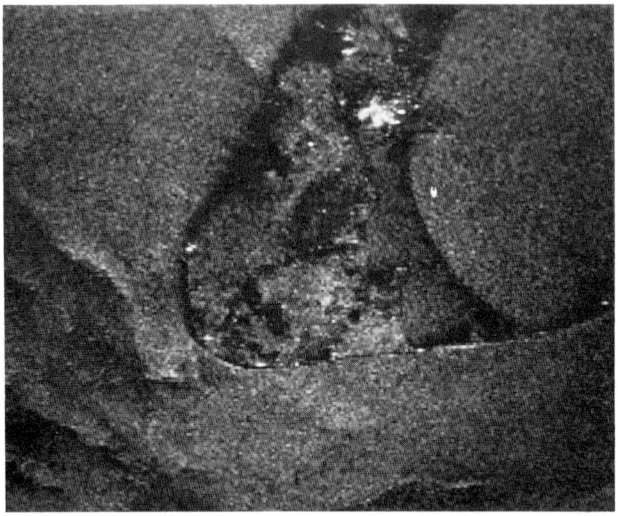

Fig. 3a

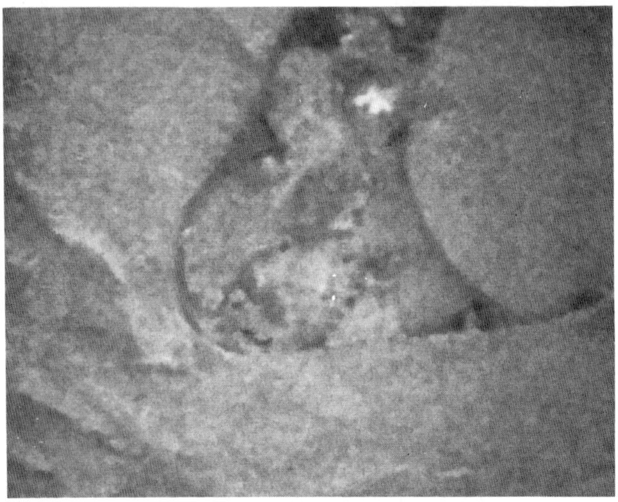

Fig. 3b

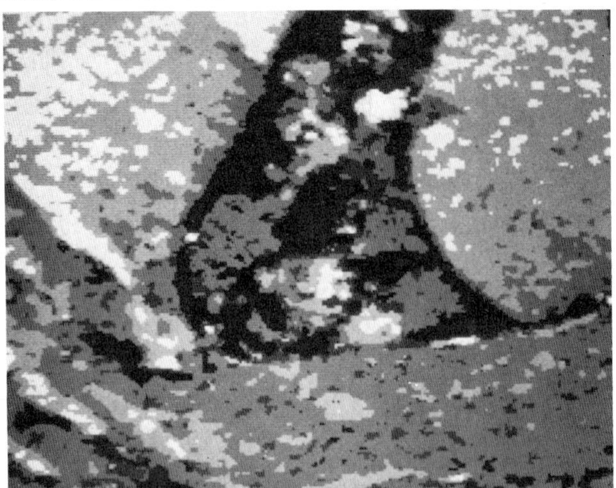

Fig. 3c

DIGITAL IMAGE PROCESSING BY MEANS OF GENERALIZED SCALE-INVARIANT FILTERS (°)

C.Braccini, G.Gambardella, A.Grattarola

Istituto di Elettrotecnica - Università di Genova
Viale F.Causa, 13 - Genova - Italy

Abstract

A 2-D linear system is said to be a generalized scale-invariant filter if its weighting function is such that the effect of a scale change of the input image is merely a (generally different) scale change of the output. This property cannot be satisfied by shift-invariant systems. After a discussion on the use of such filters in the framework of scale-independent pattern recognition, the most general class of generalized scale-invariant shift-variant 2-D systems is presented and two different implementation techniques are described, one based on the use of the Mellin transform and the other one exploiting suitable coordinate transformations in conjunction with linear shift-invariant systems.

1. Introduction

One of the research fields that strongly pushes (and takes advantage from) the development of digital image processing techniques is pattern recognition and detection. This field, on its turn, is pushed forward by some remarkable practical (mainly industrial) interests, such as those concerned with the area of robotic vision. Here, however, several research topics still have to be investigated before reliable and effective algorithms can be put to work in "the real

---

(°) This work has been partially supported by the Consiglio Nazionale delle Ricerche (C.N.R.).

NATO ASI Series, Vol. F1
Issues in Acoustic Signal/Image Processing and Recognition
Edited by C. H. Chen

world" (such as the industrial environment). Among the above topics, or "open" problems, a fundamental one is the automatic recognition of solid objects arbitrarily located with respect to the recognition system (i.e. objects "seen" through an unknown perspective). A similar though simpler problem (since it is somehow contained in the previous one) is the recognition of two-dimensional objects (i.e., "patterns", in the traditional sense) independently from their scalings, rotations and translations. This problem has recently been the object of several works and some different solutions have been suggested corresponding to different image processing techniques, such as the optical, digital and hybrid (i.e. optical and digital) techniques [1÷3].

In order to go more deeply into the above problem (which is, as we will see, the object of this paper), let us assume, as further simplifications, that no rotations and translations but only arbitrary scalings may affect the patterns to be recognized and that the input to the recognition system can be written as

$$r(x,y) = f_j(ax, ay) + n(x,y) \tag{1}$$

where $f_j(x,y)$ is one among N patterns, $n(x,y)$ is the noise and "a" is a real, unknown, scale factor.

The problem to be solved is deciding, on the base of the observation $r(x,y)$, which among the N patterns $\{f_j(x,y)\}$ is present in it (notice that the patterns have a fixed position in the input plane).

We know that under the hypothesis that $n(x,y)$ is an independent white noise, and "a" a constant factor (e.g. a=1), the problem can be solved - according to the decision theory - by means of a set of matched filters (matched, of course, to the patterns), or, which is the same, by means of the correlation functions between the input $r(x,y)$ and all the possible patterns $\{f_j(x,y)\}$. Such correlation functions should be evaluated only at the origin of the correlation space (which, on the other hand, is coincident with the output space of the matched filters) before being compared with each other.

The above classical solution extends to the case when "a", in Eq. (1), is a random variable; in such a case, however, as many matched filters (or correlators) are required for each pattern, as the possible values of "a". In practice such a number can be kept low enough, by properly quantizing "a" in its range. As an alternative, one may perform on r(x,y) a coordinate mapping before using the matched filters. If, for instance, one maps (x,y) into (x',y'), where x'= lg x and y'= lg y, then r(x,y) becomes

$$r'(x',y') = f'_j\,(x'+a',\ y'+a') + n'(x',y')$$

so that in the new plane, i.e. (x',y'), only one matched filter can be used for each pattern, provided that the outputs of the filters are now evaluated along the whole axis x'=y' (and not merely at the origin), looking for their maximum or, more rigorously, looking for the output values corresponding to the maximum conditional probabilities. Such output values must then be properly compared with each other in order to decide which of the N patterns is present in r(x,y). The position of the "winning" output value on the axis x'=y' will furtherly give (with maximum probability) the value of the scaling factor "a".

A further alternative to the above solution, which also allows using a single matched filter for each pattern to be recognized, is that [3] of using a set of filters with impulse responses of the type:

$$h_j\,(x,y) = \sum_m f_j\,(m\,\Delta_a x,\ m\,\Delta_a y)\ b(-x + m\,\Delta_a,\ -y + m\,\Delta_a) \qquad (2)$$

where m is an integer, Δ_a is a quantization interval of the scale factor "a", and b(x,y) is an auxiliary white noise whose autocorrelation function is $\delta\,(x,y)$. It can be easily checked that, if the input r(x, y) is multiplied by b(x,y), then the output of the above filter is approximately given by:

$$\sum_m \left[\iint r(x,y) f_j\,(m\,\Delta_a x,\ m\Delta_a y)\,dx\,dy\right]\ \delta(\xi - m\,\Delta_a, \eta - m\,\Delta_a)$$

which consists of as many values (along the axis $\xi = \eta$ of the output plane) as are the values of m, i.e. the quantized levels of "a". Such output values are compared with each other and used exactly in the same way as the output values of the matched filters mentioned above (i.e. within the previous type of solution).

It is worth noticing that the latter type of solution, i.e. based on the filters of Eq. (2), can be extended to a class of pattern recog nition problems wider than that represented by Eq. (1).

We have so far mentioned two different solutions of the pattern recognition problem represented by Eq. (1): they are both based on the use of the matched filters and they are both characterized by the use of a single matched filter for each pattern to be recognized.

The alternative approach to such techniques, equally aimed to solve the pattern recognition problem represented by Eq. (1), is the linguistic approach (as opposed to the signal theory approach considered above) [4]. This latter implies that the patterns $\{f_j(x,y)\}$ are symbolically represented by means of a sequence of geometric features (such as edges) characterizing them: the recognition process takes place by properly comparing the stored representation of the patterns with the representation that can be extracted from the input. It is worth noticing that this latter representation, because of the geometric nature of the above mentioned features, can be made to be independent of the size of the pattern to be recognized (i.e. independent of the scale factor "a").

Let us point out now that, whatever approach is chosen (i.e. the one based on the matched filters or the linguistic one), the performance of the recognition system is highly dependent on the degree of "fidelity" between the patterns stored within the recognition system and the patterns belonging to the input. If this latter is directly acquired from the real world, as part of a scene that is analyzed, then the acquired patterns come out to be strongly dependent on the luminance conditions of the environment (i.e. the scene) and on some other variability factors.

As a consequence, the above mentioned "degree of fidelity", between the stored and acquired patterns, may turn out to be not good enough.

In order to avoid (or at least reduce) such undesirable effects, it is convenient to base the recognition process on "contour patterns" rather than "pictorial patterns". Such a choice would also improve the speed of the recognition process whenever this latter is based on digital signal processing.

If the patterns to be recognized must be "contours", the need arises of extracting them from the input before performing the true rec ognition process. In other words, the input must be pre-processed in order to perform a "contour extraction" or, at least, a "contour enhancement" operation.

If, however, we want still the above mentioned (scale independent) recognition techniques to hold and to be applied, it is necessary that the pre-processing (aimed to extract, or enhance, the contours) be of the "form-invariant type", i.e. such that the input of Eq. (1) turns out to be transformed into a new input which still can be written in the form of Eq. (1). In other words, the new input must consist of a "pre-processed" noise added to a "pre-processed" signal which still appears to be a randomly scaled replica of one single signal. In formal terms this can be restated by saying that the pre-processing must be such as to transform $r(x,y)=f_j(ax, ay) + n(x,y)$ into $r_p(x,y) = \alpha g_j(\beta x, \beta y) + n_p(x,y)$, where α and β are real functions of the scale factor "a".

The problem that we face and solve in this work is the search and the analysis (also from the implementation point of view) of the linear pre-processing techniques which satisfy the above stated form-invariance property. We have shown, in fact, that such techniques do exist and actually consist of a well defined class of linear-shift-variant (LSV) filters, which will be referred to as generalized form--invariant filters.

The generalized form-invariant filters will be described in the next section of this paper, while the following two sections are devoted to two different implementation techniques of such filters. The first technique (section 3) is based on the Mellin transform, while the second one (section 4) is based on the linear-shift-invariant (LSI) filters in conjunction with proper coordinate mappings.

2. The generalized scale-invariant filters

We define a 2-D linear filter to be a generalized scale-invariant filter if, being f(x,y) and g(x,y) any couple of input and output signals, the output corresponding to the scaled version of the input

$f(a_1x,\ a_2y)$ turns out to be $\alpha\, g(\beta_1x,\ \beta_2y)$, where α, β_1 and β_2 are real funtions of the real numbers a_1 and a_2. When, in particular, $\beta_1=a_1$ and $\beta_2=a_2$, the filter is referred to as simply scale-invariant.

Based on the general input-output relationship of a 2-D linear filter, the equation defining a generalized scale-invariant filter is the following:

$$\iint_{-\infty}^{+\infty} f(a_1\xi,\ a_2\eta)w(\xi,\eta;\ x,y)d\xi\, d\eta =$$
$$= \alpha \iint_{-\infty}^{+\infty} f(\xi,\eta)w(\xi,\eta;\ \beta_1x,\ \beta_2y)d\xi\, d\eta \qquad (3)$$

If in the above equation the change of variables $a_1\xi = \xi'$ and $a_2\xi = \xi'$ is performed, it is easily seen that the equation is satisfied, for any input $f(x,y)$, if and only if

$$w(\xi,\eta;\ \beta_1x,\ \beta_2y) = \frac{1}{a_1\cdot a_2\cdot\alpha}\ w\left(\frac{\xi}{a_1},\frac{\eta}{a_2};\ x,y\right)$$

We have shown elsewhere [5] that the general solution of this equation is:

$$w(\xi,\eta;\ x,y) = y^{-\gamma}\, w_o\left(\frac{\xi}{x^{\sigma_1}},\frac{\eta}{y^{\sigma_2}},\frac{x}{y^{\rho}}\right)$$
$$\beta_1= a_1^{1/\sigma_1}\ ;\ \beta_2= a_2^{1/\sigma_2}\ ;\ \beta_1= \beta_2^{\rho} \qquad (4)$$
$$\alpha = a_1^{-1}\cdot a_2^{-1+\gamma/\sigma_2}\ ;\ x \neq 0,\ y\neq 0$$

where γ, σ_1, and σ_2 are arbitrary real numbers and $w_o(x,y,z)$ is any regular function of its variables. The value of ρ, as it is apparent from the above solution, is determined by the choice of σ_1 and σ_2.

We shall point out, now, some particular cases of Eq. (4) which are physically interesting.

A first case is characterized by setting the two output scaling factors, β_1 and β_2, identical to the two corresponding input factors a_1 and a_2. To this end one must choose $\sigma_1 = \sigma_2 =1$ which also implies $\rho = (\lg a_1)/\lg a_2$. By further setting $a_1 = a_2 = a$, which implies $\rho = 1$, Eq. (4) gives rise to a class of functions which contains those weighting functions (actually point-spread functions) that are shift-variant merely in the sense that they linearly expand with the distance r of the point (x,y) from the origin (0,0).

In such a case the input-output relationship of the filter can be written as:

$$g(x,y) = \iint_{-\infty}^{+\infty} f(\xi,\eta) w_r\left(\frac{x-\xi}{r}, \frac{y-\eta}{r}\right) d\xi\, d\eta \tag{5}$$

It is worth mentioning that, by properly choosing the function $w_r(\ )$, Eq. (5) represents the processing ideally performed by the peripheral visual system in man [6].

We shall now point out that for a large subclass of Eq. (4) the input-output relationship of the form-invariant systems can be written, in a suitable complex domain, under the form of Mellin transform products.

By recalling, in fact, that the Mellin transform of a (causal) signal f(x,y) is

$$\tilde{F}(s_1, s_2) = \int_0^\infty\int_0^\infty f(x,y) x^{s_1-1} y^{s_2-1}\, dxdy$$

it can be easily checked that if the weighting functions of Eq. (4) are restricted to the subclass

$$w(\xi,\eta;\, x,y) = \xi^{-1}\, \eta^{-1}\, w_1\, (x/\xi\ ,\ y/\eta) \tag{6}$$

then the input-output relationship characterizing the system can be written as:

$$\tilde{G}(s_1, s_2) = \tilde{F}(s_1, s_2)\, \tilde{W}_o(s_1, s_2) \tag{7}$$

This equation shows that, for the above systems, the Mellin transform plays exactly the same role that is played by the Laplace transform in the case of the 2-D shift-invariant systems.

It can similarly be shown that if, instead of Eq. (6), the following larger subclass of Eq. (4) is chosen:

$$w(\xi,\eta;\, x,y) = x^{-\delta_1}\, y^{-\delta_2}\, \xi^{-\gamma_1}\, \eta^{-\gamma_2}\, w_o\left(\frac{x}{\xi^{\tau_1}}, \frac{y}{\eta^{\tau_2}}\right) \tag{8}$$

then still an algebric equation in the (s_1, s_2) domain can be found, relating the input and output of the filters. In this case, in fact, we have:

$$\tilde{G}(s_1, s_2) = = \tilde{F}[\tau_1(s_1 - \delta_1) - \gamma_1 + 1, \tau_2(s_2 - \delta_2) - \gamma_2 + 1] \cdot \tilde{W}_o(s_1 - \delta_1, s_2 - \delta_2) \tag{9}$$

This algebraic equation, however, is not as simple, or straightforward, as Eq. (7) and is not equivalent to the Laplace based relationships holding for the LSI 2-D filters.

3. The Mellin - based implementation

We have shown, in the previous section, that the input-output relationship of a form-invariant system belonging to Eq. (8) - which is a subclass of Eq. (4) - can be put in the form of Eq. (9), which directly suggests an implementation technique, provided that we are able to digitally compute both the direct and the inverse 2-D Mellin Transform.

Since these latters can be written as the cascade of two 1-D Mellin Transforms, the problem reduces to computing the direct and inverse 1-D Mellin Transform. The solution to this latter problem has already been given in the literature (see, e.g., [7]): it is based on a change of variables which allows computing the Mellin Transform by means of the Fourier Transform (and, therefore, by means of the FFT). Let us write, in fact, the 1-D Mellin Transform:

$$\tilde{F}(s) = \int_0^\infty f(x)x^{s-1}\, dx \tag{10}$$

It is easily seen that, by means of the change of variables $x = \Delta x e^u$, the above equation becomes:

$$\tilde{f}(s) = (\Delta x)^s \int_{-\infty}^{+\infty} f(\Delta x\, e^u)\, e^{su}\, du \tag{11}$$

which is indeed a Fourier Transform, since $s = \alpha + j\omega$ and α can be taken to be constant. Of course, in order to apply the well known FFT algorithms, the function to be transformed in Eq. (11) must be available in sampled form, which implies that the samples of f(x) must be repositioned according to a logarithmic mapping, which on its turn implies a remarkable increase of the number of samples if one does not want to loose information.

Similarly to the LSI system case, where the computation of the output by means of the Fourier Transforms is convenient (with respect to the time convolution) only if the impulse responses are long enough, also for form-invariant LSV filters the computation of the output through the Mellin Transforms turns out to be convenient (with respect to the direct processing in the input domain) only if the size of the weighting functions w( ) is sufficiently large in the average. We may further comment that, because of the increase in the number of samples implied by the above outlined algorithm, the size of the functions w( ) which makes convenient the use of the Mellin Transforms is larger, in the average, than the size of the impulse responses which makes convenient the use of the Fourier Transforms.

Let us point out now that the Mellin Transforms can also be used for inverting a form-invariant LSV system. In fact, if one wants to recover the input f( ), knowing both the weighting function w( ) and the output g( ), the following equation can be considered (as easily derived from Eq. (7) or Eq. (9)):

$$f(x,y) = MT^{-1}\{\tilde{F}(s_1, s_2)\} = MT^{-1}\left\{\frac{\tilde{G}(s_1, s_2)}{\tilde{W}_o(s_1, s_2)}\right\} \qquad (12)$$

It must be noticed, however, that the above equation is extremely sensitive to noise. It is equivalent, in fact, to the inversion of a 2-D Fredholm integral equation of the first kind whose "ill conditioning" problems are well known (see, e.g., [8]). In order to face such problems, Eq. (12) must be dealt with by using some regularization techniques. We have been able, in fact, to solve Eq. (12) in some particular cases [9] by ignoring the contributes of $\tilde{G}/\tilde{W}_o$ originated by the values of $\tilde{W}_o$ below a certain threshold which has been fixed on the base of the noise level.

A different approach to the implementation of the form-invariant LSV systems, still based however on the use of the Mellin Transform, can be taken if the systems belong to the subclass defined by Eq. (5). As we will see, this approach is based on the use of a combined

Fourier-Mellin transform. Let us further assume, in fact, that the weighting functions $w_r(\;)$ of Eq. (5) are circularly symmetrical, that is:

$$w_r\left(\frac{x-\xi}{r},\frac{y-\eta}{r}\right) = w_o\left[\frac{(x-\xi)^2+(y-\eta)^2}{r^2}\right]$$
$$r = (x^2+y^2)^{1/2} \tag{13}$$

We perform now a mapping from the cartesian coordinates $(x,y;\xi,\eta)$ to the polar coordinates $(r,\theta;\rho,\varphi)$ according to the equations:

$$\begin{cases} x = r\cos\theta \\ y = r\sin\theta \end{cases} \qquad \begin{cases} \xi = \rho\cos\varphi \\ \eta = \rho\sin\varphi \end{cases}$$

so that we can write

$$w_o\left[\frac{(x-\xi)^2+(y-\eta)^2}{r^2}\right] = w_o\left[1+\frac{\rho^2}{r^2} - 2\frac{\rho}{r}\cos(\theta-\varphi)\right] = w_1\left(\frac{\rho}{r},\theta-\varphi\right)$$

As a consequence, Eq. (5) can be written as

$$g_1(r,\theta) = \int_0^{\infty}\int_0^{2\pi} f_1(\rho,\varphi)w_1\left(\frac{\rho}{r},\theta-\varphi\right)\rho\,d\rho\,d\varphi \tag{14}$$

where $g_1(r,\theta) = g(r\cos\theta,\ r\sin\theta)$ and $f_1(\rho,\varphi) = f(\rho\cos\varphi,\rho\sin\varphi)$.

If Eq. (14) is now Fourier transformed along the θ axis, and Mellin transformed along the r axis, the input-output relationship characterizing our system becomes:

$$\tilde{G}_1(s_r,\omega_\theta) = \tilde{F}_1(s_r+2,\omega_\theta)\,\tilde{W}_1(-s_r,\omega_\theta) \tag{15}$$

where ω_θ is the Fourier variable and $s_r = \alpha + j\omega_r$ is the Mellin variable.

Eq. (15) shows how the Fourier and the Mellin Transforms can be jointly used in order to find the output of a form invariant system characterized by Eq. (13).

As we have seen, this implementation approach (to the system) is based on a change of coordinates (from the cartesian to the polar coordinates), which is costly from the computational point of view, since it requires a non structured access to the data matrices. In addition to this, the implementation of Eq. (15) requires, of course, the two mentioned 1-D Fourier and Mellin Transforms.

It is worth pointing out that the systems that we are considering here (i.e. those characterized by Eq. (13)) are particularly interesting from the physical point of view. It has been shown, e.g., that the peripheral visual system in man can be modeled by such systems [6].

Like the systems characterized by Eq. (8) (see above), also the systems defined by Eq. (14) can be inverted through the transformed domain (i.e. through Eq. (15)).

Some experiments in this sense are in progress. As far as the "ill conditioning" problems are concerned, the experiments seem to prove what one would expect from the theoretical analysis, that is that such problems are less severe in the case of Eq. (15) than they are, as already mentioned, in the case of Eq. (12).

4. Implementation via linear shift-invariant systems

The possibility of implementing LSV systems by means of LSI systems in conjunction with suitable coordinate mappings has been investigated in the past, e.g. in [7, 10] with reference to the restoration of shift-variant image degradations (due to optical defects or to motion blur), or in [11] with reference to signal detection based on shift-variant correlators.

Mathematically, the problem can be stated as follows: find the one-to-one coordinate mappings h_1 and h_2

$$\begin{aligned} \xi' &= h_1(\xi, \eta) \qquad & x' &= h_1(x, y) \\ \eta' &= h_2(\xi, \eta) \qquad & y' &= h_2(x, y) \end{aligned} \tag{16}$$

such that the usual input-output relationship for an LSI system having a suitable impulse response w_1

$$g_1(x', y') = \iint_{-\infty}^{+\infty} f_1(\xi', \eta') w_1(x'-\xi'; y'-\eta') d\xi' \, d\eta' \qquad (17)$$

can be written as

$$g(x,y) = g_1(x', y') =$$

$$= \iint f(\xi,\eta) w_1[h_1(x,y)-h_1(\xi,\eta),\ h_2(x,y)-h_2(\xi,\eta)] \, J(\xi,\eta) d\xi \, d\eta$$

(where the integration limits are derived by inverting the mapping (16)) and then identified with the LSV relationship

$$g(x,y) = \iint_{-\infty}^{+\infty} f(\xi,\eta) \; w(\xi,\eta; x,y) d\xi \, d\eta \qquad (18)$$

being $f_1(\xi',\eta') = f(\xi,\eta)$, $g_1(x',y') = g(x,y)$ and J the giacobian of the mapping (16).

Some mathematical aspects of the problem are discussed in the quoted references. The convenience of using Eq. (17) is apparent, since for its implementation are available all the techniques already developed for the LSI systems, including the fast convolution methods based on the FFT algorithms. The main drawback implicit in Eq. (17) is due to the implementation of the mapping of Eq. (16), which poses severe approximation problems when dealing with digital images that are known over a uniform grid and must be interpolated according to the law of Eq. (16).

Coming now to the class of LSV systems that is the object of this paper, we will show that the method presented above can be used to implement two relevant subclasses of the one defined in Eq. (4), namely the subclasses considered in the previous section.

Let us start with the subclass of 2-D systems characterized by Eq. (6). To such systems, in fact, it is easy to extend the result, valid in the 1-D case [12], stating that a mapping of the form $\xi' = \lg \xi$ preceeding a LSI system with impulse response $w_1(\cdot)$ and an inverse mapping $x = e^{x'}$ is globally equivalent to a LSV system with weighting function $\xi^{-1} w(x/\xi) = \xi^{-1} w_1(\lg \frac{x}{\xi})$. In the 2-D case, the simple mapping

$$\xi' = \lg \xi \qquad x' = \lg x$$
$$\eta' = \lg \eta \qquad y' = \lg y \tag{19}$$

leads to the identification of Eq.s (17) and (18) in the following terms: the LSI impulse response $w_1(x'-\xi', y'-\eta')$ corresponds to the LSV weighting function $\xi^{-1}\eta^{-1} w(x/\xi, y/\eta)$, that guarantees output scaling factors identical to the input scaling factors. A change in the scaling factors (i.e. values of τ_1 and τ_2 in Eq. (8) different from unity) can be obtained if the mappings of Eq. (19) are modified by constant factors.

The subclass of form-invariant systems discussed so far coincides essentially with the subclass for which the use of the Mellin transform was illustrated in the previous section. It is worth noticing that the two approaches are equivalent. If, in fact, the cascade of the mappings (19) and of the LSI system $w_1(\ )$ is analyzed in the frequency domain, where

$$G_1(\omega_1', \omega_2') = F_1(\omega_1', \omega_2') W_1(\omega_1', \omega_2') \tag{20}$$

it can be easily shown that Eq. (20) is equivalent to

$$\tilde{G}(s_1, s_2) = \tilde{F}(s_1, s_2)\ \tilde{W}(s_1, s_2) \tag{21}$$

for $s_1 = -j\omega_1$, $s_2 = -j\omega_2$. Considering, e.g., the left-hand side term of Eq. (21), it can be rewritten, through the change of variables $x' = \lg x$, $y' = \lg y$, as

$$\tilde{G}(s_1, s_2) = \int_0^\infty\int_0^\infty g(x,y)x^{s_1-1}\ y^{s_2-1}\ dxdy =$$
$$= \int_{-\infty}^{\infty}\int_{-\infty}^{\infty} g_1(x', y')e^{-j\omega_1' x'}\ e^{-j\omega_2' x'}\ dx'dy'$$

which coincides with the left-hand side of Eq. (20).

The other subclass of form-invariant systems that can be implemented by means of the technique object of this section is the subclass characterized by weighting functions having size linearly increasing with the distance from the origin and being shift-invariant with respect to the angular coordinate. In other words, letting

$$x = r\cos\vartheta \qquad \xi = \rho\cos\varphi$$
$$y = r\sin\vartheta \qquad \eta = \rho\sin\varphi$$

such weighting functions are of the form

$$\rho^{-1}w(\rho/r,\ \vartheta - \varphi) \tag{22}$$

Therefore, if all the processing is performed in the polar coordinates domain, it is clear that by a simple mapping of the radial coordinates

$$\rho' = \lg \rho$$
$$r' = \lg r$$

the LSV system (22) can be implemented by a suitable LSI system.

If the cascade of the mappings and of such a LSI system is analyzed in the spatial (radial and angular) frequencies domain, an equation similar to Eq. (21) is obtained, where, however, all the transforms must be interpreted as mixed Mellin (with respect to the radial coordinate) and Fourier (with respect to the angular coordinate) transforms.

As already stated, the main problem connected with this procedure, either in the spatial or frequency domain, is the coordinate transformation from the cartesian grid (where usually digital images are available) to the polar plane and the corresponding approximation errors.

It is interesting to notice that the weighting functions (22) include those describing several physical phenomena of great interest in the field of signal processing, like (as already seen) the behaviour of the receptive fields in the peripheral visual system of man, some optical degradations [7] and the reconstruction of objects from line integrals [13].

References

[1] D.Casasent, D.Psaltis, "New optical transforms for pattern recognition", Proc. of the IEEE, vol. 65, n. 1, Jan. 1977, pp. 77-84.

[2] C.Braccini, G.Gambardella, "Linear shift-variant filtering for form-invariant processing of linearly scaled signals", Signal Processing, vol. 4, n. 2-3, Apr. 1982, pp. 209-214.

[3] J.Leger, S.Lee, "Hybrid optical processor for pattern recognition and classification using a generalized set of pattern functions", Applied Optics, vol. 21, n. 2, Jan. 1982, pp. 274-287.

[4] K.Fu, Syntactic methods in pattern recognition, Academic Press, 1974.

[5] C.Braccini, G.Gambardella, "Form-invariant linear filtering: theory and applications", submitted to IEEE Trans. on ASSP.

[6] C.Braccini, G.Gambardella, G.Sandini, "A signal theory approach to the space and frequency variant filtering performed by the human visual system", Signal Processing, vol. 3, n. 3, July 1981, pp. 231-240.

[7] G.Robbins, T.Huang, "Inverse filtering for linear shift-variant imaging systems", Proc. of the IEEE, vol. 60, n. 7, July 1972, pp. 862-872.

[8] J.McWhirther, E.Pike, "On the numerical inversion of the Laplace transform and similar Fredholm integral equations of the first kind", J. Phys., A: Math. Gen., vol. 11, n. 9, 1978, pp. 1729-1745.

[9] C.Braccini, C.Burlando, S.Gaglio, G.Gambardella, A.Grattarola, "Image reconstruction through discrete Mellin Transforms", Proc. of the Int. Conf. on Digital Signal Processing, Florence, Sept. 1981, pp. 370-376.

[10] A.Sawchuck, "Space-variant image motion degradation and restoration", Proc. of the IEEE, vol. 60, n. 7, July 1972, pp. 854-861.

[11] D.Casasent, D.Psaltis, "Deformation invariant, space-variant optical pattern recognition", in Progress in Optics, E.Wolf, Ed., vol. XVI, North-Holland, 1978, pp. 291-356.

[12] C.Braccini, G.Gambardella, "Signal detection and estimation techniques based on linear shift-variant filtering and short-space frequency variant spectral analysis", Proc. of the Workshop Italy-U.S.A. on Digital Signal Processing, Portovenere, Italy, Aug. 1981, pp. 157-170.

[13] E.Hansen, "Image reconstruction from projections using circular harmonic expansion", Ph. D. Dissertation, Dept. of El. Eng., Stanford Univ. Dec. 1979.

REPORT OF DISCUSSION SESSION ON UNRESOLVED ISSUES AND FUTURE DIRECTIONS

C.H. CHEN

This report briefly describes a sequence of comments and suggestions made during the illuminating discussion session. I am responsible for any inaccuracy of the report.

1. C.H. Chen suggested that multipath ranging using pattern recognition and image processing is a topic that requires much further study.
2. J. Kittler presented various problems on edge detection, with hardware implementation being one difficult problem.
   G. Easterbrook pointed out the importance of iterative algorithms, such as the relaxation method, in image analysis.
   J. Morlet commented that edge detection is close to line extraction in seismograms.
3. T. Kooij raised the question, "What are the participants thoughts on a set of test signals for 1-D, 2-D, 3-D processing?" The responses were unanimous that common sets of test signals are much needed. Some test signals are now available in speech and images. However no test signal is available in sonar area.
4. J.C. Hassab presented the issues of solution quality and underlying process identification. He pointed out that a problem in underlying process is how to describe a phenomenon, i.e. what class does it belong to. If we miss a class, we have problem with estimation (bias in estimation). For example when the source is moving at constant course it is in a class. Then the source makes maneuver and the the class changes. Can we determine which class it is from observation? The solution quality is a very neglected subject. We need to know how much confidence we have on the solution.
   B. Ursin raised the issue of choosing a model and selecting proper model order. There is no good solution to this problem yet.
   A quick identification of underlying process is needed.
   T. Kooij suggested the use of hypothesis testing.
5. L.J. Lloyd suggested two areas for future study: 1. Improved modelling of the source parameters and how they appear in the observed acoustic signals to improve the extraction of information. 2. Methods of adaptive processing applied to observation time (integration time) to avoid smearing of observed data.
6. V. Cappellini suggested the need for a single and fast nonlinear processing adjusted to the local properties of 2-D data or digitized image. Presently two operations must be used, one to filter the random noise and the other to remove the spike noise.
7. J. Morlet presented two issues of particular importance in seismic signal processing: 1. Multiparameter representation in 3D space and high speed scanning of 2D projections. 2. High accuracy of array element locations (for multipath processing).

NATO ASI Series, Vol. F1
Issues in Acoustic Signal/Image Processing and Recognition
Edited by C.H. Chen

8. J.N. Maksym reported that in AI approach there are many unresolved issues, among them: 1. Knowledge representation questions of what to represent, what data structure to use, and how to use the knowledge for control. 2. Self-acquisition of knowledge (learning). 3. More work on quantification of inferences. The mutual information idea in information theory can be of help. 4. Feedback from top-levels. Does it just some processing, or does it do more than this? Etc.

9. G. Easterbrook pointed out that there are currently many models of visual perception but very few digital simulations to provide new ideas for filters. The issue now is how to link the information from the visual perceptinn field to signal processing, image processing and pattern recognition.
   J.N. Maksym commented that researchers in machine vision look for operations in the past. Now they look back to human visual processing from the point of view of what knowledge is being processed and how it is to be processed.

10. R. Klemm raised the question. "How far can propagation models be used to improve the detection/estimatinn/classification performance of real-time sonar systems?". For example, if you have a dispersive field, what kind of signal is to be expected at certain time. Propagation paths are fuzzy and often do not match delays unamibiguously.

LIST OF PARTICIPANTS

Canada

Joseph N. Maksym
A. N. Venetsanopoulos

France

Herve Gautier
Genevieve Jourdain
J. Y. Jourdain
Jean P. Morlet

F.R. Germany

Richard Klemm
U. Rupe

Italy

Jorg Bummerstide
Vito Cappellini
Valerio R. Cimagalli
Sandro Fossi
Giovanni Garibotto
A. Grattarola
Lewis J. Lloyd
Even B. Lunde
Davide Marandino
Gianni Orlandi
Ruggero Reggiannini

Norway

Jens M. Hovem
Bjørn Ursin

Portugal

José M. F. Moura

U.S.A.

C. H. Chen
Joseph C. Hassab
Theo Kooij
Richard L. Moose
Kent Scarbrough

U.K.

Tariq Durrani
J. W. R.(Roy) Griffiths
J. Kittler
G. V. Werrey-Easterbrook